Toolik Lake

Developments in Hydrobiology 78

Series editor
H. J. Dumont

Toolik Lake

Ecology of an Aquatic Ecosystem in Arctic Alaska

Edited by
W. J. O'Brien

Reprinted from Hydrobiologia, vol. 240 (1992)

SPRINGER-SCIENCE+BUSINESS MEDIA, B.V.

Library of Congress Cataloging-in-Publication Data

Toolik Lake : ecology of an aquatic ecosystem in arctic Alaska / edited by W.J. O'Brien.
p. cm. -- (Developments in hydrobiology ; 78)
ISBN 978-94-010-5206-1 ISBN 978-94-011-2720-2 (eBook)
DOI 10.1007/978-94-011-2720-2
1. Lake ecology--Alaska. 2. Lake ecology--Arctic regions. 3. Lake ecology--Alaska--Toolik Lake. 4. Limnology--Alaska--Toolik Lake. 5. Limnology--Arctic regions. 6. Toolik Lake (Alaska) I. O'Brien, W. J. (William John), 1942- . II. Series.
QH105.A4T66 1992
574.5'26322'09798--dc20 92-26746

ISBN 978-94-010-5206-1

Printed on acid-free paper

Contents

Preface

The limnological study of Toolik Lake began in the summer of 1975. This research was an outgrowth of the arctic IBP project which had focused mainly on small arctic pond ecosystems on the Alaskan arctic coastal plain. It was thought desirable to study larger, deeper lakes which contained fish to further generalizations developed during the IBP study. Toolik lake became a feasible study site with the construction of the Dalton Highway, the road built to construct the trans-Alaska oil pipeline which passes within a kilometer of the lake.

Toolik Lake (68° 38′ N, 149° 36′ W) is a multiple basin kettle lake formed about 12,000 years BP. It is a dimictic lake with a maximum depth of 25 m but a mean depth of 7 m. The lake is in a morainal formation 20 km north of the Brooks Mountains at 720 m elevation. It is a very oligotrophic lake with an annual primary productivity of only 15 g $C/m^2/yr$. Ice-out is generally in mid to late June and the lake refreezes in early to late September. Much of the specific limnology of the lake is detailed in the current volume.

Initial research on Toolik Lake and the surrounding lakes and ponds focused on process studies such as annual primary productivity of the lake or the vertical migration patterns of the resident zooplankton. In 1978 the scope of the research expanded to include the study of the Kuparuk River, a fourth-order stream which crosses the Dalton highway 5–6 km north of Toolik Lake. In 1983 the philosophy of the research changed with the funding of a more integrated program. The fundamental question posed was whether arctic lake and stream ecosystems are regulated from the bottom up by nutrient availability or from the top down by the density and activity of top predators. In the Kuparuk River ecosystem this involved adding nutrients to the stream while in the lakes it involved deploying a series of limnocorrals to which nutrients and fish were added. These limnocorrals proved excellent means of studying nutrient–phytoplankton interactions, but they proved less than ideal in studying the ecosystem functions of fish and benthic invertebrates in lake ecosystems (see manuscript in this volume). Hence in 1985 began some whole lake manipulations and continued fertilizations of the Kuparuk River.

In 1987 the Toolik Lake area was designated an LTER (Long-Term Ecological Research) site, one of 18 such sites throughout the United States, Puerto Rico, and the Antarctic. The research theme for the arctic LTER also focuses on the regulation of arctic ecosystems, whether regulation comes from the top down or bottom up. The arctic LTER also incorporates a terrestrial component as well as a lake and stream component.

W. J. O'Brien

Hydrobiologia **240**: 1–14, 1992.
W.J. O'Brien (ed.), Toolik Lake – Ecology of an Aquatic Ecosystem in Arctic Alaska.

The biogeochemistry and zoogeography of lakes and rivers in arctic Alaska

George W. Kling, W. John O'Brien, Michael C. Miller & Anne E. Hershey
Department of Biology, University of Michigan, Ann Arbor, MI 48109, USA; Department of Systematics and Ecology, University of Kansas, Lawrence, KS 66045, USA; Department of Biological Sciences, University of Cincinnati, Cincinnati, OH 45221, USA; Department of Biological Sciences, University of Minnesota-Duluth, Duluth, MN 55812, USA

Abstract

Water samples from 45 lakes and 8 rivers in arctic Alaska were analyzed for major anions, cations, nutrients, chlorophyll, zooplankton, and benthos. The waters were dilute (conductivities of 30 to 843 μS cm^{-1}), and their composition varied from Na-Ca-Cl waters near the Arctic Ocean to Ca-Mg-HCO_3 waters further inland. Sea salt input in precipitation was important in determining the chemistry of coastal lakes, partly because of low groundwater flow and less time for water to react with shallow unfrozen soils. Further inland, variations in water chemistry among sites were related mainly to differences in bedrock, the age of associated glacial drift, and the input of wind blown sediment. Variations in zooplankton species composition among the lakes were related more to latitude, lake morphometery, and biotic interactions than to water chemistry. The presence of fish as predators mostly determined the overall size structure of the zooplankton community. The chironomid taxa identified have been previously reported from the Neararctic, except for *Corynocera oliveri* which is a new record. The abundance of the widely distributed chironomid *Procladius* appears to be controlled by sculpin predation.

Introduction

The most recent general reviews of arctic limnology are by Hobbie (1980, 1984). Since then, there have been further surveys or reviews on water chemistry (Welch & Legault, 1986; Lock *et al.*, 1989), on zooplankton (Haney & Buchanan, 1987; Luecke & O'Brien, 1983), and on benthos (Hershey, 1985; Miller & Stout, 1989; Maciolek, 1989; Hershey, 1990). Despite this recent research effort, there are large areas of arctic Alaska and Canada that remain unexplored limnologically.

In this study we present chemical, zooplankton, and benthos data for 45 lakes and 8 rivers along the oil pipeline in arctic Alaska, and rain chemistry data from the Toolik Lake area in the foothills of the Brooks Range. About 30 of the sites were investigated for the first time. Samples were collected in June and July, mostly in 1988 and 1990.

Site description

All study sites are located within several kilometers of the Dalton Highway on the North Slope of Alaska (Fig. 1). There is continuous permafrost under the entire area, and this north-south transect includes sites in the Brooks Range (> 900 m elevation), the foothills province (75–900 m), and the coastal plain (< 75 m). All lakes studied are natural, with the exception of sites 14, 48, 54, and 55, which are gravel pits or artificially

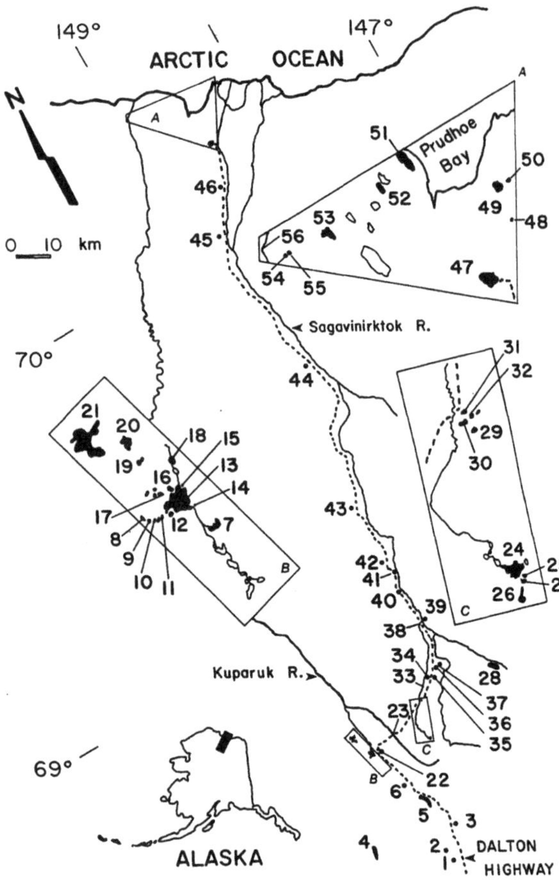

Fig. 1. Location of study sites on the North Slope of Alaska. Site numbers correspond to the numbers and names in Table 1. Enlarged areas are (A) Prudhoe Bay, (B) Toolik, and (C) Oksrukuyik drainage.

dammed (Table 1). Lakes near the coast are less than 2 m deep, and are formed by the action of wind and thermokarst processes. Most lakes at elevations greater than 145 m are morraine-dammed or kettle ponds. The surficial geology of the region is summarized by Brown & Berg (1980) and Brown & Kreig (1983). Briefly, the coastal plain is underlain by Quaternary eolian, alluvial, and marine sediments. In addition to such recent sediments, the foothills province contains older Cretaceous and Tertiary sandstones, conglomerates, and siltstones. Most of these rocks are mantled by extensive deposits of glacial drift of varying age. In the Brooks Range, the Lisburne limestone and dolomite group, and shales of Triassic to Pennsylvanian age are common.

Table 1. Location and morphometric data for study lakes and rivers. Site numbers correspond to sites in Figure 1.

Site	Latitude (°.'N)	Longitude (°.'E)	Elevation (m)	Area (ha)	Max. depth (m)
Lakes					
4 Itkillik	68.23	149.55	681	464	7.5
5 Galbraith	68.28	149.30	802	422	6.0
6 Island	68.32	149.28	881	62.0	6.0
7 I8	68.37	149.35	747	17.6	10.0
8 S13	68.38	149.40	754	1.0	5.4
9 S12	68.38	149.39	754	0.3	3.8
10 S11	68.38	149.39	750	0.4	13.1
11 S6	68.38	149.39	731	1.1	8.5
12 S3	68.38	149.38	719	4.0	5.0
13 Toolik	68.38	149.36	719	150	22.0
14 Dam	68.38	149.36	719	0.1	1.0
15 Camp	68.38	149.36	720	0.2	0.7
16 N1	68.38	149.37	731	4.4	14
17 N2	68.38	149.38	724	1.8	10.7
18 NE2	68.38	149.37	724	6.3	4.0
19 NE12	68.39	149.37	693	8.5	14
20 NE14	68.41	149.37	699	24.0	–
21 Itigaknit	68.42	149.40	663	153	6.0
22 E1	68.38	149.33	771	2.9	11
24 Campsite	68.36	149.11	864	79.5	12
25 O1	68.36	149.11	876	2.7	16.2
26 O2	68.35	149.12	892	17.1	>12.4
27 O3	68.36	149.10	876	3.0	4.7
28 Elusive	68.39	148.30	525	300	17.2
29 Moose	68.40	149.06	792	3.7	21.1
30 Hanging	68.41	149.06	785	6.1	10.5
31 Foggy	68.41	149.05	770	3.1	17
32 F4	68.41	149.04	754	2.3	3.6
34 George	68.44	148.58	597	7.3	2.0
35 Anne	68.44	148.56	556	8.5	2.0
36 Sag2	68.44	148.53	526	1.7	4.4
37 Sag1	68.45	148.52	510	18.7	2.8
42 Charles	69.02	148.51	319	10.3	2.1
44 William	69.35	148.38	145	35.5	1.3
45 Windy	69.58	148.44	57	28.6	1.9
46 Silhouette	70.05	148.32	24	56.8	1.7
47 Colleen	70.13	148.28	15	231	1.2
48 Sag-c pit	70.14	148.15	10	1.0	21.0
49 Bern	70.17	148.18	6	90.0	0.5
50 Dune	70.18	148.17	6	1.5	0.3
51 Maxine	70.22	148.30	2	180	0.55
52 Carolyn	70.21	148.35	2	66.2	0.8
53 Africa	70.20	148.48	4	91.2	0.8
54 Borrow pit1	70.20	148.56	3	1.0	–
55 Borrow pit2	70.20	148.56	3	1.0	–
Rivers					
1 West Atigun	68.16	149.28	914		
2 Atigun	68.16	149.27	914		
3 Mt. Roche	68.23	149.19	869		
23 Kuparuk	68.38	149.24	731		
33 Oksrukuyik	68.52	148.51	411		
38 Oksrukuyik	68.43	149.01	628		
39 Sagavinirktok	68.52	148.50	404		
40 Alexa	68.57	148.52	360		
41 Sagcut	69.01	148.49	343		
43 Happy Valley	69.09	148.50	290		
56 Kuparuk	70.20	149.00	3		

The climate of the southern foothills and the coastal plain is summarized in Table 2. There is a decreasing air temperature gradient moving northward toward the coastal plain. Total precipitation is higher in the mountains, but overall there are no distinct patterns across the North Slope. About 40% of the precipitation is unfrozen on the coastal plain; moving southward the thaw season lengthens and the unfrozen precipitation can reach 75% of the total in the Brooks Range (Haugen, 1982). Most lakes on the coastal plain freeze in late September and thaw in late June or early July. In the warmer foothills, lakes usually freeze in early October and thaw in late June. Deeper lakes tend to freeze later in the fall and retain their ice cover longer in the spring than do shallow lakes; for example, Hobbie (1984) reported ice cover on the 57 m deep Lake Schrader in late July and early August.

Methods

Temperature, conductivity, and pH were measured in the field. Water samples were filtered through glass fiber filters (Whatman GF/C or Gelman A/E, 1.0 μm pore size) or membrane filters (0.45 μm pore size) and stored in polyethylene bottles for no longer than 4 months. Alkalinity was determined by potentiometric titration and the titration curves were analyzed by the method of Gran (Stumm & Morgan, 1981). In some samples, dissolved inorganic carbon (DIC) concentrations were determined by acidifying samples with phosphoric acid and purging with nitrogen gas into a vacuum line. CO_2 was collected by cryogenic distillation and gas pressure measured by an electronic manometric gauge. Alkalinity was then calculated from DIC concentrations and pH using the carbonate species relations in Stumm & Morgan (1981) and Plummer & Busenberg (1982). SO_4^{2-} and Cl^- were analyzed using a Dionex ion chromatograph. Water samples for cation analyses were preserved in the field with HCl; Ca^{2+}, Mg^{2+}, and Mn^{2+} were analyzed by flame atomic absorption spectroscopy, and Na^+ and K^+ by flame emission. Soluble reactive phosphate (SRP), $NO_3^- + NO_2^-$, and NH_4^+ were determined on a Technicon Autoanalyzer. The mean deviation from cation-anion balance was 4.8% ($N = 50$, $SE = 0.9$). Waters with ion imbalances greater than 10% were usually deficient in anions; alkalinity in these samples was calculated from DIC concentrations rather than measured directly. Mineral equilibria calculations were made using Güntelberg activity coefficients.

The climate station at Toolik Lake consisted of a Campbell Scientific data logger that recorded air temperature (Fenwall thermistor), wind speed at 5 m height (MET-ONE 014A anemometer), and solar radiation (LiCor 200SB pyranometer). Climate data are reported for the year from June 1989 to June 1990. Precipitation samples were collected using an Aerochem Metrics wet-dry collector.

Chlorophyll *a* (CHLa) was extracted in 90% methanol and concentrations were calculated using equations in Wetzel & Likens (1979). Zooplankton were collected using vertical or oblique hauls of a 100 or 335 μm mesh net. Nonparametric Spearman Rank correlations were made between zooplankton occurrence and physical-chemical variables.

One benthic sample was taken with an Eckman grab in a subset of the lakes surveyed (Ta-

Table 2. Comparison of climate in the southern foothills at Toolik Lake and Galbraith and in the coastal plain at Barrow and Prudhoe. Mean annual values and the range of monthly values are given, except for precipitation from Prudhoe, Galbraith, and Toolik where the range of annual sums is given. Toolik precipitation data are from Toolik River MS117 site; Barrow data are from Hobbie (1980); Prudhoe and Galbraith data are from Haugen (1982). All precipitation data are from Wyoming snow gauges except for Barrow.

Site	Temperature (°C)	Precipitation (mm)	Insolation ($J/cm^2/d$)	Wind (m/sec)
Galbraith	−8.8	210		
	(−30.5 to 12.7)	(188 to 254)		
Toolik	−9.0	309	838	3.2
	(−31.7 to 12.4)	(194 to 406)	(0 to 2168)	(2.1 to 3.7)
Prudhoe	−11.3	203		
	(−31.9 to 9.4)	(183 to 223)		
Barrow	−12.4	108	891	6.1
	(−28.3 to 4.1)	(2.8 to 22.8)	(0 to 2330)	(5.6 to 6.8)

Table 3. Physical and chemical data for study lakes and rivers: Time (Alaska standard, hr), Temp (C), Cond (conductivity at Temp given, $\mu S\ cm^{-1}$), Secchi (secchi depth, m). All ions are reported in $\mu mol\ L^{-1}$. Chlorophyll *a* concentrations are reported in $\mu g\ L^{-1}$ for lakes and $\mu g\ cm^{-2}$ for rivers

Site	Date	Time	Temp	Cond	pH	HCO_3^-	SO_4^{2-}	Cl^-	Ca^{2+}	Mg^{2+}	Na^+	K^+	Mn^{2+}	SRP	NO_3^-	NH_4^+	Chla	Secchi
Lakes																		
4 Itkillik	10 Jul 90	2000	17.0	224.0	8.35	2158	102	8	1078	284	24	14	0.1	–	–	–	–	7.5
5 Galbraith	6 Jul 88	1617	17.5	176.0	7.80	1693	61	9	863	155	26	11	0.1	0.10	3.13	0.79	–	–
6 Island	14 Jul 88	1050	16.0	158.2	7.90	1259	274	29	477	219	111	13	0.1	–	–	–	–	6.0
7 I8	18 Jul 90	1053	14.0	46.0	7.06	445	10	8	217	57	20	9	0.1	0.04	0.17	3.50	2.9	3.4
8 S13	1 Jul 90	1520	20.0	145.0	7.90	–	10	5	744	119	13	7	–	–	–	–	–	–
9 S12	1 jul 90	1450	18.5	256.0	7.93	–	64	5	1262	238	25	5	–	–	–	–	–	–
10 S11	1 jul 90	1430	18.0	192.0	8.06	–	56	5	958	243	26	12	–	–	–	–	–	–
11 S6	14 Jul 88	1530	19.5	165.6	8.46	1405	85	5	1023	206	33	10	0.1	–	–	–	–	–
12 S3	22 Jul 86	–	17.0	–	7.49	560	–	–	–	–	–	–	–	–	–	–	–	–
13 Toolik	19 Jul 90	1000	14.0	73.2	7.29	525	12	9	257	55	22	11	0.4	0.1	0.1	0.1	1.0	5.5
14 Dam	23 Jun 90	–	10.5	108.0	7.45	1020	–	–	–	–	–	–	–	–	–	–	–	1.0
15 Camp Pond	13 Jul 88	1400	20.5	45.8	8.25	381	5	5	137	27	15	8	0.4	1.56	1.56	0.93	148	0.7
16 N1	14 Jul 88	1400	20.0	118.4	8.22	1149	13	6	591	113	17	12	0.1	0.23	0.23	0.13	–	–
17 N2	28 Jun 88	1000	14.9	110.4	9.95	1282	25	6	571	104	22	12	0.4	0.13	0.04	0.5	1.5	5.8
18 NE2	22 Jul 86	–	15.5	–	7.23	520	–	–	–	–	–	–	–	–	–	–	–	–
19 NE12	29 Jun 88	1455	15.5	126.0	7.90	1210	20	2	629	115	12	11	0.2	0.33	0.33	0.24	1.7	–
20 NE14	29 Jun 88	1330	15.0	158.5	7.91	1444	58	7	761	175	22	13	0.4	0.17	0.17	0.24	0.7	–
21 Itigaknit	29 Jun 88	1100	17.5	98.8	7.53	988	8	6	489	92	14	16	0.1	–	–	–	–	–
22 E1	22 Jul 86	–	14.5	39.5	6.85	540	24	8	324	60	24	11	0.4	–	–	–	5.4	6.0
24 Campsite	11 Jul 90	1420	13.4	42.4	7.60	381	7	4	175	34	15	11	0.4	0.19	0.27	0.64	–	6.5
25 O1	11 Jul 90	1530	15.5	31.7	7.51	251	4	3	112	25	10	11	0.4	0.20	0.44	0.86	–	4.0
26 O2	11 jul 90	1700	13.9	30.1	7.47	270	5	3	155	25	10	11	0.1	0.41	0.24	1.03	–	4.5
27 O3	11 Jul 90	1745	17.0	48.8	7.57	459	4	2	210	42	12	12	0.1	0.36	0.32	0.72	–	4.0
28 Elusive	11 Jul 90	1330	14.6	213.0	8.32	2003	232	9	1085	218	28	11	0.1	0.20	0.47	0.46	0.3	7.0
29 Moose	8 Jul 90	1400	16.0	99.0	8.16	1153	49	6	467	99	24	12	0.1	0.30	0.21	0.89	1.2	8.4
30 Hanging	8 Jul 90	1300	15.5	129	8.24	1289	43	5	739	106	24	12	0.4	0.24	0.12	0.53	6.4	9.0
31 Foggy	8 Jul 90	1145	16.0	139	8.18	1302	94	6	691	120	27	12	0.1	0.30	0.11	0.63	1.3	7.0
32 F4	8 Jul 90	1540	15.0	96.5	7.84	911	33	7	467	84	22	12	0.6	0.27	0.24	1.17	7.6	2.5
34 George	7 Jul 88	1240	19.0	166.0	8.44	1510	4	8	494	106	27	11	0.4	–	–	–	2.9	2.0
35 Anne	7 Jul 88	1030	19.5	112.0	8.16	937	39	82	541	90	36	12	0.1	0.04	1.25	3.00	2.4	2.0
26 Sag2	13 Jul 90	1200	18.0	87.5	7.90	862	5	5	444	51	14	11	0.2	0.21	0.43	0.62	2.0	3.0
37 Sag1	13 Jul 90	1100	17.0	49.3	8.38	510	2	8	250	31	14	12	0.1	0.21	0.34	0.73	2.2	2.5
42 Charles	8 Jul 88	1310	18.8	36.9	7.29	338	5	5	155	33	34	7	0.4	0.01	12.4	0.83	–	2.1
44 William	11 Jul 88	1100	17.0	157.0	8.10	1041	5	22	579	71	20	10	0.1	0.05	8.12	2.01	–	1.3
45 Windy	8 Jul 88	1645	14.0	121.3	7.35	1275	5	74	554	93	61	13	0.2	0.01	0.62	0.50	–	1.9
46 Silhouette	8 Jul 88	1745	13.0	151.7	7.72	1688	15	106	694	211	119	21	0.7	0.02	0.45	0.52	–	1.7
47 Colleen	10 Jul 88	1521	11.0	272.0	7.90	1309	8	987	791	192	444	15	0.1	0.01	4.14	0.64	–	1.2
48 Sag-c pit	10 Jul 88	945	6.0	216.0	8.40	1762	137	269	831	216	235	17	0.1	0.02	1.16	0.49	–	–
49 Bern	10 Jul 88	1215	9.5	232.0	8.16	1223	8	772	756	106	487	13	0.1	–	0.48	1.01	–	0.5
50 Dune	10 Jul 88	1145	9.2	518.0	7.80	1749	206	2274	1000	346	2031	63	0.1	0.01	0.54	0.98	–	0.3
51 Maxine	9 Jul 88	1445	9.8	843.0	8.09	1317	38	5249	1190	494	3654	77	0.2	0.02	4.77	0.81	–	0.55
52 Carolyn	9 Jul 88	1600	11.0	371.0	7.75	1621	30	1443	876	219	1109	32	0.1	0.19	0.61	0.78	–	0.8
53 Africa	9 Jul 88	1700	11.0	364.0	7.70	1150	7	1729	808	222	1027	26	0.1	0.02	1.63	0.36	–	0.8
54 Borrow pit1	9 Jul 88	1300	9.4	–	–	1174	23	468	579	146	278	24	0.2	0.13	12.1	0.68	–	–
55 Borrow pit2	9 Jul 88	1330	5.6	–	–	1322	32	530	634	168	283	25	0.4	0.04	24.4	0.71	–	–
Rivers																		
1 West Atigun	6 Jul 88	1150	9.5	147.0	7.48	895	211	8	394	237	47	11	0.1	0.05	11.0	1.11	0.2	
2 Atigun	6 Jul 88	1250	11.0	166.0	8.02	1105	184	22	524	227	144	12	0.1	0.05	12.9	0.50	0.1	
3 Mt. Roche	6 Jul 88	1430	10.5	133.0	8.10	958	102	5	619	113	43	14	0.1	0.05	7.27	0.53	0.5	
23 Kuparuk	3 Jul 90	1000	12.0	29.2	7.10	200	38	10	95	35	33	8	0.1	0.01	0.37	1.28	9.4	
33 Oksrukuyik	7 Jul 88	1430	13.9	182.0	8.30	1743	65	12	906	105	38	12	0.2	–	–	–	0.2	
38 Oksrukuyik	7 Jul 88	1700	15.0	37.9	7.28	298	37	2	125	46	28	10	0.1	0.03	0.31	1.62	0.4	
39 Sagavinirktok	7 Jul 88	1540	15.0	148.0	8.21	1163	27	5	324	189	10	2	0.6	–	–	–	0.2	
40 Alexa	10 Jul 88	–	15.0	–	7.43	1120	3	5	519	56	23	3	0.4	0.10	0.69	33.5	–	
41 Sagcut	10 Jul 88	–	15.0	–	8.07	1434	156	10	636	230	57	9	0.1	0.02	17.1	1.36	–	
43 Happy Valley	10 Jul 88	–	15.0	–	7.63	426	3	5	212	39	18	3	0.1	0.07	4.08	0.89	–	
56 Kuparuk	9 Jul 88	1130	11.2	131.4	8.14	1193	9	12	511	87	70	10	0.2	0.09	2.52	0.98	–	

ble 7). Samples were washed through a 200 μm mesh net. Residues were preserved with 95% ethanol. Chironomidae were sorted from residues and identified to genus or species. For several foothills lakes, more extensive surveys were conducted independently using diver cores or an Ekman grab (Hershey, 1985, Goyke & Hershey, this issue, Hershey, in press). Taxa whose estimated abundance exceeded 50 mg m^{-2} were reported as abundant, other taxa were reported as present or absent.

Results and discussion

Physics

The temperature of surface waters decreased from south to north along with air temperature. July is usually the warmest month on the North Slope, and midday water temperatures at this time of year ranged from 14 to 22 °C in the foothills and from 6 to 14 °C near the coast (Table 3).

In most of the shallow lakes light penetrated to the bottom and secchi depth was equal to maximum depth. In deeper lakes, secchi depth ranged from 2.5 to 9.0 m, and was similar to values in other arctic lakes (Hobbie, 1984). The variability in light transmission reflects the amount of suspended inorganic and organic material as well as the dissolved organic material (Hobbie, 1984; Miller *et al.*, 1986).

Chemistry

The rainfall at Toolik had a mean pH of 5.21 and conductivity of 2.9 $\mu S\ cm^{-1}$ in 1989 and 1990 (Table 4). The dominant ions in this acid, dilute water were Ca^{2+}, HCO_3^-, and NH_4^+. The most dilute rain reported for Alaska is at the subarctic Poker Flats site in taiga forest (Table 4). The enrichment in Ca and Mg of rain at Toolik compared to Poker Flats may be related to the proximity of limestone and dolomite outcrops in the Brooks Range, about 20 km south of Toolik. Further enrichment of ions is seen in rain at Saqvaqjuac on the northern coast of Canada. Here the effects of sea salt are most noticable and there are much higher concentrations of Na, Mg, Cl, and SO_4 compared to the rain at Toolik. There is less information on the chemical composition of snow in the Arctic, although it appears that the proportions and concentrations of ions in snow are similar to those in rain (Hobbie, 1984; Cornwell, this issue). Even less information exists on the chemistry of dry deposition. One sample from Toolik showed enrichment over rain in most ions, especially K, SO_4, and nutrients (Table 4).

In general the surface waters were also dilute (Table 3). Conductivity ranged from 30 to 843 $\mu S\ cm^{-1}$, which is similar but slightly higher than previously reported values for the Arctic (17 to 322 $\mu S\ cm^{-1}$; Rodhe, 1949; Kalff, 1968; Craig & McCart, 1975; Hobbie, 1984). The higher conductivity waters in our study were found within 2 km of the Arctic Ocean. There are also some unusual meromictic lakes along the coast on islands in the Canadian high arctic that contain trapped seawater and have very high ionic strengths (Jeffries *et al.*, 1984; Ouellet *et al.*, 1989). Perennial groundwater springs may have conductivities up to 4546 $\mu S\ cm^{-1}$, although most spring waters fall in the conductivity range of surface waters (Craig & McCart, 1975). The pH of all waters studied here was much less acidic than the pH of rain, and ranged from 6.85 to 8.46 (mean = 7.68).

The surface waters fall into two distinct groups of chemical composition. Within 20 km of the coast, but excluding rivers that drain larger areas, waters had Cl followed by HCO_3 as the dominant anions, and Na and Ca as the dominant cations. Further inland the waters were dominated by HCO_3, and Ca and Mg. The sea is the major influence on the relative proportions of ions in coastal waters. Ion ratios in these lakes are similar to seawater and sea spray, except for enrichment in Ca and HCO_3 (Table 5). This enrichment may be due to eolian input of Ca-rich silt from the nearby Sagavanirktok River delta (Walker & Webber, 1979). Because permafrost acts as a seal and prevents subsurface drainage, these lakes are similar to other coastal lakes that are perched or

Table 4. Rain chemistry (μmol L^{-1}; μeq L^{-1} for alkalinity; μS cm^{-1} for conductivity) for (1) Toolik Lake site (wet only) with volume weighted means (VWM): (2) Saqvaqjuac, Northwest Territories (VWM, bulk precipitation, 60 °N 90 °W, Welch & Legault 1986); (3) Poker Flats Alaska site (VWM, wet only, 67 °N 147 °W, Galloway *et al.*, 1982); (4) Sag River site (VWM, bulk precipitation, 68 °N 148 °N, 1986). Dry deposition values are from Toolik Lake site on 2 Sep 89 and from Sag River site on 7 Jul 86. Sag River site data are from G. Shaver (pers. comm.). Symbol '<' indicates values below detection; blank spaces indicate missing values. The relatively high alkalinity value at pH 4.34 (Toolik, 24 Jul 90) probably indicates the presence of organic acids with pK values less than 4.4.

Date	(μS cm^{-1})	(μmol L^{-1})										
	Cond	pH	Ca	Mg	Na	K	Cl	SO_4	Alk	PO_4	NO_3	NH_4
Toolik												
19 Jul 89	0.5	5.67	4.3	0.1	0.8	0.3	1.1	<		1.1	0.9	0.6
27 Jul 89	1.2	5.12	2.6	<	0.4	0.8	1.7	0.8	9.5	<	2.1	<
8 Aug 89	1.6	5.73	3.0	0.2	0.4	0.3	1.0	1.7	9.8	<	1.4	<
1 Sep 89	3.0	6.19	13.3	1.2	3.3	2.2	5.9	3.1		<	3.0	<
25 Jun 90	4.3	4.98	2.1	0.2	0.3	0.1	1.5	4.9	13.1	<	2.0	<
5 Jul 90	5.4	4.53	3.4	0.3	<	0.4	3.8	3.9	15.5	<	3.9	<
14 Jul 90	3.0	5.75	0.6	0.2	1.4	2.2	3.4	1.9		<	4.3	14.9
24 Jul 90	23.0	4.34	7.6	1.0	6.3	4.8	20.9	14.3	53.4	<	30.1	4.4
4 Aug 90	3.3	5.44	3.1	0.5	4.5	7.8	13.6	1.2		<	3.2	7.7
24 Aug 90	3.3	5.73	0.8	0.2	0.6	1.8	2.7	1.7	14.3	<	5.2	18.3
VWM =	2.9	5.21	4.2	0.4	1.1	1.0	2.7	2.7	12.2	1.1	2.7	10.7
Saqvaqjuac	8.9	4.80	4.7	2.9	21.7	1.3	25.1	8.5			0.4	1.9
Poker Flats		4.96	0.05	0.1	1.0	0.3	2.6	3.6			1.9	1.1
Sag River			0.9		2.8	12.2						
Dry Deposition												
2 Sep 89	46.3	7.11	15.2	12.0	21.7	36.0	24.5	37.4		21.9	30.4	28.0
7 Jul 86			2.7		15.1	8.8				2.0	4.5	2.4

isolated from large groundwater inputs; the chemistry of such coastal lakes is determined mainly by atmospheric inputs (Kling, 1986).

In contrast to the input of sea salt to lakes near the coast, atmospheric precipitation had little influence on the chemistry of inland lakes and rivers. In the inland waters there was substantial enrichment of all cations and HCO_3, and slight enrichment of SO_4 compared to the proportions of these ions found in rain (Table 5). This enrichment is due to weathering of bedrock, and is particularly evident in the lakes that are known to lie in or near the Lisburne limestone-dolomite formation (Table 5). Rock weathering plus the solution of evaporites explains most of the variation in surface water chemistry of the Mackenzie River basin in arctic Canada (Reeder *et al.*, 1972) and in other parts of the world (Meybeck, 1987). What is unusual in many of the waters studied here is the strong enrichment of ions in comparison to the nearby Mackenzie or to the average composition of the world's rivers. For example, Ca, Mg, and HCO_3 are enriched by at least an order of magnitude in the North Slope rivers versus the Mackenzie (Table 5). This difference may be due in part to the influence of the Brooks Range in Alaska. The Mackenzie River drains a basin of low to moderate relief, composed primarily of muskeg tundra as well as part of the Canadian Precambrian shield (Reeder *et al.*, 1972). The Sagavinirktok River, on the other hand, drains an area of relatively high relief, composed in large part of exposed rock in the Brooks Range. The headwaters of the Kuparuk River, although geographically adjacent to the Sagavanirktok, are isolated from such exposed rock, and the chem-

Table 5. Average molar ion ratios in seawater, Toolik site rain, lakes less than 20 km from the coast, lakes more than 20 km from the coast, lakes associated with limestone outcrops, rivers, lakes lying in old glacial till (sites 7, 13, 15, 22) and in new glacial till (sites 11, 16, 17), the Kuparuk River (site 23), the Sagavinirktok River (site 39), the Mackenzie River (Reeder *et al.*, 1972), and the average for world rivers (Meybeck, 1987).

Site	N	Na:Cl	Ca:Cl	Mg:Cl	K:Cl	SO_4:Cl	Alk:Cl
Seawater		0.86	0.02	0.10	0.02	0.05	0.004
Coastal lakes	9	0.67 ± 0.05	1.0 ± 0.3	0.25 ± 0.07	0.03 ± 0.01	0.08 ± 0.05	1.9 ± 0.6
Toolik rain	10	0.41	1.6	0.15	0.37	1.0	4.5
Inland lakes	30	3.4 ± 0.3	85 ± 14	16 ± 3	2.0 ± 0.3	4.4 ± 0.9	142 ± 24
Rivers	11	5.8 ± 1.0	60 ± 10	17 ± 3	1.4 ± 0.4	9.8 ± 2.7	132 ± 20
Limestone lakes	3	3.0 ± 0.1	117 ± 11	26 ± 5	1.4 ± 0.2	15 ± 6	227 ± 24
Lakes new till	3	4.4 ± 1.1	133 ± 36	26 ± 8	2.0 ± 0.0	7.8 ± 4.7	229 ± 27
Lakes old till	4	2.7 ± 0.2	31 ± 3	6.5 ± 0.5	1.3 ± 0.1	1.6 ± 0.5	64 ± 5
Sagavinirktok		2.0	65	38	0.40	5.4	233
Kuparuk		3.3	9.5	3.5	0.80	3.8	20
Mackenzie		1.2	3.3	1.7	0.11	1.5	7.3
World rivers		1.8	3.9	1.5	0.37	1.0	9.9

istry of the Kuparuk is similar to that of the Mackenzie (Table 5). Craig & McCart (1975) and Lock *et al.* (1989) also found higher conductivity streams in the Brooks Mountains than in the foothills province.

Another influence on surface water chemistry is the age of glacial drift in a drainage basin and thus the extent of primary rock exposure, soil development, and weathering. Much of the foothills province on the North Slope was heavily glaciated at some time in the Pleistocene, and Hamilton (1982) has presented a glacial sequence with at least five major stages. The glaciers carried rocks northward out of the Brooks Range to the foothills, and the influence of the age of glacial drift is seen clearly in the chemistry of several lakes around Toolik. Toolik Lake is a complex kettle formed by melting ice blocks in a terminal morraine of the relatively young Itkillik II glaciation (24000 to 15000 yr BP). Most of the drainage basin of Toolik Lake, and the land surface to the immediate east of Toolik, consists of glacial drift from the older Itkillik I glaciation (100000 yr BP). Lakes lying in the younger drift to the west have higher conductivities and are strongly enriched in Ca, Mg, and HCO_3 when compared to lakes less than 1 km to the east but surrounded by the older more weathered drift (Table 5; Fig. 1). This relationship between water chemistry and the age of the land surface can be extended by examining the Kuparuk River further to the east. The upper Kuparuk basin (area = 143 km^2; site 23) is covered by drift from the much older Sagavanirktok glaciation (mid-Pleistocene; Hamilton, 1982). The water draining this basin is the most dilute and has the lowest ion enrichments over rainfall of any surface water we studied (Table 3, 5). In addition, weathering rates in the Kuparuk basin as measured by the export of cations are as low as for any arctic or temperate zone site yet examined (Cornwell, this issue).

Our mineral equilibria calculations indicate that the product of Ca^{2+} and CO_3^{2-} concentrations was from 1.3 to 4.0 times the value theoretically required for calcite ($CaCO_3$) precipitation at eight sites (4, 11, 28, 30, 33, 34, 48, 51). It is unlikely, however, that such mineral precipitation exerts a strong influence on the overall major ion composition of surface waters on the North Slope. This is because saturation indices are only approximations of actual chemical equilibria, and a slightly positive index value does not demonstrate mineral precipitation (Nordstrom and Ball, 1989). In addition, our field measurements were made during midday when photosynthesis is maximum and thus pH and CO_3^{2-} concentrations are highest and precipitation is most likely. Finally, the amount of Ca as a percentage of total cations does not decrease as Ca concentrations increase across all inland sites, as would be expected if Ca is selectively removed by calcite precipitation.

Concentrations of dissolved nutrients were in general low; the mean values were 0.16 μM for SRP, 3.3 μM for NO_3^-, and 1.7 μM for NH_4^+. These low concentrations are expected because algal productivity is often limited by phosphorus alone or by co-limitation of phosphorus plus nitrogen in arctic surface waters (Hobbie, 1980; Miller *et al.*, 1986; Whalen & Cornwell, 1985). In addition to this biotic control, several studies have described the inorganic adsorption of phosphorus on iron hydroxides in some arctic lakes (Prentki *et al.*, 1980; Cornwell, 1987; Sugai & Kipphut, this issue). This mechanism may ultimately limit the regeneration of phosphorus from sediments, although it is unclear how widespread the phenomenon is in arctic surface waters. A third potential control on dissolved nutrient concentrations is the mineral precipitation of PO_4 with calcium. Mineral equilibria calculations on the 41 sites where Ca^{2+} and SRP data were available indicate that all waters were supersaturated with respect to hydroxyapatite [$Ca_6(PO_4)_3OH$, $pK_{sp} = 55.9$] by 5 to 15 orders of magnitude. Clearly the SRP concentrations are not controlled by hydroxyapatite formation in these waters. There are, however, reports of more amorphous mineral forms, such as β-tricalcium phosphate [β-$Ca_3(PO_4)_2$, $pK_{sp} = 22.0$], that control concentrations of dissolved PO_4 (Ferguson *et al.*, 1971; Snoeyink & Jenkins, 1980). Such amorphous mineral precipitation may play a role in controlling levels of PO_4 in arctic waters – we found 13 lakes that were supersaturated by up to 12 times with respect to β-tricalcium phosphate (sites 5, 15–17, 19, 20, 28–32, 36, and 52). At present the relative importance of these biotic and abiotic controls on nutrient concentrations, as well as the seasonal variation of each control, is unknown in arctic freshwaters.

Seasonal variations in surface water chemistry are related mainly to dilution by snowmelt and runoff, and to concentration by evaporation and by exclusion from ice. These mechanisms can increase or decrease chemical concentrations of major ions in arctic lakes and rivers by up to 30 fold (Howard & Prescott, 1973; Schindler *et al.*, 1974; Prentki *et al.*, 1980; Cornwell, this issue). It is unlikely, however, that such seasonal variations strongly affected the major chemical groupings described above because all lakes were sampled during a short time span in late June and early July. Year to year variations of alkalinity in the foothills lakes Toolik and N2 are reported by Kling *et al.* (this issue); maximum concentration changes over periods of 5 to 14 years were about 2 fold, although typically the variability was much less than that.

Zooplankton

We identified 11 species of zooplankton in samples from these lakes (Table 6). *Daphnia middendorphiana* was the most common cladoceran and *Diaptomus pribilofensis* was the most common copepod found. All species are Holarctic in distribution (Roen, 1962; Tash & Armitage, 1967; O'Brien, 1975; Moore, 1978; Haney & Buchanan, 1987).

The most striking overall pattern of species distribution is the paucity of cladocerans in the coastal plain lakes. It is possible that overwintering embryos had not yet emerged at the time of our sampling in early July, although Stross *et al.* (1980) found that *Daphnia middendorphiana* embryos hatched at ice-out and produced a brood of young by mid-July in the tundra ponds at Barrow, further north than our lakes. Low temperatures and short growing seasons associated with high latitudes are thought to decrease the diversity and limit the distribution of cladocerans in the Arctic (Poulsen, 1940; Moore, 1978; Hebert & Hann, 1986), which is consistent with our findings. Haney & Buchanan (1987) also suggest that photoperiod and light transmission in water may limit the distribution of *Daphnia* in the Arctic.

We found no significant correlations with any

Table 6. Zooplankton species composition in the study lakes; '×' indicates presence and '–' indicates absence.

Lake	*Daphnia middendorphian*	*Daphnia longiremis*	*Bosmina longirostris*	*Holopedium gibberum*	*Polyphemus pediculus*	*Heterocope septentrionalis*	*Diaptomus pribilofensis*	*Cyclops scutifer*	*Cyclopoid*	*Chaoborus*	*Brachinecta*	*Polyartemiella*
5 Galbraith	×	–	–	–	–	×	×	–	–	–	–	–
6 Island	–	×	–	–	–	–	×	×	–	–	–	–
7 I8	–	–	×	×	–	×	×	×	–	–	–	–
8 S13	×	–	–	–	×	–	×	×	–	×	×	–
9 S12	×	–	–	–	–	×	×	–	–	×	×	–
10 S11	×	–	–	–	–	×	×	–	×	×	–	–
11 S6	–	×	×	–	–	×	×	–	×	–	–	–
12 S3	×	–	×	×	×	×	×	×	×	–	–	–
13 Toolik	×	×	×	×	×	×	×	×	–	–	–	–
14 Dam	×	–	–	–	–	×	×	×	–	–	–	–
15 Camp	×	–	–	–	–	×	×	×	–	–	–	–
16 N1	×	–	–	–	–	×	×	×	–	–	–	–
17 N2	–	×	×	–	×	–	×	×	–	–	–	–
18 NE2	–	×	×	×	×	×	×	×	–	–	–	–
19 NE12	×	–	–	–	–	×	×	×	–	–	–	–
20 NE14	×	–	×	–	–	×	×	–	–	–	–	–
21 Itigaknit	×	–	–	–	–	×	×	×	–	–	–	–
22 E1	–	×	×	–	–	–	×	×	–	–	–	–
24 Campsite	×	×	–	–	×	×	×	×	–	–	–	–
25 O1	–	–	–	×	–	×	×	–	–	–	–	–
26 O2	–	–	–	×	–	×	×	×	–	–	–	–
27 O3	–	×	–	×	–	–	×	×	–	–	–	–
28 Elusive	×	×	–	–	–	×	–	×	–	–	–	–
29 Moose	×	–	×	–	×	–	×	×	–	–	–	–
30 Hanging	×	×	–	–	–	–	×	×	–	–	–	–
31 Foggy	–	×	×	–	–	–	×	×	–	–	–	–
32 F4	×	–	–	–	–	×	×	–	×	–	–	–
34 George	×	–	–	–	–	×	×	–	×	–	–	–
35 Anne	×	–	–	–	–	×	×	–	×	×	–	–
36 Sag2	×	–	–	–	–	×	×	–	–	–	–	–
37 Sag1	×	–	–	–	–	×	×	–	–	×	–	×
42 Charles	×	–	–	×	–	×	×	×	–	×	–	–
44 William	–	–	–	–	–	×	–	–	–	–	–	×
45 Windy	×	–	–	–	–	×	–	–	×	–	–	–
46 Silhouette	×	–	–	–	–	–	–	–	×	–	×	–
47 Colleen	×	–	–	–	–	×	×	×	–	–	×	–
48 Sag-c pit	×	–	–	–	–	–	×	–	–	–	–	–
49 Bern	×	–	–	–	–	–	×	×	–	–	×	–
50 Dune	–	–	–	–	–	–	–	–	×	–	–	–
51 Maxine	×	–	–	–	–	×	×	×	–	–	×	–
52 Carolyn	×	–	–	–	–	–	×	–	–	–	×	–
53 Africa	×	–	–	–	–	×	–	–	–	–	×	–

species and chlorophyll a concentration in the lake, although our data consist only of single time points and chlorophyll concentrations may be quite variable seasonally. Similarly, there were very few significant correlations between individual species and chemical variables. The presence of *Holopedium gibberum* was negatively correlated with conductivity and the dominant ions contributing to ionic strength: HCO_3^-, Ca^{2+}, and Mg^{2+} (Spearman Rank, $P<0.001$ for all). We know of no field studies or experiments demonstrating the intolerance of *Holopedium* to high conductivity. Some rotifers are intolerant of waters with conductivities above about 400 μS cm^{-1} (less than our study lakes; Green, 1986), but some copepods appear to be limited only by much more concentrated waters (LaBabera & Kilham, 1974). It is also possible that these correlations are spurious and result from the collinearity of latitude and ionic strength; *Holopedium* is absent in coastal lakes, which as a group have much higher conductivities than inland lakes.

O'Brien *et al.* (1979) summarized the zooplankton community structure in arctic lakes and ponds. In general, lakes without planktivorous fish contain mainly large-bodied zooplankton species, while lakes with fish contain mainly small-bodied species. The presence of fish is related to lake depth, with lakes shallower than 3 to 4 m generally lacking fish. In the present study we found that, for example, presence of the large-bodied *Brachinecta* was negatively correlated with depth ($r = -0.49$, $P<0.002$), while the small-bodied *Daphnia longiremis* was positively correlated with depth ($r = 0.44$, $P<0.005$). There are some exceptions, however, in that several of the lakes in the Toolik area contain either fish or no fish as well as both large and small-bodied zooplankton. The exclusion of large-bodied zooplankton by fish in these lakes appears related to the densities of predator and prey, and to the efficiency of predation (O'Brien *et al.*, 1979). Our survey results have extended these general conclusions to about 20 new lakes on the North Slope (Table 6). O'Brien *et al.* (1979) found Toolik Lake to be especially anomalous because it contained both large and small-bodied cladocerans and copepods in the presence of planktivorous fish. In addition to Toolik, we have identified two other foothills sites containing fish and with a similar planktonic zooplankton community: lakes Elusive and Campsite. These three lakes have quite different chemistries, but they are all relatively large, relatively deep, or both. It is unclear at present how these common characteristics influence zooplankton community structure, although predator avoidance in habitats made complex by depth or area may play a role.

Benthos

The distribution of Chironomidae among lakes surveyed is presented in Table 7. Because only a single Ekman sample was collected for many of the lakes, these data represent the most abundant taxa and provide only a general indication of chironomid diversity and distribution.

Stictochironomus rosenschoeldi, *Heterotrissocladius spp.*, and *Procladius spp.* were the most widely distributed taxa. *Stictochironomus* and *Heterotrissocladius* usually co-occurred and were abundant when present in foothills lakes (Table 7). However, *Heterotrissocladius* was found in only the largest of the coastal plain lakes (#47, Lake Colleen), whereas *Stictochironomus* appeared to be even more widespread on the coastal plain than in the foothills (Table 7). *Stictochironomus* is associated with a wider range of trophic conditions than *Heterotrissocladius*; it is indicative of ultraoligotrophic to mesotrophic conditions, whereas *Heterotrissocladius* is generally associated with ultraoligotrophic to oligotrophic conditions (Saether, 1979; Pinder & Reiss, 1983; Cranston *et al.*, 1983).

Procladius is widely distributed across lentic habitats worldwide (Fittkau & Roback, 1983). Goyke & Hershey (this issue) suggest that slimy sculpin, a benthic feeding fish, are important in determining *Procladius* abundance. For example, in foothills lakes predatory chironomids (primarily *Procladius*) were significantly less abundant in lakes with sculpin than in lakes without sculpin. In the present survey *Procladius* appeared to be

Table 7. Chironomidae from North Slope lakes. Lake numbers correspond to sites in Table 1. Asterisks indicate data from Hershey (1985b, in press) and Goyke & Hershey (in press). 'A' indicates abundance >50 mg m^{-2}, and 'X' indicates abundance <50 mg m^{-2}.

Lake	5	6	7*	11	12	13*	16	17*	19*	20	21	22	34	43	45	47	49	51	52	53
Chironomini																				
Chironomus					A			X						X					A	
Cryptochironomus													X			X	X	A		
Pagastiella											A		X					X		
Parachironomus						X														
Phaenopsectra								X								A				
Polypedilum			X			X		X	X											
Stictochironomus rosenschoeldi	A		A			A	A	X	A	X		X			A	X	A	A	X	A
Tanytarsini																				
Cladotanytarsus			X										A							
Constempellina brevicosa						X			X											X
Corynocera ambigua				A			X	X			X			A						
Corynocera oliveri														X		X			A	
Micropsectra						X		X				X								
Paratanytarsus		X	X			A	X	X	X			X			X					
Stempellinella	X		X			X		X	X			X								
Tanytarsus			A	X	X	A		X			X	X			X					X
Tanypodinae																				
Procladius	A	X	A	X	X	X	X	X	A		X	A	A	X	A	X	X	A	A	A
Ablabesmyia						X		X				X				X				
Arctopelopia		X				X														
Prodiamesinae																				
Monodiamesa bathyphila			A			A		A	A	X										
Diamesinae																				
Protanypus saetheri						X	X	X	X			X								
Orthocladiinae																				
Abiskomyia virgo						X														
Corynoneura						X														
Cricotopus			X			X		X				X								
Eukiefferiella							X					X								
Heterotrissocladius	A		A			A	A	X	A			A				X				
Hydrobaenus																X				
Mesocricotopus	X																			
Limnophyes																				X
Paracladius quadrinodosus			A			X		X												
Parakiefferiella	X		X	X		X		X	X			A							X	
Psectrocladius			X			X		A	A						X		X		X	X
Theinemanniella												X								
Zalutschia			X			X	X		X											

abundant (>50 mg/m^2) in six of the nine lakes either known to lack sculpin (lake 22) or presumed to lack sculpin due to shallow depth (lakes numbered >34). By comparison, *Procladius* was abundant in only three of the ten deeper lakes in the foothills (Table 7), and sculpin occur in at least nine of these (Hanson *et al.*, this issue; lake 5 has not been surveyed for sculpin).

Of the less frequently encountered taxa, most appeared in both foothills and coastal plain lakes (Table 7). The distribution of these taxa is Holarctic, and they are generally associated with oligotrophic conditions (Saether, 1979, 1983; Cranston *et al.*, 1983; Fittkau & Roback, 1983; Pinder & Reiss, 1983). However, *Corynocera oliveri* has been previously reported only in the Palaearctic (Pinder & Reiss, 1983) except as a subfossil (Walker & Mathewes, 1988). *Eukiefferiella* and *Thienemanniella* are typically lotic, thus their occurrence in foothills lakes E1 and N1 (*Eukiefferiella* only) is unusual.

Summary

We examined the biogeochemistry of 45 lakes and 8 rivers on the coastal plain and the foothills region of arctic Alaska. Water chemistry was controlled by three major factors: (1) rainfall and sea spray strongly affected the concentrations and ratios of major ions within 2 km of the Arctic Ocean, and dust blown from the Sagavinirktok delta enriched some lakes in Ca^{2+} and HCO_3^- relative to the proportions found in sea water; (2) proximity to the Brooks Range mountains and especially to limestone and dolomite outcrops produced very distinctive calcium and magnesium rich waters in several lakes and rivers; and (3) increasing age and thus extent of weathering of glacial drift surrounding lakes in the foothills region determines the overall ionic strength of surface waters – the most dilute lakes and rivers lie in the oldest terrain. Nutrient concentrations are controlled mostly by biological uptake, although adsorption of P on iron hydroxides and precipitation of amorphous calcium phosphate minerals may also be important. We identified 11 species of zooplankton and 32 species of chironomids from the study lakes. These taxa are in general widely distributed throughout the Neararctic. The effects of lower temperatures with increasing latitude seem to explain more of the variation in zooplankton species distribution than does water chemistry. Biotic interactions with zooplanktivorous fish are also a major determinant of zooplankton community composition and size structure. Although large-bodied zooplankton are nearly always eliminated in lakes with fish, we have identified three lakes that are exceptions to this generalization; it is at present unclear how the common factors of large area, great depth, or both shared by these lakes affect the zooplankton community. Fish predation may determine the abundance of the common chironomid *Procladius* in these lakes.

Acknowledgements

We thank G. Kipphut, B. Moller, C. Bauman, B. Peterson, W. Rowe, K. Foreman, K. Regan, M. Dornblaser, J. Mcrae, and J. Laundre for help in the field and lab. This research was supported by the National Science Foundation grants BSR8702328, DPP8722015, and DPP8320544, and by the A. W. Mellon Foundation.

References

Brown, J. & R. A. Kreig (Eds.), 1983. Guidebook to permafrost and related features along the Elliott and Dalton Highways, Fox to Prudhoe Bay, Alaska. Fourth International Conference on Permafrost, July 18–22, 1983, University of Alaska, Fairbanks. 230 pp.

Brown, J. & R. L. Berg (Eds.), 1980. Environmental engineering and ecological baseline investigations along the Yukon River-Prudhow Bay Haul Road. U.S. Army Cold Regions Research & Engineering Lab Report 8019. 203 pp.

Cornwell, J. C., 1987. Phosphorus cycling in arctic lake sediment: adsorption and authigenic minerals. Arch. Hydrobiol. 109: 161–179.

Cornwell, J. C., 1992. Cation export from Alaskan arctic watersheds. Hydrobiologia 240: 15–22.

Craig, P. C. & P. J. McCart, 1975. Classification of stream types in Beaufort Sea drainages between Prudhoe Bay, Alaska, and the MacKenzie delta, N.W.T., Canada. Arctic Alpine Res. 7: 183–198.

Cranston, P. S., D. R. Oliver & O. A. Saether, 1983. The larvae of Orthocladiinae (Diptera: Chironomidae) of the Holarctic region – Keys and diagnoses. (Wiederholm, ed.) Ent. Scand. Suppl. 19: 149–291.

Ferguson, J. F., D. Jenkins & A. B. Menar, 1971. Chemical processes for phosphate removal. Wat. Res. 5: 369–381.

Fittkau, E. J. & S. Roback, 1983. The larvae of Tanypodinae (Diptera: Chironomidae) of the Holarctic region – Keys and diagnoses. (Wiederholm, ed.) Ent. Scand. Suppl. 19: 33–110.

Galloway, J. N., G. E. Likens, W. C. Keene & J. M. Miller, 1982. The compostition of precipitation in remote ares of the world. J. Geophys. Res. 87: 8771–8786.

Green, J., 1986. Associations of zooplankton in six crater lakes in Arizona, Mexico and New Mexico. J. Zool., London 208: 135–159.
Haney, J. & C. Buchanan, 1987. Distribution and biogeography of *Daphnia* in the arctic. Mem. Ist. Ital. Idrobiol. 45: 77–105.
Haugen, R. K., 1982. Climate of remote areas in north-central Alaska: 1975–1979 summary. U.S. Army Cold Regions Research and Engineering Lab Report 82-35. 114 pp.
Hebert, P. D. N. & B. J. Hann, 1986. Patterns in the composition of Arctic tundra pond microcrustacean communities. Can. J. Fish. Aquat. Sci. 43: 1416–1425.
Hershey, A. E., 1985. Littoral chironomid communities in an arctic Alaskan Lake. Holarctic Ecology 8: 39–48.
Hershey, A. E., 1990. Snail populations in arctic lakes: competition mediated by predation? Oecologia 82: 26–32.
Hobbie, J. E. (Editor), 1980. Limnology of Tundra Ponds. Dowden, Hutchinson & Ross, Stroudsburg, PA. 515 pp.
Hobbie, J. E., 1984. Polar limnology, pp. 63–105 *In* F. B. Taub (ed.), Lakes and Reservoirs. Elsevier, Amsterdam.
Howard, H.H. & G.W. Prescott, 1973. Seasonal variation of chemical parameters in Alaskan tundra lakes. American Midland Naturalist 90: 154–164.
Jeffries, M. O., H. R. Krouse, M. A. Shakur & S. A. Harris, 1984. Isotope geochemistry of stratified Lake 'A', Ellesmere Island, N.W.T., Canada. Can. J. Earth Sci. 21: 1008–1017.
Kalff, J., 1968. Some physical and chemical characteristics of arctic fresh waters in Alaska and northwestern Canada. J. Fish. res. Bd Canada 25: 2575–2587.
Kling, G. W., 1986. The physicochemistry of some dune ponds on the Outer Banks, North Carolina. Hydrobiologia 134: 3–10.
LaBarbera, M. C. & P. Kilham, 1974. The chemical ecology of copepod distribution in the lakes of East and Central Africa. Limnol. Oceanogr. 19: 459–465.
Lock, M. A., T. E. Ford, D. M. Fiebig, M. C. Miller, M. Hullar, M. Kaufman, J. R. Vestal, B. J. Peterson & J. E. Hobbie, 1989. A biogeochemical survey of rivers and streams in the mountains and foot-hills province of arctic Alaska. Arch. Hydrobiol. 114: 499–521.
Maciolek, J. A., 1989. Tundra ponds of the Yukon Delta, Alaska, and their macroinvertebrate communities. Hydrobiologia 172: 193–206.
Meybeck, M., 1987. Global chemical weathering of surficial rocks estimated from river dissolved loads. Am. J. Science 287: 401–428.
Miller, M. C., G. R. Hater, P. Spatt, P. Westlake & D. Yeakel, 1986. Primary production and its control in Toolik Lake, Alaska. Arch. Hydrobiol. Suppl. 74: 97–131.
Miller, M. C. & J. R. Stout, 1989. Variability of macroinvertebrate community composition in an arctic and subarctic stream. Hydrobiologia 172: 111–128.
Moore, J. W., 1978. Composition and structure of zooplankton communities in eighteen arctic and subarctic lakes. Int. Revue ges. Hydrobiol. 63: 545–565.
Nordstrom, D. K. & J. W. Ball, 1989. Mineral saturation states in natural waters and their sensitivity to thermodynamic and analytic errors. Sci. Géol., Bull. 42: 269–280.
O'Brien, W. J., 1975. Some aspects of the limnology of the ponds and lakes of the Noatak drainage basin, Alaska. Verh. int. Ver. Limnol. 19: 472–479.
O'Brien, W. J., C. Buchanan & J. F. Haney, 1979. Arctic zooplanton community structure: exceptions to some general rules. Arctic 32: 237–247.
Oliver, D. R., 1983. The larvae of Diamesinae (Diptera: Chironomidae) of the Holarctic region – Keys and diagnoses. (Wiederholm, ed.) Ent. Scand. Suppl. 19: 115–138.
Ouellet, M., M. Dickman, M. Bisson & P. Pagé, 1989. Physico-chemical characteristics and origin of hypoersaline meromictic Lake Garrow in the Canadian High Arctic. Hydrobiologia 172: 215–234.
Pinder, L. C. & F. Reiss, 1983. The larvae of Chironominae (Diptera: Chironomidae) of the Holarctic region – Keys and diagnoses. (Wiederholm, ed.) Ent. Scand. Suppl. 19: 293–435.
Plummer, L. N. & E. Busenberg, 1982. The solubilities of calcite, aragonite and vaterite in CO_2-H_2O solutions between 0 and 90 °C, and an evaluation of the aqueous model for the system $CaCO_3$-CO_2-H_2O. Geochim. Cosmochim. Acta 46: 1011–1040.
Poulsen, E. M. 1940, The zoology of East Greenland. Freshwater entomostraca. Medd. Gronl. 121: 1–71.
Prentki, R. T., M. C. Miller, R. J. Barsdate, V. Alexander, J. Kelley & P. Coyne, 1980. Chemistry, pp. 76-178 *In* J. E. Hobbie (ed.), Limnology of Tundra Ponds. Dowden Hutchinson & Ross, Stroudsburg.
Reeder, S. W., B. Hitchon & A. A. Levinson, 1972. Hydrogeochemistry of the surface waters of the Mackenzie River drainage basin, Canada – I. Factors controlling inorganic composition. Geochim. Cosmochim. Acta 36: 825–865.
Rodhe, W., 1949. The ionic composition of lake waters. Verh. int. Ver. Limnol. 10: 377–386.
Roen, U., 1962. Studies on freshwater entomostraca in Greenland. II. Localities, ecology and geographical distribution of species. Medd. Gronl. 170: 1–249.
Saether, O. A., 1979. Chironomid communities as water quality indicators. Holarctic Ecology 2: 65–74.
Saether, O. A., 1983. The larvae of Prodiamesinae (Diptera: Chironomidae) of the Holarctic region – Keys and diagnoses. (Wiederholm, ed.) Ent. Scand. Suppl. 19: 141–147.
Schindler, D. W., H. E. Welch, J. Kalff, G. J. Brunskill & N. Kritsch, 1974. Physical and chemical limnology of Char Lake, Cornwallis Island (75°N Lat.). J. Fish. Res. Bd Can. 31: 585-607.
Snoeyink, V. L. & D. Jenkins, 1980. Water Chemistry. John Wiley & Sons, New York, 463 pp.
Stross, R. G., M. C. Miller & R. J. Daley, 1980. Zooplankton: communities, life cycles, and production. pp. 251–296 *In* J. E. Hobbie (ed.), Limnology of Tundra Ponds. Dowden, Hutchinson & Ross, Inc., Stroudsburg.
Stumm, W. & J. J. Morgan, 1981. Aquatic Chemistry, 2nd edition. John Wiley & Sons, New York, 780 pp.

Tash, J. C. & K. B. Armitage, 1967. Ecology of zooplankton of the Cape Thompson area, Alaska. Ecology 48: 129–138.

Walker, D. A. & P. J. Webber, 1979. Relationships of soil acidity and air temperature to the wind and vegetation of Prudhoe Bay, Alaska. Arctic 32: 224–236.

Walker, I. R. & R. W. Mathewes, 1988. Late-quaternary fossil Chironomidae (Diptera) from Hippa Lake, Queen Charlotte Islands, British Columbia, with special reference to Corynocera Zett. Can. Ent. 120: 739–751.

Welch, H. E., 1976. Ecology of Chironomidae (Diptera) in a Polar Lake. J. Fish. Res. Bd Can. 33: 227–247.

Welch, H. E. & J. A. Legault, 1986. Precipitation chemistry and chemical limnology of fertilized and natural lakes at Saqvaqjuac, N.W.T. Can. J. Fish. aquat. Sci. 43: 1104–1134.

Wetzel, R. G. & G. E. Likens, 1979. Limnological Analyses. W. B. Saunders, Philidelphia. 357 pp.

Whalen, S. C. & J. C. Cornwell, 1985. Nitrogen, phosphorus and organic carbon cycling in an arctic lake. Can. J. Fish. aquat. Sci. 42: 797–808.

Whalen, S. C. & V. Alexander, 1986. Seasonal inorganic carbon and nitrogen transport by phytoplankton in an arctic lake. Can. J. Fish aquat. Sci. 43: 1177–1186.

Hydrobiologia **240**: 15–22, 1992.
W.J. O'Brien (ed.), Toolik Lake – Ecology of an Aquatic Ecosystem in Arctic Alaska.

Cation export from Alaskan arctic watersheds

Jeffrey C. Cornwell
University of Maryland CEES, Horn Point Environmental Laboratory, Cambridge, MD 21613-0775, USA

Key words: arctic streams, cations, watershed fluxes, mass balances

Abstract

The concentrations and stream fluxes of Na, K, Ca and Mg were determined in 1978, 1980 and 1981 in the Toolik Lake watershed and in 1978 in the upper Kuparuk River watershed. The annual export of K was strongly influenced by high K concentrations at the initiation of spring melt. Potassium concentrations were positively correlated with concentrations of particulate and dissolved organic carbon, particulate and dissolved phosphorus, and particulate, dissolved organic and ammonium nitrogen, suggesting a common origin in the decomposition of tundra plants. Calcium and Mg stream concentrations were generally highest in July and August when the depth of the soil active layer reached a maximum of ~ 0.5 m. Precipitation-derived Na and K accounted for up to half of the Na and K stream export while chemical weathering supplied $>85\%$ of exported Ca and Mg. Cation export rates in these arctic foothill watersheds are among the lowest recorded.

Introduction

Elemental mass balances have proven to be of great use in understanding the biogeochemical processes occurring in terrestrial and aquatic ecosystems (e.g. Likens *et al.*, 1977; Likens, 1984). The composition and flux of ions in surface and ground waters depends upon the forms and reactivity of geological substrates (Johnson *et al.*, 1968; Bricker *et al.*, 1967; Velbel, 1985), the rates of atmospheric inputs of water and ions, the climate and the structure and maturity of terrestrial ecosystems (Vitousek & Reiners, 1975; Likens *et al.*, 1977). In the North American arctic, detailed studies of watershed elemental mass balances are limited to an Alaskan coastal plain site at Barrow, Alaska (Barsdate & Alexander, 1974; Hobbie, 1980; Chapin *et al.*, 1978), the Char Lake study (Schindler *et al.*, 1974), the Saqvaqjuac Lakes region (Welch & Legault, 1986), and Toolik Lake, Alaska (Cornwell, 1983; Whalen & Cornwell, 1985). Arctic ecosystems, both terrestrial and aquatic, are characterized by short growing seasons, low nutrient concentrations and fluxes, and low primary production. In comparison to temperate and tropical watersheds, it may be hypothesized that dissolved ion export in arctic watersheds is low because of the low primary weathering rates which arise from low precipitation and temperatures. To test this hypothesis, the major features of stream cycles of Ca, Mg, Na and K were determined in two arctic watersheds in the foothills of the Brooks Range in arctic Alaska.

Study area

This study includes measurements from two distinct systems, the inflow of Toolik Lake and the

Kuparuk River (Fig. 1). Both watersheds consist of glacial moraines and outwash left after the retreat of the Itkillik II glaciation 12000 to 14000 yr ago (Hamilton & Porter, 1975). The soils in the arctic foothill region are poorly drained with a peaty surface layer underlain by silt loams and silty-clay loams (Brown & Berg, 1980). The vegetation is mostly of the tussock tundra variety, with a high percentage of ground cover. The mean monthly temperature is −10 °C, with mean monthly temperatures less than −25 °C in December through March and over 10 °C during July and August. The mean yearly precipitation is approximately 20 cm yr^{-1}, with roughly equivalent amounts of frozen and unfrozen precipitation. Overall, the climate is somewhat more continental than the arctic coastal plain, with a thaw season of 100 to 140 days (Brown & Berg, 1980). A gravel Haul Road constructed for pipeline construction and the Alyeska oil pipeline pass through each watershed, with major construction activity in the region initiated in the mid-1970's. The major watershed of Toolik Lake (area = 65 km^2) was sampled at the main inlet stream and includes a series of small lakes and ponds. The Kuparuk River watershed was sampled where the river passes beneath the Haul Road at a point where the upstream watershed covers 143 km^2, an area equivalent to <2% of its total watershed area as the river enters the Arctic Ocean (Peterson *et al.*, 1986).

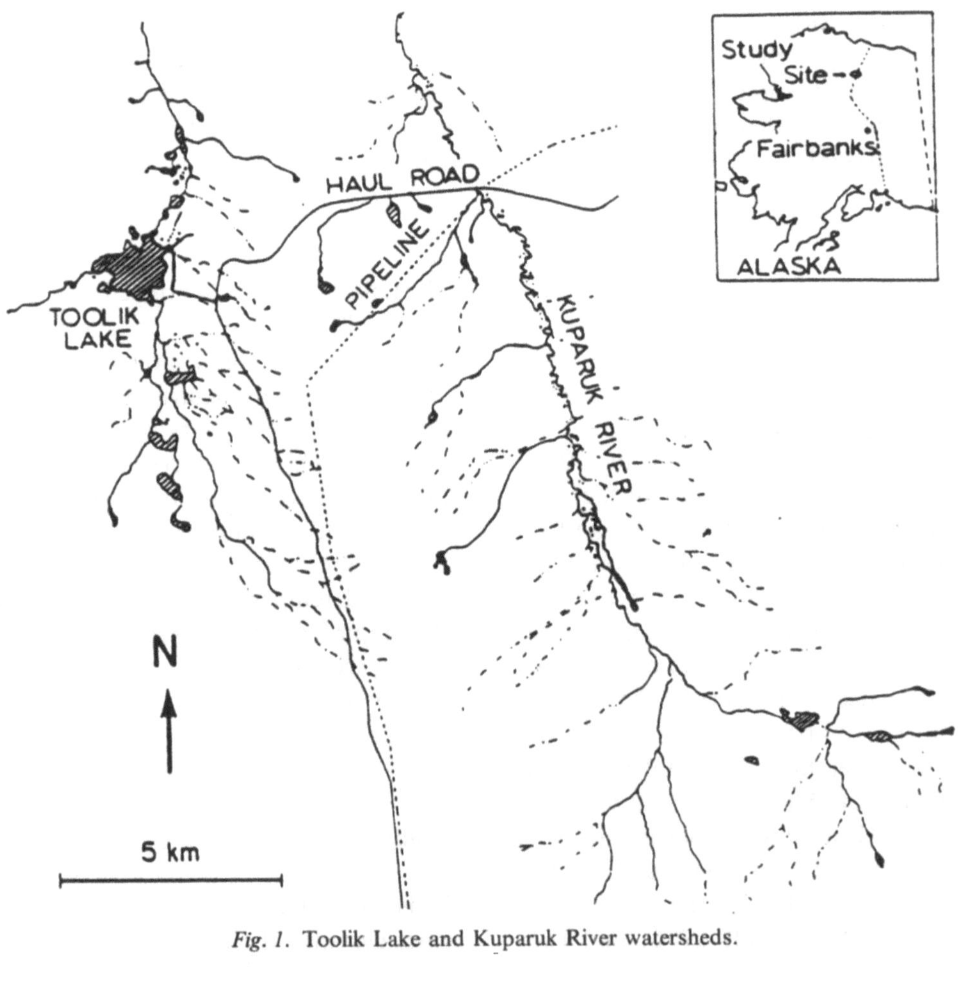

Fig. 1. Toolik Lake and Kuparuk River watersheds.

Methods

Water flow was measured in the Toolik inlet and the Kuparuk River either by the use of a Gurley pigmy current meter on a cross stream transect or by Rhodamine dye dilution (Peterson *et al.*, 1986). At the time of flow measurement, a 4 l poly bottle of water was collected and immediately returned to the laboratory. Analyses of P, N, Fe, Mn, Si, organic carbon, water color and suspended sediments are presented in Cornwell (1983). A sample of rain water for cation analysis was collected in August 1980. Measurements of the dissolved cation concentration in snow were made in April 1978 and May 1980 by collection of samples from the surface of Toolik Lake; no samples from fresh snowfall were analyzed. Snow samples were kept frozen until transported to the Fairbanks laboratory.

Samples for major cations were vacuum filtered through Gelman A/E glass fiber filters through which > 100 ml of sample was passed before collection of solutes. Samples were stored at room temperature in acid washed polyethylene bottles and preserved with nitric acid. The analysis of dissolved cations was made on a Perkin-Elmer model 503 atomic absorption spectrophotometer, with lanthanum added to suppress ionic interferences during the analysis of Ca and Mg (USEPA, 1984).

Daily stream export rates were calculated as the product of the stream flow and the cation concentration. Annual fluxes were estimated as the sum of fluxes for intervals between sampling dates, with the interval flux calculated as the product of the mean flux of adjacent sampling dates and the time interval. Missing data was assumed to be equal to the mean of bracketing values or in the case of unmeasured late season export, equal to the previous value. The most complete data set is found for the Toolik Inlet in 1980; the 1978 fluxes are poorly represented in late summer and the 1981 data is missing the first two weeks of flow because of logistic difficulties. As a consequence, the 1980 export rates present the best estimates.

Results

Patterns of stream flow varied considerably from year to year in the Toolik Inlet (Figs. 2, 3, 4). In 1980, the measurement of stream flow and stream chemistry was initiated within several hours of the initial stream flow. In 1981, logistical difficulties prevented sampling during the first 10 days of flow. Stream flow in the Kuparuk in 1978 generally followed the pattern of the Toolik inlet in the same year except for an absence of an early season peak (Fig. 5). A short thaw in April 1978 may have caused Kuparuk River flow prior to our sampling.

Potassium concentrations were highest at the initiation of stream flow during the spring melt, with extremely low concentrations found at all other times. The importance of this is seen in the 1980 Toolik data which show that ~33% of K fluxes occurred during the first 10% of the flow season. In that same time period, 8–9% of Ca,

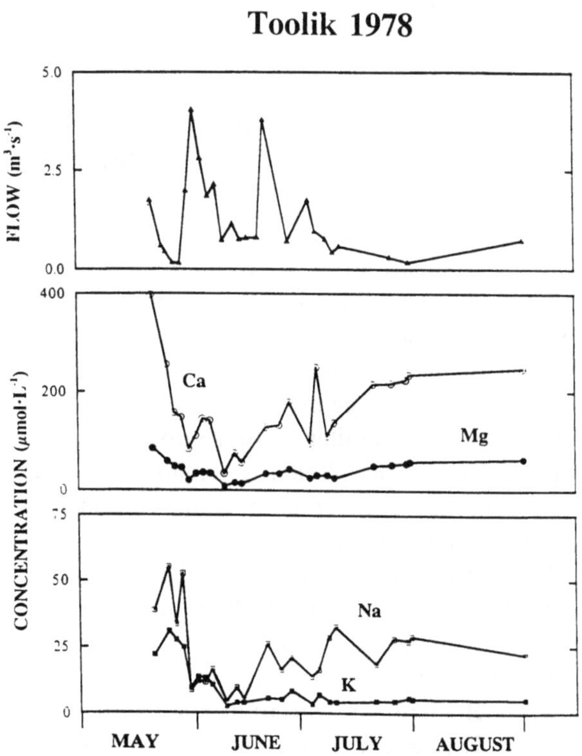

Fig. 2. 1978 Toolik Lake main inlet stream flow and major cation concentrations.

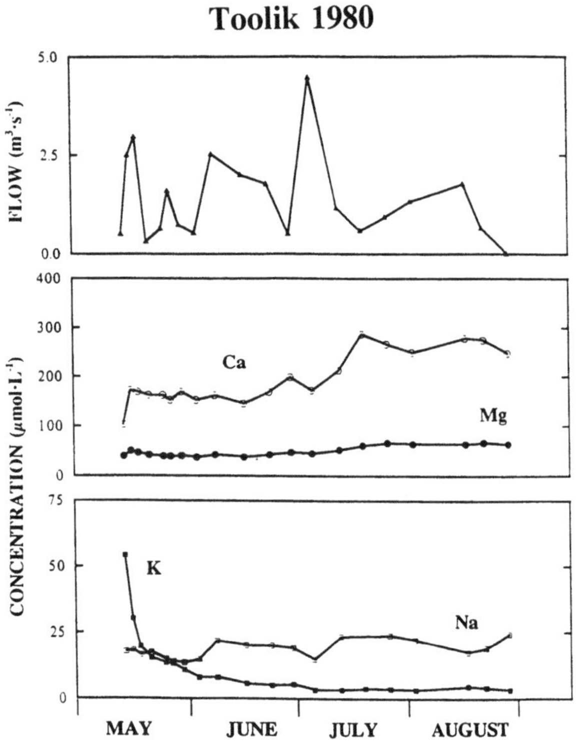

Fig. 3. 1980 Toolik Lake main inlet stream flow and major cation concentrations.

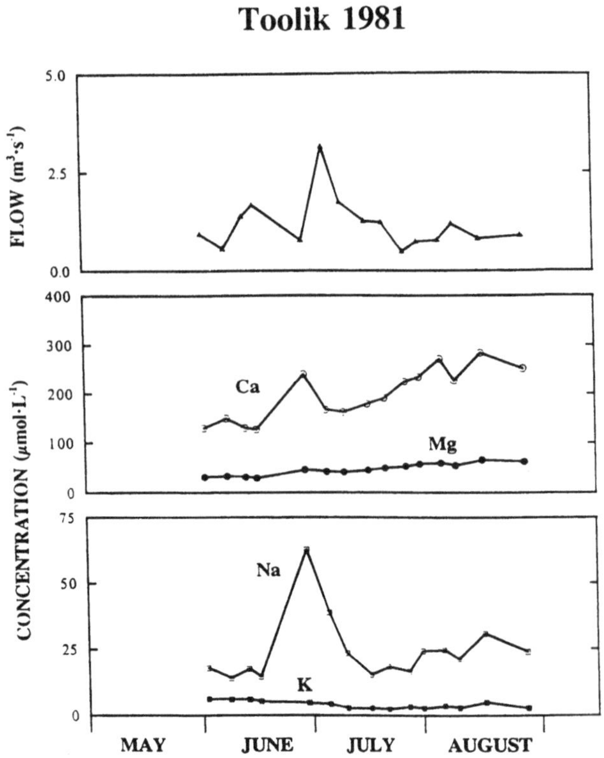

Fig. 4. 1981 Toolik Lake main inlet stream flow and major cation concentrations.

Mg and Na fluxes occurred. Incomplete early season sampling for K in 1978 and 1981 lead to the underestimation of fluxes; the other cations are less affected. Early season peaks in Na, Ca and Mg concentrations are found in some years, followed by a mid-season minimum and a subsequent gradual increase in concentration through the rest of the season. The seasonal patterns of concentration change in the Toolik inlet and Kuparuk River are similar to those observed in the Saqvaqjuac region (Welch & Legault, 1986), except that in the Toolik region, K does not increase significantly in the late summer.

Concentration differences of cations in snow, rain and streams between well-studied arctic systems do not show any consistent pattern (Table 1). Cation concentrations in snow and rain have coefficients of variation than can exceed $\pm 50\%$. Higher concentrations of Na in the three other arctic study sites result from the input of coastally-derived sea salts. The Toolik and Kuparuk systems had by far the lowest measured concentrations of K in the North American arctic. The concentrations of Ca and Mg in the Toolik region were at the lower end of those found in these arctic studies. Much higher atmospheric concentrations of Ca and Mg are found in the Char Lake region. The concentrations of Na and K were similar for the Toolik and Kuparuk watersheds, but the Kuparuk watershed has much lower concentrations of Ca and Mg. The standard deviation of Na, Ca and Mg in different years in the Toolik Lake inlet was $< \pm 21\%$, with K concentrations changing by a factor of 2 or more. Low concentrations of K in 1981 (relative to 1978 and 1980) may be attributed to the incomplete sampling in early spring, when the highest concentrations are expected.

The watershed flux of cations (Table 2) in the Toolik and Kuparuk watersheds represent minimum estimates in 1978 and 1981 because of incomplete early season sampling and paucity of

Table 1. Mean concentrations of major cations in North American arctic watersheds. The Barrow data are from Barsdate and Prentki (unpublished), the Char Lake data are from Schindler *et al.* (1974) and the data from six Saqvaqjuac watersheds are from Welch and Legault (1986). The annual stream data from the Toolik and Saqvaqjuac regions are flow weighted, with the Saqvaqjuac stream data representing the mean (± S.D.) of annual mean concentrations from 1 to 4 jears of stream monitoring. Char Lake mean stream values are not flow weighted and represent the mean of all samples collected from three of Char Lake's inflow stream. Saqvaqjuac rain values are the mean of 5 years of volume-weighted summer precipitation data; variability is higher within a given year than that calculated from the annual mean concentrations. N represents the number of either (1) discrete samples used for the calculation of sample statistics or (2) the number of volume weighted annual means (Saqvaqjuac only) used for calculation.

Location		Concentration (μmol l^{-1})				
		N	Na	K	Ca	Mg
Barrow, AK	Snow	3	155 ± 102	6 ± 3	60 ± 14	29 ± 19
	Rain	5	36 ± 42	3 ± 2	45 ± 23	14 ± 9
Char Lake, N.W.T.	Snow	8	42 ± 53	5 ± 5	52 ± 36	24 ± 17
	Rain	5	35 ± 43	5 ± 3	488 ± 74	30 ± 9
	Stream	23	365 ± 635	23 ± 32	838 ± 369	273 ± 180
Saqvaqjuac, N.W.T.	Snow	5	12 ± 24	<1	3 ± 9	2 ± 3
	Rain		46 ± 33	<2	6 ± 2	4 ± 1
	Meadow	4	144 ± 23	19 ± 3	249 ± 111	41 ± 17
	P & N	3	106 ± 31	15 ± 0	79 ± 5	30 ± 7
	Far	3	98 ± 36	9 ± 3	75 ± 44	29 ± 10
	Spring	3	204 ± 31	19 ± 3	106 ± 26	36 ± 9
	Jade	2	101 ± 46	12 ± 2	101 ± 58	28 ± 8
	Saqvaqjuac	1	76	10	141	33
Toolik region	Snow 1978	5	12 ± 2	1.4 ± 0.7	15 ± 20	3 ± 2
	Snow 1980	5	10 ± 5	1.5 ± 0.6	30 ± 9	6 ± 2
	Rain 1980	1	3	1.3	6	3
	Inlet 1 1978	23	21	9	167	41
	Inlet 1 1980	20	19	7	200	50
	Inlet 1 1981	15	28	4	208	48
	Kuparuk 1978	17	26	6	61	24

late season data. In the main inlet, the reasonable agreement between these years and the very complete 1980 flow (a high flow year) suggest that with the possible exception of K, the fluxes of cations are reasonable. The watershed fluxes of Na and K in the Toolik inlet and particularly in the Kuparuk River, were lower than has been documented for arctic (Table 2) and temperate zone ecosystems (Likens *et al.*, 1974), with only the Caribou-Poker Flats site in subarctic Alaska showing similar low fluxes. The Ca and Mg fluxes in the Toolik region were quite low relative to many temperate systems but were similar to those found in Hubbard Brook and are typical of arctic and subarctic systems. The Ca and Mg fluxes in the Kuparuk River watershed ar the lowest on record.

A rough estimate for precipitation inputs of cations may be estimated based on the single rain and 10 snow samples. Considering the variance found in other arctic systems, such estimates might be high or low by a factor of two. Sodium fluxes are ~ 1.4–2.4 mmol m^{-2} yr^{-1}, K fluxes are ~ 0.2–0.4 m^{-2} yr^{-1}, Ca fluxes are 4-6 mmol m^{-2} yr^{-1} and Mg fluxes are 0.7–1.2 m^{-2} yr^{-1}. Even if these data are poor estimates of the 'true' atmospheric flux, it is clear that soils are the main source of stream cations (Table 2), rather than direct atmospheric inputs. In particular, <15% of Ca and Mg flux can be accounted for by at-

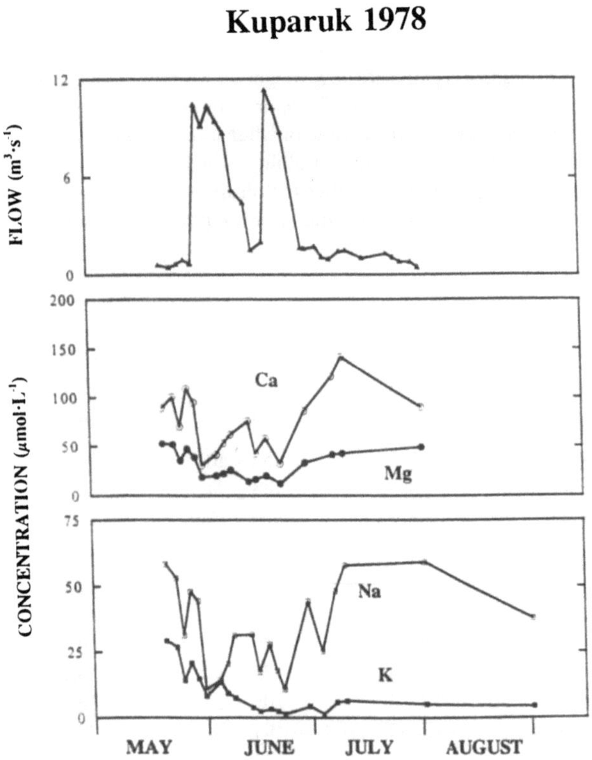

Fig. 5. 1978 Kuparuk River stream flow and major cation concentrations.

mospheric inputs while up to half of the Na stream output could have an atmospheric origin.

Discussion

The fluxes of major ions in the Toolik inlet are reasonably consistent for the three years of study. The primary exception to this pattern is the low 1981 K flux, a reflection of incomplete spring sampling. The higher concentrations and fluxes of K at the outset of stream flow are found in both the Toolik Inlet and the Kuparuk River and may result from: (1) enhanced K concentrations in snow, (2) freeze out or cryoconcentration of solutes in soils or below ice in streams and ponds or (3) the leaching of K from plant litter at the tundra surface. Mechanism (1) is not supported by K concentration data in the spring snow pack and mechanisms (2) is unlikely because cryoconcentration must result in a tenfold increase in K concentration from late summer values, requiring that $>90\%$ of lake or pond water freeze within the watershed and that water must be undiluted by surface runoff when snow melt is at its peak. Such an extreme concentration process is not evident for Na, Ca and Mg, suggesting that this mechanism is unlikely.

During spring flow, higher concentrations of K coincide with higher concentrations of nutrient elements. Potassium was positively correlated ($p<0.01$; $df\geq 15$) with total dissolved P ($r=0.95$),

Table 2. Watershed cation export rates for selected North American watersheds. The Caribou-Poker Flats (central Alaska) data are recalculated from Hawkins & Glover (1982), the MacKenzie data are from Brunskill *et al.* (1973) and the Hubbard Brook data are from Likens *et al.* (1974).

Location		Watershed export (mmol m^{-2} yr^{-1})			
		Na	K	Ca	Mg
Char Lake, N.W.T.		42	12	97	32
Caribou-Poker Flats		6	2	37	15
Saqvaqjuac Lakes	low	99	9	75	29
	high	204	19	249	41
MacKenzie R. region	low	16	2	42	24
	high	55	9	309	156
Hubbard Brook		32	5	32	24
Toolik Lake	1978[a]	4	1	28	7
		5	2	56	14
	1981[a]	5	1	35	8
Kuparuk River	1978	4	1	9	4

[a] Annual fluxes are minimal estimates because of incomplete spring sampling.

dissolved organic N ($r = 0.76$), DOC ($r = 0.87$) and particulate C, N and P (Cornwell, 1983; Whalen and Cornwell, 1985; Cornwell, unpublished data), supporting the hypothesis that the decomposition of terrestrial organic matter is the source of early season K.

The seasonal pattern of Na, Ca and Mg concentration within each watershed are similar within a given year, but the overall pattern can vary between years. Annual patterns of calcium and Mg concentrations are similar in each system. In the Kuparuk River and Toolik Inlet in 1978 (but not 1980) the spring concentrations of Na, Ca and Mg follow the same pattern as K, though the relative dynamic range is considerably greater for K. Higher concentrations of Na, Ca and Mg in July and August are found in both the Kuparuk and Toolik Inlet in all years. Solar insolation through the summer months increases the depth of the soil active layer to a maximum of ~0.5 m in August (Brown & Berg, 1980). As thaw depths increase, weathering of cations from deeper, less weathered mineral soils proceeds and cation concentrations increase. An additional mechanism for cation release may involve the release of cations which have been cryoconcentrated in soils during the fall and winter by soil water freezing which proceeds from both the soil surface and from below. The exclusion of ions from ice would result in the accumulation of cations in intermediate soil depths from which they would be released when thaw depths reach the cation-enriched soil horizon.

Examination of concentration-flow relationships reveal no significant correlations. High cation concentrations are found during early spring flow and with the exception of K, also in the latter part of summer (July and August) when flows are quite variable. Minor decreases in cation concentrations are observed when flows increase (especially Figs. 3 and 5), but these are insufficient to influence the overall seasonal concentration-flow pattern.

The flux of cations measured in the Toolik region has three potential sources: (1) atmospheric precipitation, (2) primary weathering of mineral soils and (3) mineralization of plant organic matter in the soil. Atmospheric fluxes, though poorly quantified in this study, are probably minor relative to the other sources. While the relative importance of primary weathering and organic decomposition has been inferred from seasonal patterns of concentration and flux, it is difficult in practice to separate each of these source. The cations utilized as plant nutrients may be derived from primary weathering, decomposition or precipitation. A stream data set cannot unequivocably identify the processes which control cation biogeochemisty in tundra ecosystems. Stream studies, such as this one, establish the net fluxes from the watershed and establish limits to the biogeochemical processes which are integrated into a net flux number.

The concentrations and fluxes of cations in arctic watersheds (Tables 1 and 2) show the same high degree of variability as found in temperate watersheds (Likens *et al.*, 1974). Atmospheric cation inputs vary between studies, and higher inputs of sea salt-derived Na are found in the coastal sites (Barrow, Char Lake and Saqvaqjuac Lakes) than in the Toolik region. The composition of bedrock varies in these arctic studies with the poorly consolidated glacial outwash in the Toolik region being derived from mostly Devonian sandstones and limestones in the Brooks Range (Brown & Berg, 1980). The bedrock in the Saqvaqjuac region is granitic and despite the absence of carbonate minerals (Welch & Legault, 1986), higher stream fluxes of Ca and Mg were found. Annual precipitation inputs to the Saqvaqjuac Lakes area are ~29 cm (Welch & Legault, 1986), approximately 50% higher than the Toolik region, possibly accounting for higher cation fluxes at the Saqvaqjuac watersheds. Vegetation can have a large influence on the composition and fluxes of elements in watersheds (Likens *et al.*, 1977), a result of the storage and release of cations in living and dead organic matter and enhanced weathering via plant-derived organic acids. Plants also influence the thaw depth of arctic tundra soils with unvegetated sites having a greater active zone (Brown & Berg, 1980). Precipitation inputs of water vary within the arctic (though they are generally low), and Welch &

Legault (1986) have clearly shown that the timing of melt water runoff and summer rain events can dramatically influence the flux rates.

Conclusions

The stream export of cations in arctic watersheds is generally low relative to temperate watersheds because of low weathering rates. Low rates of weathering occur because of a short thaw season, shallow thaw depths, low temperatures and low inputs of precipitation. The large differences in cation flux found within different arctic regions are consistent with the large degree of variability found in temperate zone watersheds and similarly arise from differences in climate, geology and the influence of biological uptake and recycling. The Toolik inlet and Kuparuk cation fluxes are among the lowest on record.

This study has shown the critical importance of early season stream fluxes of cations, particulary K, in mass balance studies. The Toolik region watershed export estimates help to provide a broader picture of arctic cation fluxes, but the limited data set does not address the year-to-year variability often found in similar temperate zone and arctic watershed studies (Likens, 1977; Schindler *et al.*, 1976; Welch & Legault, 1986). Detailed multi-year studies of precipitation inputs, weathering and the influence of terrestrial biological processes on ion fluxes are required for a better understanding of processes influencing tundra elemental mass balances.

Acknowledgements

This research was part of the Arctic Lake Process Study funded by the National Science Foundation, with additional support from a State of Alaska fellowship through the University of Alaska Institute of Marine Science. I thank Steve Whalen and Bruce Peterson for assistance in sample collection and Mike Williams and two anonymous reviewers for significantly improving this paper.

References

Barsdate, R. J. & V. Alexander, 1975. The nitrogen balance of arctic tundra: pathways, rates and environmental implications. J. envir. Qual. 4: 111–117.

Bricker, O. P., A. E. Godfrey & E. T. Cleaves, 1967. Mineral-water interaction during the chemical weathering of silicates. In Trace Inorganics in Water. Adv. Chem. Series 73: 128–142. American Chemical Society, Washington, D.C.

Brown, J. & R. L. Berg, 1980. Environmental engineering and ecological baseline investigations along the Yukon River-Prudhoe Bay Haul Road. CRREL Report 80–19.

Brunskill, G. J., D. M. Rosenberg, N. B. Snow, G. L. Vascatto & R. W. Wagemann, 1973. Ecological studies of aquatic systems in the MacKenzie-Porcupine drainages in relation to proposed pipeline and highway developments. Canada Task Force N. Oil Dev. Environ. Soc. Comm. Vol. II. Report 73–41. 345 pp.

Chapin, F. S. III & R. J. Barsdate, 1978. Phosphorus cycling in Alaskan coastal tundra: a hypothesis for the regulation of nutrient cycling. Oikos 31: 189–199.

Cornwell, J. C., 1983. The geochemistry of manganese, iron and phosphorus in an arctic lake. Ph.D. Dissertation, University of Alaska, Fairbanks. 238 pp.

Cornwell, J. C., 1991. Manganese and iron mass balances in an arctic lake. Can. J. Fish. aquat. Sci. (submitted).

Hamilton, T. D. & S. C. Porter, 1975. Itkillik glaciation in the Brooks Range, northern Alaska. Quat. Res. 5: 471–497.

Hawkins, D. B. & D. M. Glover, 1982. Hydrogeochemistry of the Caribou-Poker Creeks Research Watershed. Report IWR-102. Institute of Water Resources, University of Alaska. Fairbanks, Alaska.

Hobbie, J. E., 1980. Polar Limnology. In F. B. Tabu (ed), Lakes and Reservoirs. Elsevier, New York.

Hobbie, J. E., T. L. Corliss & B. J. Peterson, 1983. Seasonal patterns of bacterial abundance in an arctic lake. Arct. Alp. Res. 15: 253–259.

Johnson, N. M., G. E. Likens, F. H. Bormann & R. S. Pierce, 1968. Rate of chemical weathering of silicate minerals in New Hampshire. Geochim. Cosmochim. Acta. 32: 531–545

Likens, G. E., 1984. Beyond the shoreline: a watershed-ecosystem approach. Verh. int. Ver. Limnol. 22: 1–22.

Likens, G. E., F. H. Bormann, R. S. Pierce, J. S. Eaton & N. M. Johnson, 1977. Biogeochemistry of a forested ecosystem. Springer-Verlag, New York. 146 pp.

Peterson, B. J., J. E. Hobbie & T. L. Corliss, 1986. Carbon flow in a tundra stream ecosystem. Can. J. Fish. aquat. Sci. 43: 1259–1270.

Schindler, D. W., R. W. Newbury, K. G. Beaty & P. Campbell, 1976. Natural water and chemical budgets for a small Precambrian lake basin in central Canada. J. Fish. Res. Bd Can. 33: 2526–2543.

Schindler, D. W., H. E. Welch, J. Kalff, G. J. Brunskill, H. Kling & N. Kritsch, 1974. Physical and chemical limnology of Char Lake, Cornwallis Island. J. Fish. Res. Bd Can. 31: 587–607.

USEPA, 1974. Methods for Chemical Analysis of Water and Wastes. USEPA, Washington, D.C.

Velbel, M. A., 1985. Geochemical mass balances and weathering rates in forested watersheds of the southern Blue Ridge. Am. J. Sci. 285: 904–930.

Vitousek, P. M. & W. H. Reiners, 1975. Ecosystem succession and nutrient retention: a hypothesis. BioScience 25: 376–381.

Welch, H. E. & J. A. Legault, 1986. Precipitation chemistry and chemical limnology of fertilized lakes at Saqvaqjuac, N.W.T. Can. J. Fish. aquat. Sci. 43: 1104–1134.

Whalen, S. C. & J. C. Cornwell, 1985. Nitrogen, phosphorus and organic carbon cycling in an arctic lake. Can. J. Fish. aquat. Sci. 42: 797–808.

Hydrobiologia **240**: 23–36, 1992.
W.J. O'Brien (ed.), Toolik Lake – Ecology of an Aquatic Ecosystem in Arctic Alaska.

The flux of CO_2 and CH_4 from lakes and rivers in arctic Alaska

George W. Kling, George W. Kipphut[1] & Michael C. Miller[2]
Department of Biology, University of Michigan, Ann Arbor, MI 48109, USA; [1] *Institute of Marine Science, University of Alaska, Fairbanks, AK 99775, USA; present address: Center for Reservoir Research, Murray State University, Murray, KY 42071, USA;* [2] *Department of Biological Sciences, University of Cincinnati, Cincinnati, OH 45221, USA*

Key words: Alaska, arctic, carbon budgets, methane, limnology

Abstract

Partial pressures of CO_2 and CH_4 were measured directly or calculated from pH and alkalinity or DIC measurements for 25 lakes and 4 rivers on the North Slope of Alaska. Nearly all waters were supersaturated with respect to atmospheric pressures of CO_2 and CH_4. Gas fluxes to the atmosphere ranged from −6.5 to 59.8 mmol m^{-2} d^{-1} for CO_2 and from 0.08 to 1.02 mmol m^{-2} d^{-1} for CH_4, and were uncorrelated with latitude or lake morphology. Seasonal trends include a buildup of CO_2 and CH_4 under ice during winter, and often an increased CO_2 flux rate in August due to partial lake turnover. Nutrient fertilization experiments resulted in decreased CO_2 release from a lake due to photosynthetic uptake, but no change in CO_2 release from a river due to the much faster water renewal time. In lakes and rivers the groundwater input of dissolved CO_2 and CH_4 is supplemented by in-lake respiration of dissolved and particulate carbon washed in from land. The release of carbon from aquatic systems to the atmosphere averaged 24 g C m^{-2} y^{-1}, and in coastal areas where up to 50% of the surface area is water, this loss equals 1/5 to 1/2 of the net carbon accumulation rates estimated for tundra.

Introduction

There are large stores of carbon in arctic environments (Bliss *et al.*, 1981). Arctic ecosystems are expected to be affected strongly by future climatic change (Lachenbruch & Marshall, 1986; Grotch, 1988; Abelson, 1989), and if these ecosystems either accumulate additional carbon or lose carbon to the atmosphere they will be important to the global carbon budget (Bolin *et al.*, 1986; Trans *et al.*, 1990; Post, 1990).

At present tundra ecosystems are considered to be small sinks for CO_2. The amount of carbon extracted from the atmosphere during photosynthesis is slightly more than that respired by plants and soils and lost to the atmosphere (Chapin *et al.*, 1980; Billings, 1987). But these estimates of carbon balance in the arctic are only for terrestrial ecosystems; there is evidence that additional carbon loss to and from aquatic ecosystems may be important. Schell & Ziemann (1983) found that the rate of carbon loss by erosion of peat into coastal lakes, and through transport by rivers to the sea, was a substantial percentage of the long-term carbon accumulation rates on land. The input of terrestrial peat can dominate carbon cycling in more inland rivers as well (Peterson *et al.*, 1986). There is also some indication that terrestrial carbon washed into water bodies is respired and lost to the atmosphere as CO_2. For example,

Coyne & Kelley (1974) found that one coastal pond and one coastal lake near Point Barrow were strongly heterotrophic, supersaturated with CO_2, and continuously released CO_2 to the atmosphere. In contrast to the terrestrial CO_2 sink, tundra ecosystems are considered to be sources of CH_4 to the atmosphere (Whalen & Reeburgh, 1990a). Several taiga and arctic lakes have also been reported to release CH_4 to the atmosphere at rates similar to those observed for the tundra (Whalen & Reeburgh, 1990b).

The CO_2 flux data from this study were described in brief by Kling *et al.* (1991); in this paper we present the details of the processes controlling annual and seasonal variation in CO_2 flux from surface waters in the Arctic, and we present new data on CH_4 concentration and flux. We conclude that because some of the carbon respired in the terrestrial environment makes its way into lakes and rivers where it is released to the atmosphere, this loss of carbon from aquatic ecosystems must be included in regional carbon budgets for the Arctic.

Site description

All sites are located on the North Slope of arctic Alaska and cover an area from Mt. Roche stream (68.23° N) in the Brooks Range foothills to Lake Maxine (70.22° N) near Prudhoe Bay on the Arctic Ocean (Fig. 1, Table 1). All sites are within several kilometers of the Dalton Highway.

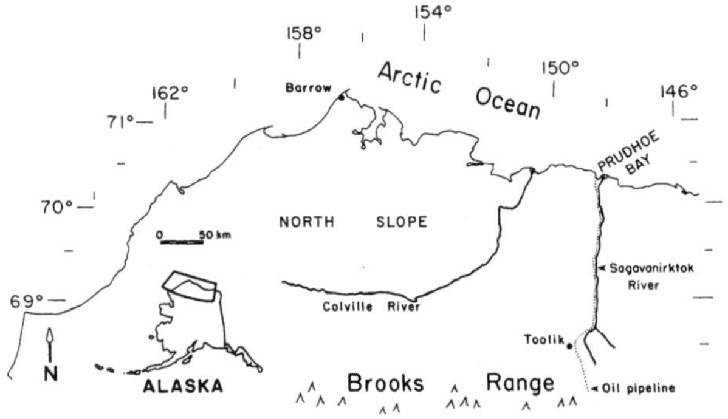

Fig. 1. Map showing the North Slope of Alaska and the location of the oil pipeline and Toolik Lake.

Samples from the Kuparuk River were taken near the oil pipeline crossing. Exact locations and morphometric data for each site are given in Kling *et al.* (1992).

Sellmann *et al.* (1975) presented a contour map of water coverage for much of the North Slope. We have integrated their map contours by planimetry to arrive at the relative proportions of land (86%) and water (14%) for arctic Alaska north of latitude 69.5° and between Point Lay (162.5° W) and Brownlow Point (146° W) (Fig. 1).

Methods

CO_2 partial pressures were measured directly or calculated given pH, temperature, and concentrations of alkalinity or dissolved inorganic carbon (DIC). Alkalinity was determined by potentiometric titration and the curves analyzed by the method of Gran (Stumm & Morgan, 1981). Once titration endpoints were known and found to be consistent over time for particular lakes, titrations were made to a specific endpoint. Water for DIC analysis was collected from just below the lake or stream surface without exposure to the atmosphere and preserved with $HgCl_2$. In the lab, samples were acidified with phosphoric acid and purged with nitrogen gas into a vacuum line. CO_2 was collected by cryogenic distillation and gas pressures measured by an electronic manometric gauge. The precision of this procedure was about $\pm 5\ \mu mol\ CO_2\ kg^{-1}$. Dissociation constants from Plummer & Busenburg (1982) and carbonate species relations from Stumm & Morgan (1981) were used for conversion of alkalinity and DIC to dissolved $CO_2(aq)$ concentrations. In some waters, gas pressures were measured directly by equilibration of water samples with an air-filled (for CO_2) or nitrogen-filled (for CH_4) headspace. After equilibration, headspace gas was analyzed on a thermal conductivity (for CO_2) or a flame ionization (for CH_4) gas chromatograph. Atmospheric concentrations of CH_4 were taken as 1.7 μatm; concentrations of CO_2 were taken as 350 μatm (Tans *et al.*, 1990, Point Barrow Sta-

Table 1. CO_2 flux data for North Slope lakes and rivers. Positive fluxes indicate evasion of CO_2 from water to atmosphere. Mean and grand mean values are given ± 1 SE. Site names and locations correspond to data in Klint *et al.* (this issue).

Lakes	DATE	*N*	Temp (°C)	P_{CO_2} (μatm)	CO_2 Flux (mmol m^{-2} d^{-1})
E1	Aug 83–Jul 86	3	–	2758 ± 1215	59.8 ± 30.2
NE14	Aug 86–Jun 88	2	–	1954 ± 904	39.7 ± 21.3
Toolik	Jun 75–Aug 89	62	–	1847 ± 238	35.0 ± 5.3
N1	Aug 83–Jul 88	3	–	1170 ± 689	20.5 ± 16.4
Island	Jul 86–Jul 88	2	–	1040 ± 6	17.8 ± 0.3
N2	Aug 83–Aug 89	44	–	825 ± 89	12.5 ± 2.2
S6	Jul 86–Jul 88	2	–	628 ± 303	7.8 ± 7.6
I8	22 July 86	1	14.0	2155	45.5
NE2	22 July 86	1	15.5	1972	41.5
Windy	8 July 88	1	14.0	1836	40.1
Dam	4 July 85	1	11.0	1591	30.9
Silhouette	8 July 88	1	13.0	1552	32.4
Galbraith	6 July 88	1	17.5	1477	29.1
Carolyn	9 July 88	1	11.0	1333	26.1
S3	22 July 86	1	17.0	1194	22.1
NE12	29 June 88	1	15.5	925	15.3
Africa	9 July 88	1	11.0	919	15.1
Charles	8 July 88	1	18.8	818	13.0
Coleen	10 July 88	1	11.0	761	11.0
Maxine	9 July 88	1	9.8	584	6.2
William	11 July 88	1	17.0	504	4.5
Bern	10 July 88	1	9.5	459	3.0
George	7 July 88	1	19.0	359	1.1
Anne	7 July 88	1	19.5	291	−0.7
Camp	13 July 88	1	20.5	93	−5.5
Rivers					
Kuparuk	Aug 83–Aug 89	21	–	812 ± 177	11.9 ± 1.5
Hershey	Aug 85–Jul 88	3	–	845 ± 182	12.8 ± 4.3
Mt. Roche	6 July 88	1	10.5	369	1.4
Oksrukuyuk	7 July 88	1	13.9	418	2.3

tion) and seasonal variations (up to 10 μatm; Coyne & Kelley, 1974) were ignored. P_{CO_2} and the concentration of CO_2 in equilibrium with the atmosphere (CO_2eq) was calculated using Henry's law and the CO_2 fugacity-pressure relationship presented in Weiss (1974). The concentration of CH_4 in equilibrium with the atmosphere was calculated using equations in Wilhelm *et al.* (1977). Corrections for vapor pressure were made using equations in Buck (1981).

CO_2 and CH_4 flux was calculated using the equation

$$Flux = (D/z) * (Aq - Eq),$$

where D is the molecular diffusion of CO_2 or CH_4 ($cm^2\ s^{-1}$) taken from Himmelblau (1964), z is the thickness of the surface boundary layer (μm), Aq is the concentration of gas at the bottom of z, and Eq is the concentration of gas in water exposed to the atmosphere at the top of z (see Broecker & Peng, 1974). Because the pH of natural waters studied was typically much less than 9.0, chemically enhanced CO_2 gas exchange was assumed to be negligible (Emerson, 1975a).

From 1988 to 1990 wind speed was measured with an RM Young anemometer at 5 m height at two stations near Toolik Lake. The mean wind speed in the ice-free months (June–September) was 3.3 ± 0.2 m sec^{-1} (± 1 SE; taken from a total of 7 monthly means of averaged daily wind speeds). This value was similar to the mean of 2.3 m sec^{-1} at 0.5 m height reported by Miller *et al.* (1986). On the coastal plain at Barrow the average wind speed is somewhat higher (June–September mean = 6.3 ± 0.2 m sec^{-1} at 10 m height, Hobbie, 1980), due in part to the greater height of measurement. Maximum wind speeds were seldom great enough to affect CO_2 saturation values due to bubble entrainment and 'wind pumping' (Smith & Jones, 1985; Merlivat & Memery, 1983).

The boundary layer *z* for all lakes was conservatively estimated as 200 μm using an empirical relation with wind speed. The regression equation used was $\log_{10}$ (z, μm) = 2.56 − 0.133∗(wind speed, m s^{-1}), $SE_{intercept} = 0.04$, $SE_{slope} = 0.007$, $r^2 = 0.71$, N = 147 (data not corrected to common reference height; values from Kanwisher, 1963; Hoover & Berkshire, 1969; Liss, 1973; Emerson, 1975b; Broecker *et al.*, 1978; Broecker *et al.*, 1980; Bower & McCorkle, 1980; Torgersen *et al.*, 1982; Merlivat & Memery, 1983; Hartman & Hammond, 1984; Jähne *et al.*, 1984; Wanninkhof *et al.*, 1985, 1987). This value of 200 μm, which translates to a CO_2 exchange coefficient at 10 °C of 2.11 cm hr^{-1}, may be somewhat too large. For example, using the CO_2 flux measurements of Coyne & Kelly (1975) for an arctic coastal lake, which were made directly in chambers and are therefore independent of the assumptions required to estimate *z*, we back calculated a boundary layer *z* equal to only 44 μm. This value is similar to recent measurements made in other lakes (Wanninkhof *et al.*, 1985). Boundary layer thickness in rivers is difficult to estimate because surface turbulence is generated by advection in addition to wind sheer. Wanninkhof *et al.* (1990) estimated a gas exchange coefficient of 21 cm hr^{-1} (boundary layer of about 25 μm) for a river; we assumed a thickness of 200 μm for both the rivers and lakes in our study. Therefore, our value of 200 μm may be conservative in some lakes and is most likely too conservative in the rivers, and thus our calculated CO_2 fluxes may underestimate actual fluxes.

We tested this flux equation in one lake by measuring wind speed and the flux of CO_2 into a floating plexiglass chamber (surface area = 0.2 m^2). The calculated flux of 8.6 mmol m^{-2} d^{-1} (estimated boundary layer of 400 μm at ambient conditions) compared very well to a measured chamber flux of 6.7 mmol m^{-2} d^{-1}.

Annual and seasonal variations in CO_2 flux as well as the effects of nutrient fertilization were estimated from weekly data taken from Lake N2. From 1985 to 1990 this lake has been divided by a plastic curtain and one half fertilized with phosphoric acid and NH_4NO_3. Each year nutrients were added continuously starting on 1 July and ending on 15 August at loading rates of about 3 mmol N m^{-2} d^{-1} and 0.25 mmol P m^{-2} d^{-1}. Chlorophyll *a* (CHLa) was extracted in 90% acetone and concentrations were calculated using equations in Wetzel & Likens (1979). Primary production was measured using 24 hr incubations with ^{14}C labeled bicarbonate (method details in Miller *et al.*, 1986). Numeric variables used for statistical analyses were log transformed to normalize distributions and to reduce heteroscedasticity.

Results and discussion

Although the 25 lakes and 4 rivers studied lie in varied terrain, there were no systematic patterns of gas flux with site location. Nor were there statistically significant linear or non-linear trends of gas flux with lake area or maximum depth.

The mean surface water P_{CO2} for all lakes was 1162 ± 134 μatm (± 1 SE), which is supersaturated by more than 3 times with respect to the atmosphere (Table 1). The flux of CO_2 from water to the atmosphere ranged from −5.5 to 59.8 mmol m^{-2} d^{-1}, and averaged 20.9 ± 3.3 (N = 25) in the lakes and 7.1 ± 3.0 mmol m^{-2} d^{-1} in the rivers (N = 4). Only two lakes showed invasion of CO_2 from the atmosphere to water (negative fluxes). These CO_2 evasion rates are similar to rates reported previously for North Slope waters: 44.3 and 12.7 mmol m^{-2} d^{-1} for a pond and a lake, respectively, located near Point Barrow (Coyne & Kelley, 1974). Hesslein *et al.* (1980)

reported smaller fluxes for an oligotrophic lake in northwestern Ontario.

The mean surface water P_{CH4} for the 9 sites examined was 272 ± 50 μatm (± 1 SE, N = 10), or more than 150 times the pressure at atmospheric saturation (Table 2). The calculated CH_4 flux to the atmosphere averaged 0.43 ± 0.08 mmol $m^{-2} d^{-1}$ (N = 10). We also measured directly a CH_4 flux of 0.13 mmol $m^{-2} d^{-1}$ using a floating plexiglass chamber in Dam Pond. CH_4 supersaturation in unmanipulated arctic and taiga surface waters was also reported by Whalen and Reeburgh (1990b), who measured fluxes ranging from 0.29 to 8.17 mmol $m^{-2} d^{-1}$ (N = 6, median = 0.52). There was a substantial buildup of CH_4 over winter at the bottom of the fertilized side of Lake N2 and in Lake N3 (Table 2). Lake N3 lies just downstream of N2 and has been affected by fertilizer addition to Lake N2. Some of the CH_4 that accumulated in these lakes was oxidized or lost to the atmosphere during ice-out and spring turnover; in Lake N2 bottom water, P_{CH4} decreased from 70 590 μatm in May to 4 100 μatm in July after ice-out. The origin of this CH_4 is due to microbial decomposition in anaerobic sediments or bottom waters; it is unlikely that large amounts of CH_4 are transported into the lake during winter when the ground is frozen. Similar CH_4 concentrations have been reported for a eutrophic lake on the Canadian shield (Rudd & Hamilton, 1978) and for oligotrophic lakes in the Canadian arctic (Welch *et al.*, 1980).

Variations in CO_2 flux

Year to year variations in CO_2 flux for Lake N2 are shown in Table 3. The results are based on weekly data taken from late June through August

Table 2. Methane pressure, concentration, and flux in lakes and in the Kuparuk River; P_{CH_4} was measured directly and the concentration of CH_4 was calculated using solubility data. Pond386 is located next to the Dalton Highway at a latitude of 69° 50' N. The values for Lake William are duplicates, and the values for Lakes N3 and N2-fert on 12 May 90 were taken under ice.

Site	Date	Depth (m)	Temp (°C)	CH_4 (μmol l^{-1})	P_{CH_4} (μatm)	CH_4 Flux (mmol $m^{-2} d^{-1}$)
Toolik	14 Jul 90	0	15	1.10	642	1.02
		23	4	0.46	203	.
N2–cont	14 Jul 90	0	15	0.57	333	0.53
		6	8	0.25	123	.
N2–fert	14 Jul 90	0	15	0.25	146	0.23
		6	8	0.28	138	.
		10	6	8.80	4097	.
	12 May 90	3	1.6	0.15	62	.
		5	2.5	0.26	110	.
		6	2.8	0.17	72	.
		7	3.0	0.13	56	.
		8	3.0	0.55	235	.
		9	3.0	165	70590	.
N1	14 Jul 90	0	15	0.09	53	0.08
		12	6	0.08	37	.
Dam	14 Jul 90	0	15	0.46	269	0.43
	15 Jul 90	0	15	0.55	321	0.51
William	5 Aug 90	0	12	0.34	185	0.29
		0	12	0.37	201	0.32
Coleen	5 Aug 90	0	8	0.41	201	0.31
Windy	5 Aug 90	0	9	0.29	146	0.23
Pond386	5 Aug 90	0	10	0.62	321	0.50
N3	12 May 90	3.5	2.5	16.7	7042	.
Kuparuk R.	5 Aug 90	0	8	0.48	236	0.36

Table 3. Year to year variations in CO_2 flux (FCO_2), DIC, CHLa, and temperature in Lake N2. Samples were taken weekly and mean values are reported ±1 SE. Temp (°C) is reported as the average value from 0 to 3 m depth to reduce variability introduced by short term fluctuations in surface temperatures. ANOVA statistics are given for the null hypothesis that the response of these variables is the same across years. Scheffe's multiple comparison of means is reported only when the overall F value is significant; different letters indicate statistically significant differences between years ($P<0.05$).

Year	Lake N2–Control			
	FCO_2 (mmol m^{-2} d^{-1})	DIC (μmol l^{-1})	CHLa (μg l^{-1})	Temp (°C)
1985	0.9 ± 1.6A	792 ± 50A	0.65 ± 0.12	13.4 ± 0.6
1986	8.9 ± 1.3AB	849 ± 47A	0.90 ± 0.14	14.7 ± 0.9
1987	15.5 ± 6.1B	1105 ± 42B	1.38 ± 0.33	14.9 ± 0.7
1988	12.5 ± 4.6B	1268 ± 52B	1.82 ± 0.53	13.4 ± 1.0
1989	15.5 ± 3.4B	1317 ± 38B	–	13.7 ± 0.4
ANOVA F	5.94	31.3	1.87	0.86
df	42	42	26	40
$P>F$	0.0008	0.0001	0.16	0.49
Year	**Lake N2–Fertilized**			
	FCO_2 (mmol m^{-2} d^{-1})	DIC (μmol l^{-1})	CHLa (μg l^{-1})	Temp (°C)
1985	−4.4 ± 1.0	782 ± 50A	5.41 ± 1.78	138 ± 0.6
1986	1.0 ± 2.1	1334 ± 43BC	4.63 ± 0.88	14.1 ± 0.6
1987	5.4 ± 4.5	1231 ± 29B	7.40 ± 1.80	14.7 ± 0.7
1988	2.0 ± 1.6	1515 ± 33D	7.28 ± 1.75	12.8 ± 0.8
1989	1.4 ± 2.0	1457 ± 24DC	–	13.2 ± 0.5
ANOVA F	1.91	72.7	0.27	1.38
df	51	51	32	43
$P>F$	0.12	0.0001	0.85	0.26

for the period 1985–1989. An analysis of variance (ANOVA) shows that the variation in CO_2 flux over time is significant ($F=5.94$, $df=42$, $P<0.0008$), and an *a posteriori* multiple comparison of means indicates that the average CO_2 flux was significantly lower in 1985 than in subsequent years (Table 3). The reasons for this difference in 1985 are unclear, although DIC concentrations were also significantly lower in 1985 and 1986 than in other years. DIC concentrations may have been altered because installation of the curtain restricted flow between the two sides of the lake and confined the major stream inflow to the control side. Insufficient data exist over these years to determine the effect of variable rain and snowfall on CO_2 flux.

Year to year variations in CO_2 flux for Toolik Lake were also small. Although the overall F-test for the hypothesis that CO_2 flux is the same among years was significant ($P>F=0.009$), Scheffe's test was unable to discriminate between any two specific years (Table 4). There were also no significant differences in DIC concentrations among all years (Table 4). The absence of a strong trend over the last 14 years suggests that such CO_2 flux is not a recent phenomenon; it is probable, but not demonstrable, that CO_2 evasion has been a characteristic of these arctic ecosystems for a minimum of several tens to several hundreds of years.

Inter-annual variability in CO_2 flux of other lakes ranged from low in Island Lake (17.6 and 18.1 mmol m^{-2} d^{-1} in July 1986 and July 1988, respectively) to high in Lake E1 (58, 8.4, and 113 mmol m^{-2} d^{-1} in August 1983, June 1985, and July 1986). Part of this high annual variation in Lake E1 may be due to the fact that sampling was done in different seasons each year.

Table 4. Year to year variations in CO_2 flux and DIC concentrations in Toolik Lake. Means (± 1 SE) are reported for N surface water samples taken weekly during eacht ice-free season (see Fig. 3). ANOVA statistics are given for the null hypothesis that the response of these variables is the same across years.

Year	TOOLIK LAKE		
	N	CO_2 Flux (mmol $m^{-2} d^{-1}$)	DIC (μmol l^{-1})
1975	4	20.5 ± 6.7	396 ± 53
1976	7	34.6 ± 19.8	501 ± 46
1977	11	66.7 ± 20.0	568 ± 56
1978	5	58.3 ± 19.7	495 ± 45
1979	10	27.9 ± 5.5	516 ± 22
1980	10	37.0 ± 10.1	557 ± 49
1983	5	12.2 ± 1.4	478 ± 34
1984	2	10.6 ± 0.6	633 ± 64
1989	8	10.7 ± 1.8	517 ± 23
ANOVA F		2.92	1.83
df		61	73
P>F		0.009	0.092

Seasonal variations in CO_2 flux were studied intensively in Lake N2 and Toolik Lake. Figure 2 illustrates the general trends in CO_2 flux, pH, and concentrations of DIC and CHLa during the summer growing season in Lake N2 over 5 years. In the control side of the lake there were no strong seasonal trends in any variable for each year or when all five years were grouped together; for example, none of the slopes of the regression of CO_2 flux on sampling date were significantly different ($P<0.05$) from zero for each year. With few exceptions, the CO_2 flux was always from the lake to the atmosphere. In the fertilized side of the lake, however, there were distinct seasonal trends as well as statistically significant differences in the response of these variables relative to the control side (Fig. 2, Table 5). Just after ice-out in mid to late June the CO_2 flux was positive and roughly equal in both sides of the lake. With the onset of nutrient enrichment the CHLa concentrations increased in the fertilized side. As the phytoplankton reduced the pool of CO_2(aq) in the fertilized side, surface water pH rose and CO_2 flux fell. After mid July the flux of CO_2 from water to atmosphere reversed and there was net invasion of CO_2 into the fertilized side of the lake (Fig. 2). This invasion of CO_2 may be somewhat underestimated due to chemically enhanced CO_2 exchange at high pH, although the relatively thin boundary layer and high DIC concentrations in Lake N2 would moderate the enhancement (see Emerson, 1975a). Maximum invasion rates during this time were similar to invasion rates in

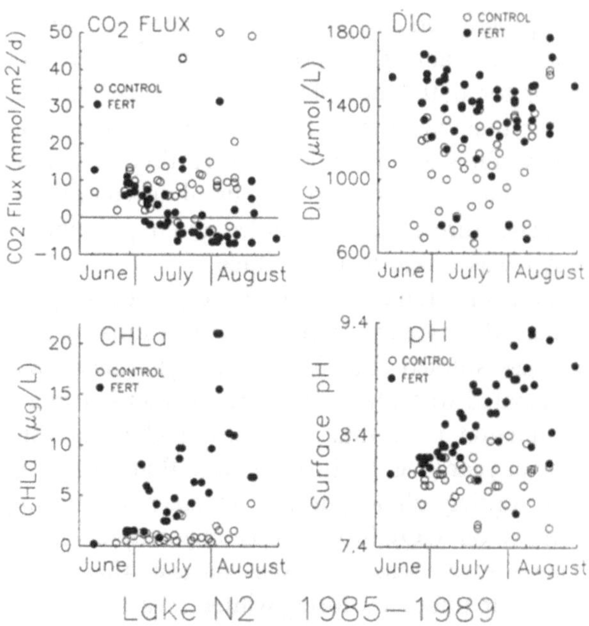

Fig. 2. Seasonal variation in CO_2 flux, DIC, CHLa, and pH for years 1985–1989 in Lake N2. All values are for surface water.

Table 5. Statistical comparison between control and fertilized sides of Lake N2 and reaches of Kuparuk River for CO_2 flux (mmol m^{-2} d^{-1}), DIC (μmol l^{-1}), P_{CO_2} (μatm), CHLa (μg l^{-1}), pH and temperature (°C). Temperature is reported as the average value from 0 to 3 m depth to reduce variability introduced by short term fluctuations in surface temperatures. All years with data are lumped together for each analysis. ANOVA statistics are given for the null hypothesis that the response of the variables is the same between control and fertilized sides.

Lake N2–Control vs. Fertilized

	CO_2 flux	DIC	CHLa	Temp	pH
Mean ± SE					
Control	12.6 ± 2.2	1120 ± 38	1.2 ± 0.2	14.0 ± 0.32	7.93 ± 0.03
Fert	1.4 ± 1.0	1351 ± 35	6.5 ± 0.9	13.6 ± 0.30	8.36 ± 0.05
ANOVA F	35.8	17.9	47.7	0.87	60.2
df	94	94	59	84	94
P>F	0.0001	0.0001	0.0001	0.35	0.0001

Kuparuk River–Control vs. Fertilized

	CO_2 flux	P_{CO_2}
Mean ± SE		
Control	12.8 ± 2.4	846 ± 102
Fert	11.0 ± 1.8	775 ± 76
ANOVA F	0.21	0.16
df	11	11
P>F	0.66	0.69

other softwater lakes (Herczeg, 1987). As summer progressed into August, the highest CO_2 evasion rates and DIC concentrations tended to occur for both sides of Lake N2. Hypolimnetic water had higher CO_2(aq) concentrations than did surface water, and therefore this tendency may be in response to partial breakdown of thermal structure as the lake cooled and hypolimnetic water was mixed to the surface.

In Toolik Lake the highest CO_2 evasion rates occurred very early in the season in most of the 9 years studied (Fig. 3). These high rates were associated with DIC concentrations and primary production rates that were also highest at this time of year (Fig. 3). It is apparent that uptake of CO_2 during photosynthesis is an unimportant control on CO_2 flux in this oligotrophic lake. These early season fluxes suggest that CO_2 builds-up under ice during winter. Evasion of this CO_2 during ice-out may be missed if sampling begins in late June or early July as it did in Lake N2. For example, P_{CO_2} values calculated under ice on 10 May 1990 were higher than maximum values recorded during the open water season in Lakes N1, N2, S6, and Toolik (Table 6). Similar increases in P_{CO2} under ice were found in arctic coastal plain lakes (Coyne & Kelley, 1974). Some of this CO_2 may be depleted by photosynthetic

Table 6. Calculated P_{CO_2} values for under–ice versus open–water samples. Under–ice samples were taken on 10 and 11 May 1990. Open–water samples are presented as a mean with the maximum value in parentheses (see Table 1).

Lake	P_{CO_2} (μatm)	
	Open–water	Under–ice
N1	1170(2548)	3913,4894
S6	628 (930)	6348
N2 control (all years)	825(2916)	8866
N2 fert (all years)	373(1581)	5547
Toolik (1989)	758(1125)	3701

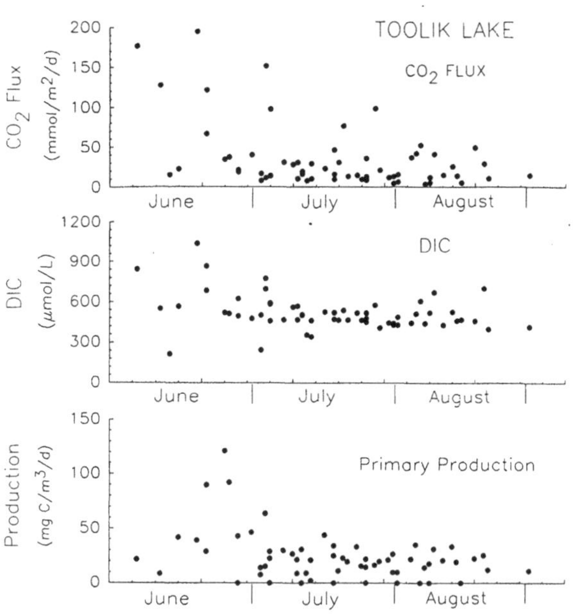

Fig. 3. Seasonal variation in CO_2 flux, DIC, and primary production in Toolik Lake during 1975–1980, 1983, 1984, and 1989. All values are for surface water. Production data for 1975–1980 are from Miller *et al.* (1986).

uptake before ice-out, because net primary production can be substantial under ice if there is little snow cover (Hobbie, 1980; Miller *et al.*, 1986). Such depletion appears to be small in these lakes; for example in Lake N2, a net production rate of 2.5 mmol C m^{-3} d^{-1} from 10 May to 10 June (rate taken from June 1988) would decrease the concentration of CO_2(aq) in surface waters from a measured 600 μmol l^{-1} to 525 μmol l^{-1}. This value of 525 μmol l^{-1} is still more than an order of magnitude higher than CO_2(aq) concentrations measured at the end of June following ice-out. Thus there must be a large evasion of CO_2 from these lakes during ice-out and lake turnover in early June. While CO_2 evasion most likely continues until freeze-up, episodes of CO_2 invasion from the atmosphere are probably rare because of deteriorating conditions for photosynthesis in late August and September (Miller *et al.*, 1986).

The seasonal response of decreased CO_2 flux with increased phytoplankton growth following fertilization in Lake N2 contrasts with a similar fertilization experiment done in the Kuparuk River. Since 1983 the Kuparuk has been fertilized with *P* and *N* and a control reach was maintained upstream (Peterson *et al.*, 1985). In 1985, 1988, and 1989 we sampled both fertilized and control reaches, and found no significant differences in CO_2 flux (Table 5). This contrast between lake and river most likely exists because the renewal time of water in the river is much faster than in the lake, and thus the amount of DIC moving through the fertilized reach during summer is too great to be changed by algal uptake.

We have no measurements of diurnal variations in CO_2 flux. Miller *et al.* (1986) reported that diurnal variations in photosynthesis in Toolik Lake were related positively to light and temperature. If the negative relationship between algal uptake of CO_2 and the flux of CO_2 from lake to atmosphere seen in lake N2 is typical during midsummer, our measurements made during the middle of the day will slightly underestimate the total daily CO_2 flux. This assumes that wind conditions remain constant or are higher at night.

Sources of CO_2

We believe that much of the CO_2 released from these lakes and rivers originates in the terrestrial environment. In most lakes of low to moderate alkalinity the invasion of CO_2 from the atmosphere supports net primary production, and the lakes are considered autotrophic (Schindler *et al.*, 1972). If CO_2 is continually released to the atmosphere, however, the lakes are heterotrophic or they are acting as conduits for terrestrially produced CO_2. In either case, there must be a net influx of carbon or a critical change in sediment conditions such that old organic carbon accumulated in the sediment is respired. It is unlikely that bottom mud is being depleted to provide the CO_2 lost from these systems, judging from sediment accumulation rates in arctic lakes (Livingstone *et al.*, 1958; Cornwell, 1985). In addition, rates of net benthic respiration, which include respiration of terrestrial detritus washed into the lake, are much less than the flux of CO_2 to the atmosphere in Lake N2 and Toolik Lake. In the control side of Lake N2 net benthic respiration is on the order of 5 mmol C $m^{-2} d^{-1}$ (Kipphut, unpubl. data). In Toolik Lake, net consumption of O_2 by the sediments averages 7.6 mmol $m^{-2} d^{-1}$ ($N = 5$; Cornwell & Kipphut, this issue), which is almost 5 times less than the CO_2 flux to the atmosphere (respiration quotient of 1; Table 1). Welch & Bergmann (1985) report an average respiration rate in ten arctic lakes of about 6 mmol O_2 $m^{-2} d^{-1}$.

On the other hand, there are several lines of evidence indicating an important influx of dissolved and particulate carbon of terrestrial origin to arctic lakes and rivers. Thermokarst and fluvial processes cause great erosion of peat into lakes, especially along the coast (Sellmann *et al.*, 1975; Schell & Ziemann, 1983), and some of this terrestrial carbon is transferred to the aquatic food chain (Schell, 1983). In the Kuparuk River, Peterson *et al.* (1986) showed that the carbon cycle was dominated by inputs of peat from the streambank and dissolved organic carbon leached from tundra.

There is also direct evidence that DIC in groundwater is moving from land to surface waters. For example, from 1985 to 1990, P_{CO2} values from 11 soil lysimeters near the Kuparuk River averaged 8315 ± 2191 μatm ($N = 16$; range 2240 to 26650 μatm). These are minimum values because CO_2 in the large headspace of the lysimeter was not measured. High groundwater P_{CO2} values were also measured in the Toolik Lake watershed in summer 1990 (up to 46500 μatm). Such high P_{CO2} values are due in part to permafrost conditions; groundwaters are confined by permafrost to shallow (50 cm) organic rich soil layers, where they can accumulate CO_2 produced by plant and microbial respiration. Movement of these CO_2-charged groundwaters into the Kuparuk River can explain the high P_{CO2} observed in the river (mean = 812 ± 177 μatm).

Our proposed mechanism of gas transport in groundwater is supported by CH_4 measurements. P_{CH4} values in two water samples from soil lysimeters near Toolik Lake were 36500 and 12600 μatm (collected 14 July 1990). These values are much higher than the values we measured in lakes and rivers, suggesting that terrestrially produced CH_4 in groundwater moves into surface waters where it is released to the atmosphere.

An independent estimate of the importance of CO_2 transport from land to water comes from balancing the inputs and outputs of CO_2(aq) in the Kuparuk River basin. On the basis of CO_2(aq) concentrations in rainwater (air-saturated water) and an average groundwater P_{CO2} of 8315 μatm in the upper Kuparuk basin, we calculate that 295×10^5 mol CO_2(aq) are input to the basin each year. Losses include a river output of 182 $\times 10^5$ mol C y^{-1}, a CO_2 flux to the atmosphere of 1.4×10^5 mol y^{-1} (areal flux from Table 1 and a river surface area of 0.12 km^2), and consumption by net primary production of 1.3×10^5 mol CO_2(aq) y^{-1} (Peterson *et al.*, 1986), leaving an excess input from the terrestrial system of 110×10^5 mol C yr^{-1}. The river output assumes an average DIC concentration of 308 μmol l^{-1} and that respiration of peat in the river is negligible (Peterson *et al.*, 1986). This output includes CO_2(aq) plus bicarbonate and therefore is conservative with respect to the contribution of CO_2 from the terrestrial landscape. Thus in this river it appears that the excess CO_2 originates mainly from soil respiration and transport by groundwater. By comparison, in lakes the groundwater input of dissolved CO_2 is supplemented by in-lake respiration of particulate and dissolved organic carbon washed in from the land, especially in sites on the coastal plain.

Tundra carbon budgets

The North Slope of Alaska is considered to be a net sink for carbon both today and in the past (Billings, 1987). This results from an excess of CO_2 uptake by plants over release of CO_2 during soil and root respiration. Chapin *et al.* (1980) assembled data from several sources and estimated this accumulation to be from 59 to 120 g C $m^{-2} y^{-1}$ on the basis of a one year budget at Point Barrow. Billings (1987) estimated the arctic tundra accumulation to be about 41 g C $m^{-2} y^{-1}$. On the basis of an eddy correlation method involving profiles of CO_2 concentration above tundra at Barrow, Coyne & Kelley (1975) estimated a net accumulation of 40 g C m^{-2} y^{-1}. If one assumes that the lower atmosphere is well-mixed on a spatial scale of one to two km^2, this eddy correlation method would include CO_2 efflux from lakes and ponds. The additional efflux of CH_4 is less than 4 g C m^{-2} y^{-1} (39 mg C m^{-2} d^{-1}, from Whalen & Reeburgh, 1990b; 100 active days per year). Even so, the above estimates imply that the inventory of soil carbon stored above permafrost could have accumulated in the unreasonable time of only several hundred years (Billings, 1987). In comparison to these budget estimates, Schell & Ziemann (1983) used ^{14}C dating of peat cores and reported a smaller long-term carbon accumulation rate of 13.3 g C m^{-2} y^{-1} on the coastal plain.

Estimates of the importance of carbon transport from terrestrial to aquatic systems vary in these published budgets. Chapin *et al.* (1980) considered this pathway of carbon to play a 'relatively minor role in the annual budgets' of tundra near Barrow, and Billings (1987) assumed (but questioned) a loss of only 2 g C m^{-2} y^{-1} to aquatic systems. In their study of peat erosion and accumulation on the coastal plain, Schell & Ziemann (1983) found instead that loss of carbon to aquatic environments was a 'major factor'. The method of ^{14}C-dating of peat cores measures a long-term net accumulation and would account for carbon losses from land to aquatic systems that the experimental measures would miss. We found that roughly 24 g C m^{-2} are lost to the atmosphere from lakes and ponds on the North Slope each year (100 active days per year; includes 0.5 g C m^{-2} lost as CH_4). The true value may be slightly higher considering that under-ice respiration in lakes is not included. In tundra rivers, some of the DOC and POC exported to the sea will be respired and lost to the atmosphere. Assuming that 50% of the surface area near the coast is water or marsh (Sellmann *et al.*, 1975), this loss through aquatic systems is up to half of the terrestrial carbon accumulation rate. Subtracting 24 g C m^{-2} y^{-1} from the terrestrial budget estimates of carbon accumulation results in a net accumulation closer to radiocarbon based estimates of long-term storage. Accounting for this loss of carbon to aquatic systems also gives more reasonable accumulation rates when considering the inventory of carbon currently in soils (Billings, 1987). Although lake density decreases away from the coast, water coverage on the entire North Slope above 69.5° latitude is still about 14% of the total surface area. The total loss of carbon from tundra lakes and rivers is about 0.02 Pg C y^{-1}, which is 7–20% of the current estimated carbon sink of 0.1–0.3 Pg C y^{-1} for arctic tundra (14% water coverage, tundra area of 5×10^6 km^2; see Post, 1990). These results support the idea that transport and eventual loss of dissolved and particulate carbon from land to water is an important process in Alaskan and probably other arctic environments.

CO_2 flux and changing climate

According to global climate models, northern latitudes will experience the largest temperature changes (4–12 °C) under a scenario of future global warming with a doubling of greenhouse gases (Grotch, 1988; Abelson, 1989; Lachenbruch & Marshall, 1986). Accumulation or loss of arctic carbon stocks depends on the balance between primary production and decomposition, although at present the effect of climate warming and the associated buildup of greenhouse gases on carbon cycling in the Arctic is unresolved. It is known, however, that in general primary production in aquatic and terrestrial arctic ecosystems is limited by nutrients rather than by temperature (Kalff & Welch, 1974; Hobbie, 1980; Miller *et al.*, 1986; Chapin *et al.*, 1980; Billings *et al.*, 1984; Shaver & Chapin, 1986). On the other hand, temperature is most limiting to decomposition or respiration and thus to the generation and release of CO_2 (Heal *et al.*, 1981; Billings *et al.*, 1982). Because of these relationships between production, respiration, and temperature, recent projections hold that large releases of CO_2 to the atmosphere will occur given a 5 °C increase in average annual temperature (Post, 1990). The release of CH_4 to the atmosphere is also expected to increase (Post, 1990; Livingston & Morrissey, 1990), although the importance of methane consumption in soils and its response to changing temperature and moisture regimes is somewhat in question (Whalen & Reeburgh, 1990a).

These increased carbon releases were predicted for terrestrial systems. our results indicate that they would be substantially augmented by CO_2 evasion from lakes and rivers. It is preliminary to assign values to future aquatic gas fluxes, because the response of land plants and the role of hydrology and nutrients remain ill-defined. The generation and transport of carbon through the landscape may be altered by changing water availability (Lachenbruch & Marshall, 1986; Post, 1990), and nutrient loss from the soil as a result of warmer temperatures (Nadelhoffer *et al.*, 1991) may stimulate primary production in receiving water bodies. Our experiments indicate that given substantial nutrient enrichment and a slow water renewal time, photosynthesis could reduce the loss of CO_2 from naturally heterotrophic Arctic waters.

Acknowledgements

S. Whalen provided the CH_4 data for Lakes N2 and N3 on 12 May 90, and pH and alkalinity data for some sites. B. Moller, S. Sugai, B. Wang, M. Cheek, M. Castro, and A. Giblin assisted in the lab and field. We thank J. Hobbie, G. Shaver, B. Peterson, D. Schell, and one anonymous reviewer for discussion and comments. Research was supported by the National Science Foundation grants BSR8702328, DPP8722015, and DPP8320544, and by the A. W. Mellon Foundation. This is The Ecosystems Center, Marine Biological Center, University of Alaska Institute of Marine Science Contribution 901.

References

Abelson, P. H., 1989. The Arctic: A key to world climate. Science 243: 873.

Billings, W. D., 1987. Carbon balance of Alaskan tundra and taiga ecosystems: past, present and future. Quat. Sci. Rev. 6: 165–177.

Billings, W. D., J. O. Luken, D. A. Mortensen & K. M. Peterson, 1982. Arctic tundra: A source or sink for atmospheric carbon dioxide in a changing environment? Oecologia 53: 7–11.

Billings, W. D., K. M. Peterson, J. O. Luken & D. A. Mortensen, 1984. Interaction of increasing atmospheric carbon dioxide and soil nitrogen on the carbon balance of tundra microcosms. Oecologia 65: 26–29.

Bliss, L. C., O. W. Heal & J. J. Moore, 1981. Tundra ecosystems: A comparative analysis. IBP Handbook 25. Cambridge University Press, Cambridge, 715 pp.

Bolin, B., B. R. Doos, J. Jager & R. A. Warrick, 1986. The greenhouse effect, climate change, and ecosystems. SCOPE 29. John Wiley & Sons, New York, 541 pp.

Bower, P. & D. McCorkle, 1980. Gas exchange, photosynthetic uptake, and carbon budget for a radiocarbon addition to a small enclosure in a stratified lake. Can. J. Fish. aquat. Sci. 37: 464–471.

Broecker, H. C., J. Peterman & W. Siems, 1978. The influence of wind on CO_2-exchange in a wind-wave tunnel, including the effects of monolayers. J. Mar. Res. 36: 595–610.

Broecker, W. S. & T.-H. Peng, 1974. Gas exchange rates between air and sea. Tellus 16: 21–35.

Broecker, W. S., T.-H. Peng, G. Mathieu, R. Heslein & T. Torgersen, 1980. Gas exchange rate measurements in natural systems. Radiocarbon 22: 676–683.

Buck, A. L., 1981. New equations for computing vapor pressure and enhancement factor. J. appl. Meteorol. 20: 1527–1532.

Chapin, F. S., III, P. C. Miller, W. D. Billings & P. I. Coyne, 1980. Carbon and nutrient budgets and their control in coastal tundra. In J. Brown, P. C. Miller, L. L. Tieszen & F. K. Bunnell (eds), An Arctic Ecosystem, the Coastal Tundra at Barrow, Alaska. IBP Handbook 12. Dowden, Hutchinson & Ross, Inc., Stroudsburg: 458–482.

Cornwell, J. C., 1985. Sediment accumulation rates in an Alaskan arctic lake using a modified ^{210}Pb technique. Can. J. Fish. aquat. Sci. 42: 809–814.

Cornwell, J. C. & G. W. Kipphut, 1992. Biogeochemistry of manganese- and iron-rich sediments in Toolik Lake, Alaska. Hydrobiologia 240: 45–59.

Coyne, P. I. & J. J. Kelley, 1974. Carbon dioxide partial pressures in arctic surface waters. Limnol. Oceanogr. 19: 928–938.

Coyne, P. I. & J. J. Kelley, 1975. CO_2 exchange over the Alaskan arctic tundra: meteorological assessment by an aerodynamic method. J. appl. Ecol. 12: 587–611.

Emerson, S., 1975a. Chemically enhanced CO_2 gas exchange in a eutrophic lake: A general model. Limnol. Oceanogr. 20: 743–753.

Emerson, S., 1975b. Gas exchange rates in small Canadian shield lakes. Limnol. Oceanogr. 20: 754–761.

Grotch, S. L., 1988. Regional intercomparison of general circulation model predictions and historical climate data.

DOE/NBB-0084 TR041. U.S. Department of Energy. Washington. 291 pp.

Hartman, B. & D. E. Hammond, 1984. Gas exchange rates across the sediment-water and air-water interfaces in south San Francisco Bay. J. Geophys. Res. 89: 3593–3603.

Heal, O. W., P. W. Flanagan, D. D. French & S. F. MacLean, Jr., 1981. Decomposition and accumulation of organic matter in tundra. In L. C. Bliss, O. W. Heal & J. J. Moore (eds), Tundra Ecosystems: A Comparative Analysis. IBP Handbook 25. Cambridge University Press, Cambridge: 587–633.

Herczeg, A. L., 1987. A stable carbon isotope study of dissolved inorganic carbon cycling in a softwater lake. Biogeochemistry 4: 231–263.

Hesslein, R. H., W. S. Broecker, P. D. Quay & D. W. Schindler, 1980. Whole-lake radiocarbon experiment in an oligotrophic lake at the Experimental Lakes Area, Northwestern Ontario. Can. J. Fish. aquat. Sci. 37: 454–463.

Hobbie, J. E., 1980. Limnology of tundra ponds. IBP Handbook 13. Dowden, Hutchinson & Ross, Inc., Stroudsburg, 514 pp.

Hoover, T. E. & P. C. Berkshire, 1969. Effects of hydration on carbon dioxide exchange across air-water interface. J. Geophys. Res. 74: 456–474.

Himmelblau, D. M., 1964. Diffusion of dissolved gases in liquids. Chem. Rev. 64: 527–550.

Jähne, B., K. H. Fischer, J. Imberger, P. Libner, W. Weiss, D. Imboden, U. Lemnin & J. M. Jaquet, 1984. Parametrization of air/lake gas exchange. In W. Brutsaert and G. H. Jirka (eds), Gas Transfer at Water Surfaces. D. Reidel, Dordrecht: 459–466.

Kalff, J. & H. E. Welch, 1974. Phytoplankton production in Char Lake, a natural polar lake, and in Meretta Lake, a polluted polar lake, Cornwallis Island, Northwest Territories. J. Fish. Res. Bd Can. 31: 621–636.

Kanwisher, J., 1963. On the exchange of gases between the atmosphere and the sea. Deep-Sea Res. 10: 195–207.

Kling, G. W., G. W. Kipphut & M. C. Miller, 1991. Arctic lakes and streams as gas conduits to the atmosphere: implications for tundra carbon budgets. Science 251: 298–301.

Kling, G. W., W. J. O'Brien, M. C. Miller & A. E. Hershey, 1992. The biogeochemistry and zoogeography of lakes and rivers in arctic Alaska. Hydrobiologia 240: 1–14.

Lachenbruch, A. H. & B. V. Marshall, 1986. Changing climate: Geothermal evidence from permafrost in the Alaskan arctic. Science 234: 689–696.

Liss, P. S., 1973. Process of gas exchange across an air-water interface. Deep-Sea Res. 20: 221–238.

Livingston, G. P. & L. A. Morrissey, 1990. An interannual comparison of arctic methane emissions: a climatic warming scenario. p. 105 In International Conference on the Role of the Polar Regions in Global Change, June 11–15, 1990, University of Alaska, Fairbanks. 230 pp.

Livingstone, D. A., K. Bryan, Jr. & R. C. Leahy, 1958. Effects of an arctic environment on the origin and development of freshwater lakes. Limnol. Oceanogr. 3: 192–214.

Merlivat, L. & L. Memery, 1983. Gas exchange across an air-water interface: Experimental results and modeling of bubble contribution to transfer. J. Geophys. Res. 88: 707–724.

Miller, M. C., G. R. Hater, P. Spatt, P. Westlake & D. Yeakel, 1986. Primary production and its control in Toolik Lake, Alaska. Arch. Hydrobiol. Suppl. 74: 97–131.

Nadelhoffer, K. J., A. E. Giblin, G. R. Shaver & J. A. Laundre, 1991. Effects of temperature and substrate quality on element mineralization in six arctic soils. Ecology 72: 242–253.

Peterson, B. J., J. E. Hobbie, A. E. Hershey, M. A. Lock, T. E. Ford, J. Robie Vestal, V. L. McKinley, M. A. J. Hullar, R. M. Ventullo & G. S. Volk, 1985. Transformation of a tundra river from heterotrophy to autotrophy by addition of phosphorus. Science 229: 1383–1386.

Peterson, B. J., J. E. Hobbie & T. L. Corliss, 1986. Carbon flow in a tundra stream ecosystem. Can. J. Fish. aquat. Sci. 43: 1259–1270.

Plummer, L. N. & E. Busenberg, 1982. The solubilities of calcite, aragonite and vaterite in CO_2-H_2O solutions between 0 and 90 °C, and an evaluation of the aqueous model for the system $CaCO_2$-CO_2-H_2O. Geochim. Cosmochim. Acta 46: 1011–1040.

Post, W. M., 1990. Report of a workshop on climate feedbacks and the role of peatlands, tundra, and boreal ecosystems in the global carbon cycle. Publ. No. 3289. Oak Ridge National Laboratory, Environmental Sciences Division, Oak Ridge, 32 pp.

Rudd, J. W. M. & R. D. Hamilton, 1978. Methane cycling in a eutrophic shield lake and its effects on whole lake metabolism. Limnol. Oceanogr. 23: 337–348.

Schell, D. M., 1983. Carbon-13 and carbon-14 abundances in Alaskan aquatic organisms: delayed production from peat in Arctic food webs. Science 219: 1068–1071.

Schell, D. M. & P. J. Ziemann, 1983. Accumulation of peat carbon in the Alaska arctic coastal plain and its role in biological productivity. pp. 1105–1110. In Permafrost, Fourth International Conference, National Academy Press, Washington. 1524 pp.

Schindler, D. W., G. J. Brunskill, S. Emerson, W. S. Broecker & T.-H. Peng, 1972. Atmospheric carbon dioxide: its role in maintaining phytoplankton standing crops. Science 177: 1192–1194.

Sellmann, P. V., J. Brown, R. I. Lewellen, H. McKim & C. Merry, 1975. The classification and geomorphic implications of thaw lakes of the arctic coastal plain, Alaska. Report 344. U.S. Army Cold Regions Research and Engineering Lab, Hanover, 24 pp.

Shaver, G. R. & F. S. Chapin, III, 1986. Effect of NPK fertilization on production and biomass of Alaskan tussock tundra. Arct. Alp. Res. 18: 261–268.

Shaver, G. R., K. J. Nadelhoffer & A. E. Giblin, 1992. Biogeochemical diversity and element transport in a heterogeneous landscape, the North Slope of Alaska. In M. Turner & R. Gardner (eds), Quantitative Methods in Landscape Ecology. Springer-Verlag. In press.

Smith, S. D. & E. P. Jones, 1985. Evidence for wind-pumping of air-sea gas exchange based on direct measurements of CO_2 fluxes. J. Geophys. Res. 90: 869–875.

Stumm, W. & J. J. Morgan, 1981. Aquatic chemistry, 2nd edition. John Wiley & Sons, New York, 780 pp.

Tans, P. P., I. Y. Fung & T. Takahashi, 1990. Observational constraints on the global atmospheric CO_2 budget. Science 247: 1431–1438.

Torgersen, T., G. Mathieu, R. H. Hesslein & W. S. Broecker, 1982. Gas exchange dependency on diffusion coefficient: direct ^{222}Rn and ^{3}He comparisons in a small lake. J. Geophys. Res. 87: 546–556.

Wanninkhof, R., J. R. Ledwell & W. S. Broecker, 1985. Gas exchange-wind speed relation measured with sulfur hexafluoride on a lake. Science 227: 1224–1226.

Wanninkhof, R., J. R. Ledwell & W. S. Broecker, 1987. Gas exchange on Mono Lake and Crowley Lake, California. J. Geophys. Res. 92: 14567–14580.

Wanninkhof, R., P. J. Mulholland & J. W. Elwood, 1990. Gas exchange rates for a first-order stream determined with deliberate and natural tracers. Water Resour. Res. 26: 1621–1630.

Weiss, R. F., 1974. Carbon dioxide in water and seawater: the solubility of a non-ideal gas. Mar. Chem. 2: 203–215.

Welch, H. E. & M. A. Bergmann, 1985. Winter respiration of lakes at Saqvaqjuac, N.W.T. Can. J. Fish. aquat. Sci. 42: 521–528.

Welch, H. E., J. W. M. Rudd & D. W. Schindler, 1980. Methane addition to an Arctic lake in winter. Limnol. Oceanogr. 25: 100–113.

Wetzel, R. G. & G. E. Likens, 1979. Limnological analyses. W. B. Saunders, Philadelphia, 357 pp.

Whalen, S. C. & W. S. Reeburgh, 1990a. Consumption of atmospheric methane by tundra soils. Nature 346: 160–162.

Whalen, S. C. & W. S. Reeburgh, 1990b. A methane flux transect along the Trans-Alaska pipeline Haul Road. Tellus B42: 237–245.

Wilhelm, E., R. Battino & R. J. Wilcock, 1977. Low-pressure solubility of gases in liquid water. Chem. Rev. 77: 219–262.

Hydrobiologia **240**: 37–44, 1992.
W.J. O'Brien (ed.), Toolik Lake – Ecology of an Aquatic Ecosystem in Arctic Alaska.

A silicon budget for an Alaskan arctic lake

Jeffrey C. Cornwell & Susan Banahan
University of Maryland System CEES, Horn Point Environmental Laboratory, P.O. Box 775, Cambridge, MD 21613, USA

Key words: arctic lakes, biogenic silica, nutrient cycles, diatoms

Abstract

The cycling of dissolved reactive silicate (DRS) and sedimentary biogenic SiO_2 has been examined in ultraoligotrophic Toolik Lake, Alaska. Watershed output of DRS (~ 7 mmol m^{-2} yr^{-1}) is similar to other arctic study sites, but a short water residence time results in the poor retention of DRS ($\sim 17\%$) within the lake. Biogenic SiO_2 concentrations in surficial sediments average 123 mg g^{-1} with the highest concentrations arising from the production of benthic diatoms in the littoral zone. Lake water DRS concentrations are highest prior to spring flow and photic zone concentrations exceed 23 μmol l^{-1} at the time of greatest primary production. Wintertime increases in watercolumn DRS concentrations and spring and summer pore water DRS flux estimates indicate that internal cycling within the lake may supply an amount of DRS equivalent to that which biologically utilized.

Introduction

The primary productivity of arctic lakes is considerably lower than for lakes in other biogeographic provinces of the world (Hobbie, 1973; Welch & Kalff, 1974; Miller *et al.*, 1986). While a number of studies have indicated that low inputs of nitrogen and phosphorus appear to limit phytoplankton productivity in arctic lakes and ponds (Schindler *et al.*, 1974a, b; Prentki *et al.*, 1980; Whalen & Cornwell, 1985), detailed considerations of silicon's role in arctic freshwater nutrient cycles are absent. Silicon is an essential nutrient for diatom production (Hutchinson, 1957; Wetzel, 1975) and exhibits a very dynamic cycle in arctic and subarctic marine ecosystems (e.g. Banahan & Goering, 1986).

In this report, we present a silicon budget for Toolik Lake, Alaska, an arctic lake with well documented productivity, nitrogen and phosphorus cycles (Whalen & Cornwell, 1985; Whalen & Alexander, 1986; Miller *et al.*, 1986; Cornwell, 1987). Our research objectives are to 1) examine the rates of sedimentary silicon recycling and burial and 2) determine the importance of benthic diatoms to the silicon balance. The measurement of 1) stream dissolved reactive Si (DRS) fluxes, 2) water column DRS concentrations and 3) pore water DRS and biogenic SiO_2 concentrations in ^{210}Pb-dated sediments provide the elements for a biogeochemical silicon budget.

Study area

Toolik Lake (68° 38′ N; 149° 38′ W) is a multiple-basin kettle lake in the northern foothills of the Brooks Range, Alaska (Fig. 1). The mean and maximum depths of this 1.5 km^2 lake are 7 and 25 m, respectively. The 65 km^2 watershed is al-

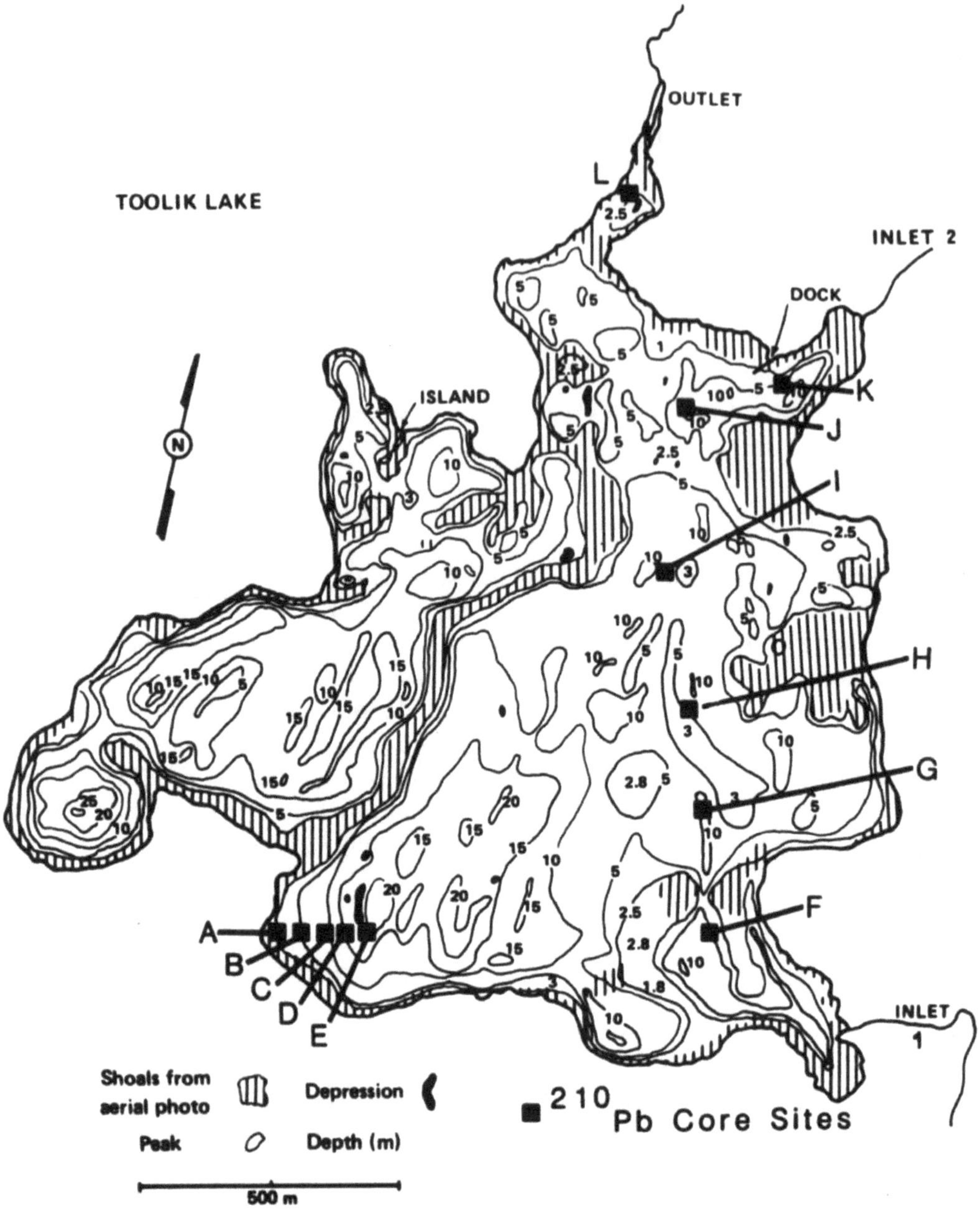

Fig. 1. Map of Toolik Lake, Alaska, showing sampling sites.

most completely covered with tussock tundra vegetation which overlies 12000 year old glacial deposits. The lake is ice free ~3 months per year, with thermal stratification during the summer. Stream flow commences in the major inlets and outlet during mid-May and ceases in mid-September. The main inlet stream drains 75% of the watershed, a second stream drains 7% and several ephemeral streams drain the remaining 18%.

The alkalinity of Toolik Lake is ~0.4 mequiv l^{-1} and Ca is the predominant cation. Photic zone chlorophyll a concentrations average 1.3 $\mu g\,l^{-1}$ during the summer, and planktonic primary production is about 10 g C $m^{-2}\,yr^{-1}$ (Whalen & Cornwell, 1985; Miller *et al.*, 1986). Nutrient retention in Toolik Lake is low because of short water residence times under the ice during spring flow and sustained primary production requires the continual turnover of organic N and

P (Whalen & Cornwell, 1985; Whalen *et al.*, 1988).

The lake bottom is covered with fine-grained muds at water depths exceeding ~2.5 m and sediment accumulation rates are extremely slow (27 g m^{-2} yr^{-1}; Cornwell, 1985). High concentrations of Mn and Fe oxides (>15% each in deepwater sediments) result from post-depositional migration of Mn and Fe (Cornwell, 1983). As a consequence of Mn and Fe enrichment, trace metals and P are also enriched in some sediment horizons (Cornwell, 1986).

Methods

Lake water was collected using Van Dorn or similar type water samplers. Stream water flow rates were determined using a Gurley meter (Whalen & Cornwell, 1985) and water was collected in 4 l plastic bottles at the time of flow measurement. Lake and stream water was filtered through Gelman A/E filters and frozen in plastic bottles until analysis. DRS was determined in stream, lake and sediment pore waters using automated colorimetry (Armstrong *et al.*, 1967). Frozen samples were thawed and kept at room temperature for three days to ensure analysis of monomeric silicic acid.

Undisturbed sediments were collected through the ice in May and June 1980 using a KB gravity corer. Sediment pore waters were obtained using Reeburgh (1967) core squeezers with squeezing generally at ~10 °C, ~6 °C warmer than *in situ* temperatures. Sediments from a parallel core were collected for ^{210}Pb dating and the analysis of major sediment components, trace metals and biogenic silica. Amorphous SiO_2, assumed to be primarily biogenic SiO_2, was analyzed using a Na_2CO_3 digestion technique with an Al correction for the dissolution of clay minerals (Eggiman *et al.*, 1980). Digests were analyzed for Si using manual colorimetry (Parsons *et al.*, 1984). Repeated analysis of a surficial sediment sample with ~114 mg g^{-1} biogenic SiO_2 gave a COV of <2% ($n = 6$). Techniques and results of other analyses are presented in Cornwell (1983).

The calculation of stream inputs to the lake were made assuming that the ungauged input to the lake (~22% of inflow) had DRS concentrations similar to the major inlet stream. Whole lake sediment budgets were calculated using the mean deposition rates and the area encompassed in a series of depth intervals (0–2.5, 2.5–5, 5–10, 10–25 m). Rocky, shallow water sediments (generally depths <2.5 m) were assumed to have no fine sediment accumulation (Cornwell, 1985).

Results

Silicon distribution

Stream concentrations of DRS in both the inlet and outlet streams are lowest in early June (Fig. 2). The inlet DRS concentrations show higher values early in the season, as do total dissolved P and N (Whalen & Cornwell, 1985). Unlike P and N, however, DRS concentrations in the main inlet increase in July and August, similar to increases in Ca and Mg (Cornwell, 1983). The pattern of temporal variation of DRS in inlet 2, a secondary inlet, was similar to inlet 1. Several small ponds upstream of Toolik Lake modify the composition of inlet water to an unknown degree, but lake May/early June minima in both the inlet and outlets may arise by dilution

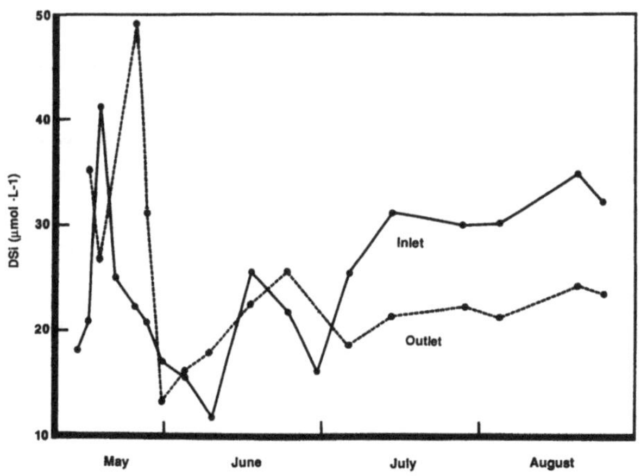

Fig. 2. Concentrations of DRS in the main inlet and outlet of Toolik Lake in 1980.

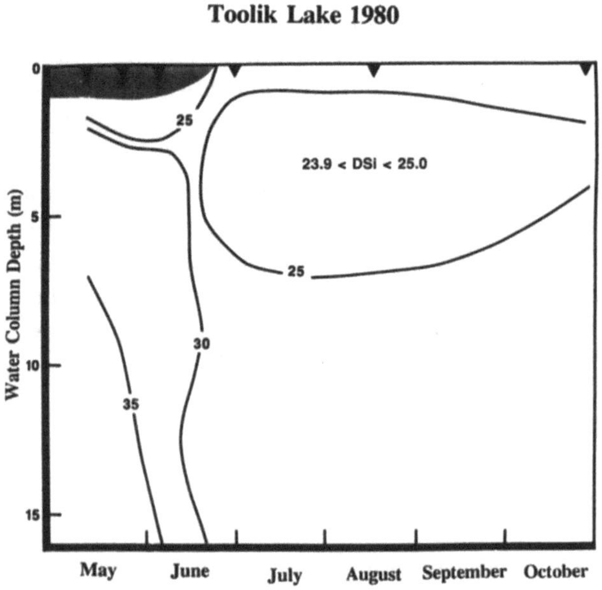

Fig. 3. Seasonal DRS isopleths from 1980 for Toolik Lake. Sampling dates are indicated by triangles at the top of the figure. Concentrations are $\mu mol\ l^{-1}$.

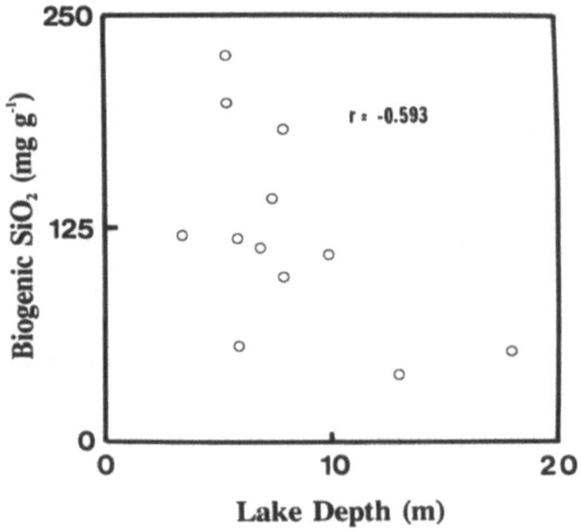

Fig. 4. Surficial sediment biogenic SiO_2 concentrations versus depth in Toolik Lake. Biogenic SiO_2 is inversely correlated with depth ($r = -0.593$, $n = 12$, $P < 0.05$).

from melting pond and lake ice. There was no apparent relationship between stream flow (Whalen & Cornwell, 1985) and DRS concentrations. Outlet concentrations of DRS are generally higher than inlet concentrations in May, reflecting the loss of regenerated Si at a time of low primary production. Lower DRS concentrations in the outlet relative to the inlet in July and August are likely a result of retention of DRS via the sedimentation of diatom-derived biogenic SiO_2. The seasonal isopleths of DRS (Fig. 3) show higher early season DRS concentrations in deeper waters. The mid-summer depletion of DRS occurred in the euphotic zone after vertical water column mixing following ice-out and prior to summer stratification.

Biogenic SiO_2 concentrations are measurable in surficial sediments throughout the lake, with the highest concentrations found at shallow water depths (Fig. 4). A weak negative correlation ($P < 0.05$) exists between lake depth and biogenic SiO_2 concentrations, but no significant correlations with other sediment parameters (organic C, metals, P, N) were found. Higher concentrations of biogenic SiO_2 may be a result of benthic primary production (specifically diatoms; Cuker, 1983). Such benthic productivity has been shown to be a major part of carbon cycling in high latitude lakes and ponds (Welch & Kalff, 1974; Stanley, 1976; Yeakel, 1977; Welch & Bergmann, 1985).

Vertical profiles of biogenic SiO_2 and DRS show that Toolik Lake sediments are important sites for the deposition and regeneration of DRS (Fig. 5). Biogenic SiO_2 concentrations decreased with depth in all cores, though profiles were not smooth for sites A, B and K. In cores A and B, nearshore physical disturbances may affect profiles (e.g. storm events, resuspension) while in core K, localized stream inputs may result in vertical heterogeneity. In cores C, D and E, most of the biogenic SiO_2 at the sediment surface appears to dissolve before permanent burial. Regeneration of biogenic SiO_2 results in the increase of pore water DRS concentrations (Calvert, 1983 and references therein), with the highest pore water DRS concentrations found at sites with the highest biogenic SiO_2 concentrations in surficial sediments.

Silicon budget calculations

Stream fluxes of DRS from the tundra were calculated using stream flow rates from Whalen

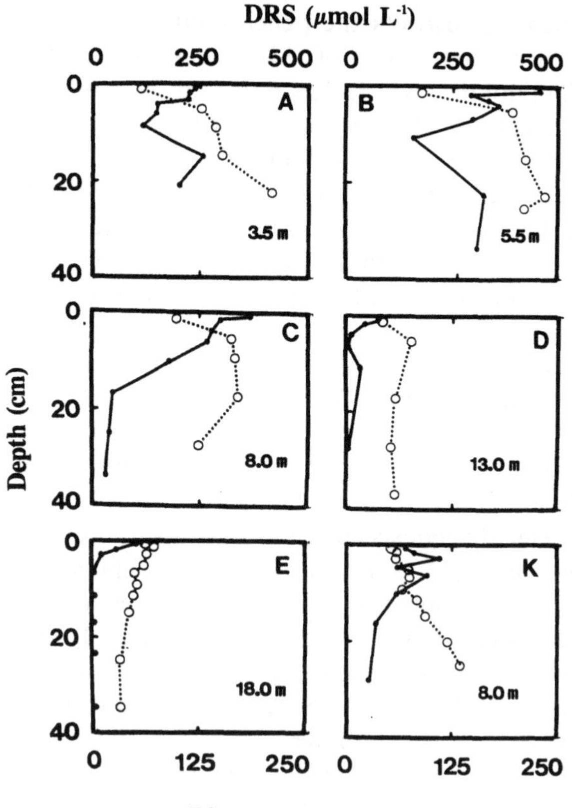

Fig. 5. Pore water DRS (dashed lines, hollow circles) and biogenic SiO_2 (solid lines, solid circles) versus core depth. Depths of overlying water (m) are indicated in the lower right corner for each core. The arrow at the top of each panel indicates the concentration of DRS in overlying water.

& Cornwell (1985). A watershed efflux of 7.4 $mmol\ m^{-2}\ yr^{-1}$ DRS compares favorably with values of 8 ± 1 $mmol\ m^{-2}\ yr^{-1}$ and 1–19 $mmol\ m^{-2}\ yr^{-1}$ measured in two other arctic watersheds by Welch & Legault (1986) and Brunskill *et al.* (1973), respectively. In contrast, the loss of DRS from the Char Lake watershed was considerably lower (~ 1 $mmol\ m^{-2}\ yr^{-1}$; Schindler *et al.*, 1974b).

While DRS was not measured in Toolik Lake precipitation, < 0.5 $mmol\ m^{-2}\ yr^{-1}$ would be contributed assuming Toolik Lake snow and rain had concentrations similar to Char Lake's. The bulk of stream DRS is ultimately derived from the weathering of silicate minerals and as in other systems (Likens *et al.*, 1977), differences between arctic sites may reflect differences in surficial geology.

Table 1. Stream and sediment silica fluxes.

	Flux ($mmol\ m^{-2}\ yr^{-1}$)
Main inlet	241
Inlet 2	29
Ungauged	~ 70
Total input	340
Outlet	282
Net retention	58
Surficial SiO_2 Burial	64
'Deep' SiO_2 Burial	178 ± 89
Water column accumulation	50–70

* All expressed per m^{-2} of lake area.

The 1980 DRS inputs to Toolik Lake (Table 1) were calculated directly from stream data for inlets 1 and 2, but estimated for the $\sim 20\%$ of water coming in from diffuse inputs (Whalen & Cornwell, 1985). The concentrations for these inputs were assumed to be equal to the weighted mean of inlet 1 DRS concentrations. Measurement of DRS in ephemeral streams on July 4 and 5, 1980 gave DRS concentrations of ($X \pm SD$) $109 \pm 37\%$ of the main inlet values. Approximately 83% of the incoming DRS exits via the outlet, with a net retention of 58 $mmol\ m^{-2}\ yr^{-1}$. Because this represents the difference between two relatively large numbers, it is unlikely that this retention value is very precise. The percentage DRS retention is somewhat lower than the Saqvaqjuac Lakes (57–90%; Welch & Legault, 1985) because Toolik Lake's greater watershed to lake area ratio results in a shorter residence time. Despite the higher supply of DRS to Toolik, the areal retention (19–67 $mmol\ m^{-2}\ yr^{-1}$) at Saqvaqjuac Lake was similar. Comparable rates of primary production (Welch & Bergmann, 1985; Miller *et al.*, 1986) suggest there may be similar biogenic silica production rates.

From solid phase biogenic SiO_2/Al ratios and Al-based ^{210}Pb deposition rates (used to compensate for sediment dilution by authigenic Mn and Fe oxyhydroxides; Cornwell, 1985), the burial of biogenic SiO_2 can be calculated for these cores. Burial rates of surficial biogenic SiO_2 range from 29 to 292 mmol m^{-2} yr^{-1} ($n = 12$), with a whole lake estimate of 64 mmol m^{-2} yr^{-1}.

The deeper, 'permanent' burial estimates were made using the six cores for which biogenic SiO_2 profiles were available (Fig. 5). Rates of biogenic SiO_2 burial in these deeper horizons range from 0 to 73 mmol m^{-2} yr^{-1}, with a whole lake estimate of 16 mmol m^{-2} yr^{-1}.

If dissolution of biogenic SiO_2 is a process occurring primarily below 1 cm of sediment, a dissolution rate of 48 mmol m^{-2} yr^{-1} is predicted from biogenic silica disappearance. A Fickian diffusion calculation may be made using the gradient between bottom water and surficial (0–1 cm) pore water DRS concentrations and a diffusion coefficient corrected for temperature (~ 4 °C; Li & Gregory, 1974) and tortuosity (Lerman, 1979; Ullman & Aller, 1982). The production of the mean DRS gradient (0.122 ± 0.061 μmol cm^{-4} ($n = 10$)), surficial porosity (~ 0.95) and a corrected diffusion coefficient (154 cm^2 yr^{-1}) yields DRS fluxes of 178 ± 89 mmol m^{-2} yr^{-1}, two to six times the dissolution estimate based on the distribution of biogenic SiO_2.

The dissolution of 'fresh' biogenic SiO_2 at the sediment-water interface is a quantitatively important process to the silicon cycle. Parker *et al.* (1977) found that diatom frustrule deposition rates in near-bottom sediment traps exceeded those estimated for surficial sediments, suggesting that dissolution of biogenic silica is important at the sediment-water interface.

Discussion

The largest terms in the Toolik Lake Si budget are for sediment-water fluxes and stream inputs and outputs. The short residence time (~ 0.5 yr in 1980) of water in Toolik Lake combined with low rates of biological utilization (low primary production) result in low rates of DRS retention. Poor retention of N and P has been observed in Toolik Lake by Whalen and Cornwell (1985), partially because the spring nutrient pulse occurs when ice restricts the penetration of light. 'Permanent' retention of Si ranges from 16 to 58 mmol m^{-2} yr^{-1}, depending on the method of calculation. This retention should correspond to the burial of biogenic SiO_2.

The 'true' rate of biological utilization of DRS in Toolik Lake may exceed our estimated rates based on the sum of 'permanent' retention of biogenic SiO_2 and sediment recycling of DRS (eg. 64–194 mmol m^{-2} yr^{-1}). Significant biogenic SiO_2 dissolution has been found to occur in the water column (e.g. Nriagu, 1978; Parker *et al.*, 1977). However, the shallow Toolik Lake water column may result in lower rates of water column dissolution.

The diagenesis and sedimentary recycling of biogenic SiO_2 appears to be redox dependent in some environments (e.g. Hammond *et al.*, 1986; Mortimer, 1941, 1942). Adsorption of DRS to iron and manganese oxyhydroxide surfaces (Calvert, 1983) may influence DRS concentrations at the sediment-water interface. A large upward flux of reduced Mn^{2+} and Fe^{2+} in Toolik Lake sediment pore waters (Cornwell, 1983) results in the formation of Mn and Fe oxyhydroxide enriched layers close to the sediment-water interface. Such high concentrations of metal oxides may result in an adsorptive barrier which could impede the diffusion of pore water DRS across the sediment–water interface. Indeed, if the early spring and fall water column concentrations of DRS are examined, a water column DRS budget extrapolated from the winter months to a whole year would suggest that 50 to 70 mmol m^{-2} yr^{-1} DRS is produced by dissolution. Enhanced biogenic SiO_2 dissolution in water and sediment during the summer (warmer temperatures, fresher diatom SiO_2) could increase this number considerably.

Benthic and planktonic primary production in Toolik Lake may be similar to each other in magnitude (Yeakel, 1977: Whalen & Alexander, 1986), suggesting a total lakewide primary production of ~ 1700 mmol C m^{-2} yr^{-1}. If silica utilization by diatoms is ~ 64–194 mmol Si m^{-2} yr^{-1}, the Si:C atom ratio would be ~ 0.04 to 0.12. The Si:C ratio of Toolik diatoms has not been determined, but the higher estimate compares favorably with the Si:C ratio of 0.13 for marine diatoms (Brezinski, 1985; Banahan, 1983).

The high concentrations of biogenic SiO_2 in Toolik Lake sediments suggest that benthic diatoms, along with *Nitella* and mosses, play an important role in lake metabolism. In the rocky littoral zone of Toolik Lake, benthic diatoms comprise the bulk of the epilithic algal community (Cuker, 1983). Senescence of diatoms and their grazing by snails could provide a continuous turnover of biogenic SiO_2 in nearshore sediments. Modest rates of primary productivity at the surface of soft sediments has been noted in another lake in the Toolik region (Barsdate & Cornwell, unpublished data) and at least a portion of biogenic SiO_2 in Toolik Lake sediments may be both formed and deposited at a single site. The preservation of biogenic SiO_2 in Toolik Lake sediments is high in the littoral zone as a result of increased biogenic SiO_2 deposition at the sediment water interface. The two deepwater stations with lower biogenic SiO_2 content should reflect a greater proportion of water column diatoms inputs. Preliminary sediment flux experiments in nearshore, stream-dominated sites, showed both negative and positive DRS fluxes from sediments, perhaps reflecting differences in diatom production and biogenic SiO_2 regeneration at a given site (Kipphut & Cornwell, unpublished data). Clearly, the productivity and Si-cycle of benthic diatoms requires more detailed research.

Conclusions

The silica cycle of an arctic lake has been outlined for the first time. Conclusions drawn on observational and quantitative data include:

1. low retention of stream DRS is a result of low primary production and short water residence times;
2. enhanced deposition of biogenic SiO_2 in littoral zone sediments points to significant benthic productivity by diatoms;
3. internal cycling of DRS can provide a significant part of biologically utilized DRS.

Acknowledgements

This research was part of the Arctic Lake Process Study funded by the National Science Foundation, with additional support to JCC from a State of Alaska Fellowship through the University of Alaska Institute of Marine Science. We thank Steve Whalen for assistance in sample collection.

References

Armstrong, F. A. J., C. R. Stearns & J. D. H. Strickland, 1967. The measurement of upwelling and subsequent biological processes by means of the Technicon Autoanalyzer and associated equipment. Deep-Sea Res. 14: 381–389.

Banahan, S., 1983. The uptake of silicic acid by diatoms in the Bering Sea. M.S. thesis, University of Alaska, Fairbanks. 121 p.

Banahan, S. & J. J. Goering, 1986. The production of biogenic silica and its accumulation on the southeastern Bering Sea shelf. Cont. Shelf Res. 5: 199–213.

Brezinski, M. A., 1985. The Si:C:N ratio of marine diatoms: interspecific variability and the effect of some environmental variables. J. Phycol. 21: 347–357.

Brunskill, G. J., D. M. Rosenberg, N. B. Snow, G. L. Vascatto & R. W. Wagemann, 1973. Ecological studies of aquatic systems in the MacKenzie-Porcupine drainages in relation to proposed pipeline and highway developments. Canada Task Force N. Oil Dev. Environ. Soc. Comm. Vol. II. Report 73–41. 345 pp.

Calvert, S. E., 1983. Sedimentary geochemistry of silicon. pp. 143–186 In: (S. R. Aston, Ed.), Silicon Geochemistry and Biogeochemistry. Academic Press, N.Y.

Cornwell, J. C., 1983. The geochemistry of manganese, iron and phosphorus in an arctic lake. Ph.D. Dissertation, University of Alaska, Fairbanks. 238 pp.

Cornwell, J. C., 1985. Sediment accumulation rates in an Alaskan arctic lake using a modified ^{210}Pb technique. Can. J. Fish. aquat. Sci. 42: 809–814.

Cornwell, J. C., 1986. Diagenetic trace metal profiles in arctic lake sediments. Envir. Sci. Technol. 20: 299–302.

Cornwell, J. C., 1987. Phosphorus cycling in arctic lake sediments: adsorption and authigenic minerals. Arch. Hydrobiol. 109: 161–179.

Cuker, B. E., 1983. Grazing and nutrient interactions in controlling the activity and composition of the epilithic algal community of an arctic lake. Limnol. Oceanogr. 28: 133–141.

Eggiman, D. W., F. T. Manheim & P. R. Betzer, 1980. Dissolution and analysis of amorphous silica in marine sediments. J. Sed. Petrol. 50: 215–225.

Hammond, D. E., J. Boucher & R. L. Rea, 1986. The effect of oxygen on silica reaction kinetics in marine sediments. Trans. Am. Geophys. Union 65: 933.

Hobbie, J. E., 1973. Arctic limnology: a review. p. 127–168. In M. E. Britton (ed.) Alaskan arctic tundra. Arct. Inst. N. Am. Tech. Pap. 25.

Hutchinson, G. E., 1957. A Treatise on Limnology. Vol. 1. John Wiley. 1015 pp.

Lerman, A., 1979. Geochemical Processes: Water and sediment Environments. Wiley, N.Y.

Li, Y.-H. & S. Gregory, 1974. Diffusion of ions in seawater and in deep-sea sediments. Geochim. Cosmochim. Acta. 38: 703–714.

Likens, G. E., F. H. Bormann, R. S. Pierce, J. S. Eaton & N. M. Johnson, 1977. Biogeochemistry of a Forested Ecosystem. Springer Verlag, NY. 138 pp.

Miller, M. C., G. R. Hater, P. Spatt, P. Westlake & D. Yeakel, 1986. Primary production and its control in Toolik Lake, Alaska. Arch. Hydrobiol./Suppl. 74: 97–131.

Mortimer, C. H., 1941. The exchange of dissolved substances between mud and water in lakes (Parts I and II). J. Ecol. 29: 280–329.

Mortimer, C. H., 1942. The exchange of dissolved substances between mud and water in lakes (Parts III and IV). J. Ecol. 30: 147–201.

Nriagu, J. O., 1978. Dissolved silica in pore waters of Lake Ontario, Erie and Superior sediments. Limnol. Oceanogr. 23: 53–67.

Parker, J. I., H. L. Conway & E. M. Yaguchi, 1977. Dissolution of diatom frustrules and recycling of amorphous silicon in Lake Michigan. J. Fish. Res. Bd Can. 34: 545–551.

Parsons, T. R., Y. Maita & C. M. Lalli, 1984. A Manual of Chemical and Biological Methods for Seawater Analysis. Pergamon Press, NY. 173 pp.

Prentki, R. T., M. C. Miller, R. J. Barsdate, V. Alexander, J. Kelley & P. Coyne, 1980. Chemistry. In J. E. Hobbie (ed), Limnology of Tundra Ponds, Barrow, Alaska. Dowden, Hutchinson and Ross, Stroudsburg.

Reeburgh, W. S., 1967. An improved interstitial water sampler. Limnol. Oceanogr. 12: 163–165.

Schindler, D. W., J. Kalff, H. E. Welch, G. J. Brunskill, H. Kling & N. Kritsch, 1974a. Eutrophication in the high arctic-Meretta Lake, Cornwallis Island. J. Fish. Res. Bd Can. 31: 647–662.

Schindler, D. W., H. E. Welch, J. Kalff, G. J. Brunskill & N. Kritsch, 1974b. Physical and chemical limnology of Char Lake, Cornwallis Island. J. Fish. Res. Bd Can. 31: 587–607.

Stanley, D. W., 1976. Productivity of epipelic algae in tundra ponds and a lake near Barrow, Alaska. Ecology 57: 1015–1024.

Ullman, W. J. & R. C. Aller, 1982. Diffusion coefficients in nearshore marine sediments. Limnol. Oceanogr. 27: 552–556.

Welch, H. E. & J. Kalff, 1974. Benthic photosynthesis and respiration in Char Lake. J. Fish. Res. Bd Can. 31: 609–620.

Welch, H. E. & M. A. Bergmann, 1985. Winter respiration of lakes at Saqvaqjuac, N.W.T. Can. J. Fish. aquat. Sci. 42: 521–528.

Welch, H. E. & J. A. Legault, 1986. Precipitation chemistry and chemical limnology of fertilized and natural lakes at Saqvaqjuac, N.W.T. Can. J. Fish. aquat. Sci. 43: 1104–1134.

Wetzel, R. G., 1975. Limnology. Saunders, Phila. 743 p.

Whalen, S. C. & V. Alexander, 1986. Seasonal inorganic carbon and nitrogen transport by phytoplankton in an arctic lake. Can. J. Fish. aquat. Sci. 43: 1177–1186.

Whalen, S. C. & J. C. Cornwell, 1985. Nitrogen, phosphorus and organic carbon cycling in an arctic lake. Can. J. Fish. aquat. Sci. 42: 797–808.

Whalen, S. C., J. C. Cornwell & V. Alexander, 1988. Comparison of chemical and biological N budgets in an Arctic lake: implications for phytoplankton production. SCOPE/UNEP Sonderband 66: 99–115.

Yeakel, J. D., 1977. Production and control of epilithic periphyton in an arctic lake. M.S. thesis, University of Cincinnati, Ohio. 53 p.

Hydrobiologia **240**: 45–59, 1992.
W.J. O'Brien (ed.), Toolik Lake – Ecology of an Aquatic Ecosystem in Arctic Alaska.

Biogeochemistry of manganese- and iron-rich sediments in Toolik Lake, Alaska

Jeffrey C. Cornwell[1] & George W. Kipphut[2,3]
[1] *University of Maryland CEES, Horn Point Environmental Laboratory, Cambridge, MD 21613-0775, USA;* [2] *Institute of Marine Science, University of Alaska-Fairbanks, Fairbanks, AK 99775-1080, USA;* [3] *Present address: Center for Reservoir Research, Murray State University, Murray, KY 42071, USA*

Key words: arctic lakes, trace metals, lake sediments, manganese, iron, phosphorus, sediment-water flux, diagenesis, Toolik Lake

Abstract

The sediments within Toolik Lake in arctic Alaska are characterized by extremely low rates of organic matter sedimentation and unusually high concentrations of iron and manganese. Pore water and solid phase measurements of iron, manganese, trace metals, carbon, nitrogen, phosphorus, and sulfur are consistent with the hypothesis that the reduction of organic matter by iron and manganese is the most important biogeochemical reaction within the sediment. Very low rates of dissolved oxygen consumption by the sediments result in an oxidizing environment at the sediment-water interface. This results in high retention of upwardly-diffusing iron and manganese and the formation of metal-enriched sediment. Phosphate in sediment pore waters is strongly adsorbed by the metal-enriched phases. Consequently, fluxes of phosphorus from the sediments to overlying waters are very small and contribute to the oligotrophic nature of the Toolik Lake aquatic system. Toolik Lake contains an unusual type of lacustrine sediment, and in many ways the sediments are similar to those found in oligotrophic oceanic environments.

Introduction

The composition of lake sediment is a reflection of processes occurring within the lake watershed, water column and sediments. Slow rates of organic matter decomposition within the sediments of oligotrophic arctic lakes are expected because of low temperatures and low rates of autochthonous carbon deposition (Hobbie, 1973; Welch & Bergmann, 1985), the latter a result of low external nutrient fluxes (Schindler *et al.*, 1974; Welch & LeGault, 1986). The rate of organic matter supply affects not only the overall rate of decomposition but also the nature of decomposition pathways in sediments (Reeburgh, 1983). In sediments with low carbon loading, such as the deep sea, the primary oxidants for organic diagenesis are O_2, NO_3^-, Mn(IV) and Fe(III) (Froelich *et al.*, 1979). Higher rates of organic loading result in the predominance of sulfate reduction and/or methanogenesis. The study of coupling between organic matter oxidation and specific oxidants is simplified in sediments with low rates of sediment metabolism. Arctic lake sediments are relatively oxidizing (Coakley & Rust, 1968) and provide an opportunity to investigate authigenic iron and manganese oxide formation in an environment other than the oligotrophic ocean.

The sediments within Toolik Lake have been studied more intensively than those of any other lacustrine site in arctic Alaska. In this paper we describe the major and minor features of the sediment biogeochemistry of Toolik Lake, with emphasis on the highly biogenic and authigenic nature of the sediments. In particular, we focus on those processes which are responsible for the formation of the Mn- and Fe-enriched sediments. Our conclusions are based on new observations, as well as our previously published work on ^{210}Pb sedimentation rates, sediment-water exchange rates, Fe, Mn and trace metal distributions, biogenic SiO_2 distributions, phosphorus diagenesis, and burial fluxes of carbon, nitrogen and phosphorus.

Study area

Many of the physical, chemical and biological features of Toolik Lake have been described previously and will be summarized only briefly here. Toolik Lake, Alaska (68° N, 149° W) was formed in the northern foothills of the Brooks Range by glacial moraines and outwash deposited after the retreat of the Itkillik II glaciation, ~12000 to 14000 yr ago (Hamilton & Porter, 1975). Toolik Lake has an area of 1.5 km^2 within a watershed of ~65 km^2. It is fed by two main streams on the lake's east side (Fig. 1), with Inlet 1 draining 75% of the watershed, Inlet 2 draining 7% and emphemeral streams draining the remainder. The complex lake morphometry results from the formation of distinct kettles by the melting of glacial ice. Five basins are separated by rocky shoals, and the lake margin generally consists of cobbles and boulders to depths of ~2.5 m. The mean depth of the lake is 7 m, with a maximum of 25 m in the northwest basin.

Ice covers Toolik Lake from late September until late June, with stream flow generally commencing in mid-May and ceasing in September. The lake is dimictic with a summer thermocline at 5–7 m depth and a maximum epilimnetic temperature of ~16 °C. Low rates of nutrient inputs (Whalen & Cornwell, 1985) result in low rates of water column primary production (Miller *et al.*, 1986) and organic sedimentation (Whalen & Cornwell, 1985).

Methods

Sediments were collected in 1977–1980 and 1983 from Toolik Lake. In Fig. 1 we have identified the locations of 1) the ^{210}Pb and trace metal cores, 2) the pore water cores and 3) other cores used for solid phase analysis. Not shown are 26 coring sites used primarily for solid phase analysis of surficial (0–1 cm depth interval) sediments.

Four different sediment collection devices were used. A KB gravity corer was slowly lowered into soft sediments to obtain undisturbed cores 25 to 45 cm in length. Metal tubes ~3 m in length were used to push 7.5 cm diameter plastic core liners into nearshore sediments. An Ekman dredge was used to collect surficial sediment samples. Cores from M and N from Limnobay (Fig. 1) were collected in plastic core liners by a SCUBA diver. Pore water was extracted from sediments using a Reeburgh (1967) style squeezer, with sediment extrusion in a N_2-filled glove bag to minimize oxidation artifacts (Bray *et al.*, 1974). Pore water analyses are described elsewhere (Cornwell, 1983, 1987a). Sediment for solid phase analysis was dried at 65 °C and ground with a ceramic mortar and pestle. A $LiBO_2$ fusion technique was used for sediment dissolution (Medlin *et al.*, 1969), followed by atomic absorption spectrophotometric metal analysis and colorimetric P analysis (Parsons *et al.*, 1984). Inorganic P, Fe and Mn associated with poorly crystalline oxides of Fe and Mn were extracted with oxalate (Williams *et al.*, 1971), followed by colorimetric inorganic P analysis (Parsons *et al.*, 1984) and Mn and Fe analysis via atomic absorption spectrophotometry (Raad *et al.*, 1969). Total carbon was analyzed using a CHN analyzer, carbonate-C was determined using Stainton's (1973) gas chromatographic technique and ^{210}Pb was analyzed fol-

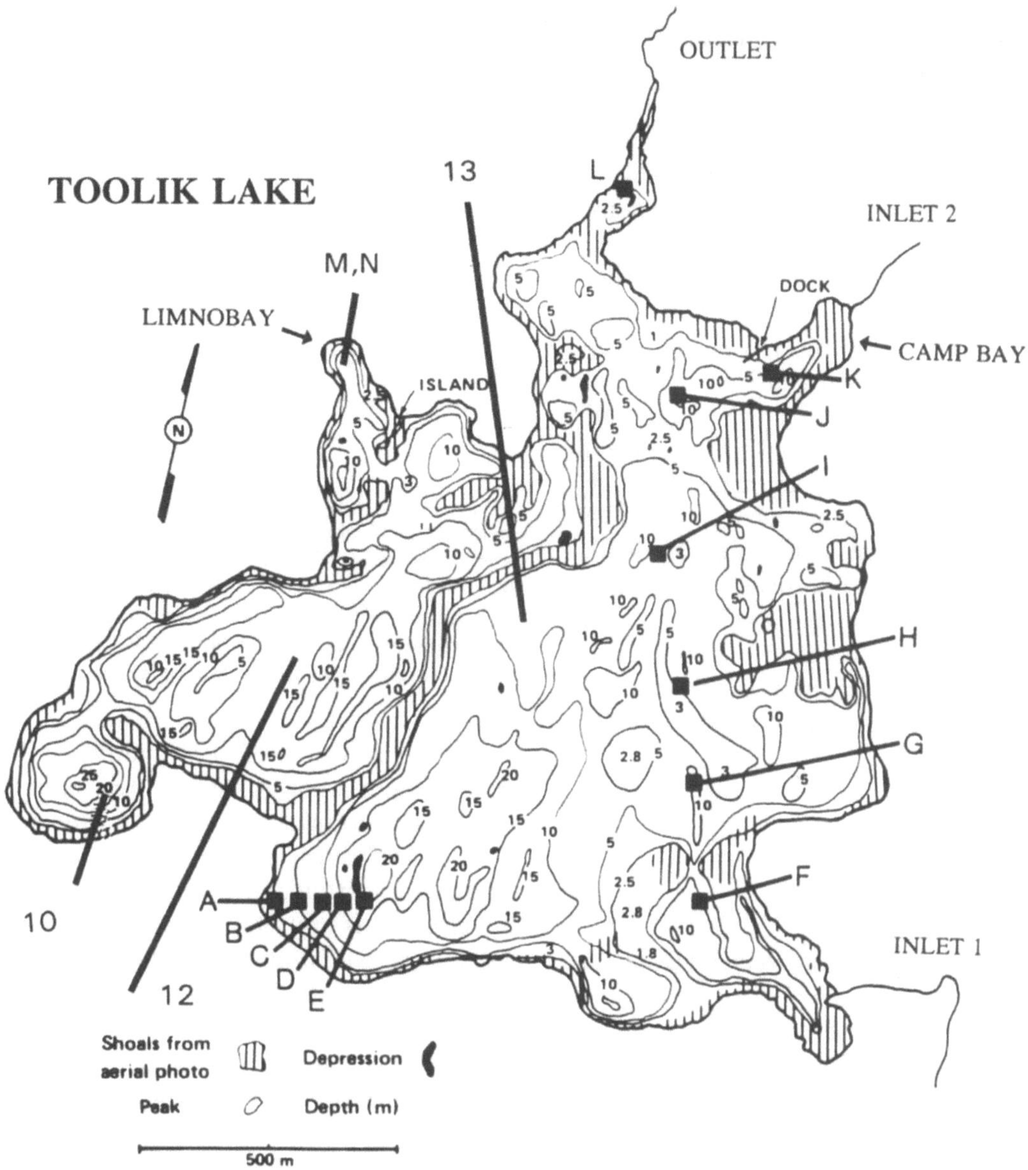

Fig. 1. Toolik Lake Map. Coring locations marked with a filled box (A through N) are ^{210}Pb dating sites, with A–E and K used for pore water studies. Core sites 10–13 and CB1 are sites for solid phase analysis. Not identified are 26 other coring sites located throughout Toolik Lake that have been used for solid phase analysis. Core MID corresponds to site E.

lowing Chanton *et al.* (1983). Dried sediment samples were analyzed for chromium-reducible S following the digestion procedure of Canfield *et al.* (1986) with a potentiometric sulfide analysis (Cornwell & Morse, 1987). Chromium-reducible sulfur is a measure of acid volatile sulfide, pyrite-S and elemental S, all of which are formed via sulfate reduction. These samples had been stored dry for several years, but Canfield *et al.* (1986) and Morse & Cornwell (1987) present evidence that drying and storage artifacts are minimal.

Oxygen fluxes between sediments and overlying waters were measured using clear plexiglass chambers emplaced upon the sediments. The chambers covered approximately 0.2 m^2 of sediment surface and enclosed 30–40 l of overlying water. Water in the chambers was mixed with a battery operated 1 rpm stirring paddle. Water samples were withdrawn with syringes attached to sampling ports on the top of the chambers. Dissolved O_2 was measured by Winkler titration.

Sedimentation rates were determined in Toolik Lake using the ^{210}Pb dating technique (Robbins, 1978). Measurements of deposition rates were undertaken to estimate burial fluxes of Fe, Mn, P, organic C, N and biogenic SiO_2. The high concentrations of authigenic Mn and Fe oxides presented difficulties for simple interpretation of ^{210}Pb profiles, a problem also noted by Gubala *et al.* (1990). The problem was minimized in high Mn and Fe sediments by normalizing the ^{210}Pb activity to Al concentrations and expressing depth on a cumulative Al scale (Cornwell, 1985); this procedure was not used in the low Mn and Fe sediments of Limnobay (cores M and N). On a whole-lake basis, this correction procedure decreased the apparent deposition rate by ~34%.

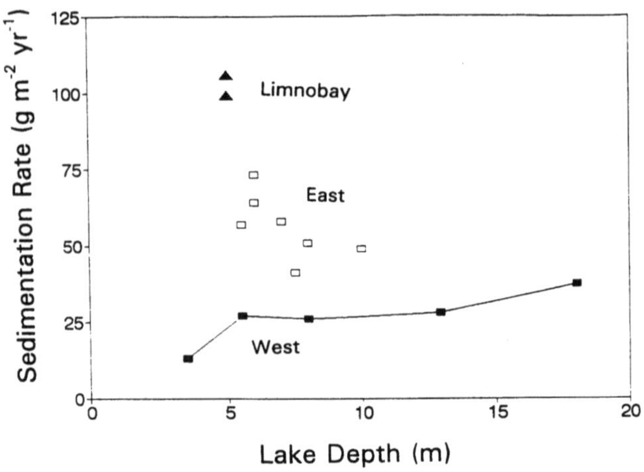

Fig. 2. Toolik Lake sedimentation rates versus overlying water depth. The Limnobay cores sites are sites M and N, the west cores were collected in a transect (cores A through E) from the lake's western shore and the east samples are located on a transect between the lake's major inlet and the outlet. The sedimentation rates from all Mn and Fe-rich sites have corrected for dilution by authigenic oxides (Cornwell, 1985).

Results and discussion

Sedimentation

Unlike a number of other small lakes (Evans & Rigler, 1980, 1983), sediment deposition rates in Toolik Lake do not have a simple positive correlation to water depth (Fig. 2). Indeed, Toolik Lake ^{210}Pb sedimentation rates are highly variable, with the only overall trend indicating lower sedimentation rates at greater water depths. The deposition rates on the east side of Toolik Lake (sites F–L on Fig. 1) are higher and more variable than those on the west side, probably reflecting the proximity to stream inputs. Two cores from a small embayment on the lake's north side (sites M and N) had the highest rates of sedimentation found in this study. From observations of the spring snowpack it is clear that this protected basin, surrounded on three sides by hills, concentrates terrestrial windborne debris.

The whole lake sedimentation rate is ~27 g m^{-2} yr^{-1} (Cornwell, 1985). Total particulate input via streams accounts for 40–50% of burial fluxes (Cornwell, 1983); most (~70–80%) of this input was organic matter. However, the sediment organic matter concentrations are <20% at all sites, leaving the source of inorganic particles uncertain. One potential source is atmospheric inputs of particulates. Summer inputs are unknown, but spring sampling of snow in the main basin showed a particulate concentration of 40 ± 27 g m^{-3} ($n = 5$). Assuming a direct snow input equivalent to ~0.1 m of water (Brown & Berg, 1980), the winter particulate input was on the order of 4 g m^{-2} yr^{-1}, ~15% of the depositional flux. With the lack of snow cover in the summer months and the increased potential for dust inputs from the nearby Dalton Highway, atmospheric inputs of dust may be higher than in the winter. The erosion of shorelines may be another significant but unmeasured source of primarily inorganic particulates. At present, lake shore erosion is clearly evident along steep banks on the northern margin of the lake.

Toolik Lake sedimentation rates are among the lowest reported for lakes anywhere, with comparable whole lake rates reported only in Char Lake (de March, 1978) in the Northwest Territories (Table 1). In that study, sedimentation was esti-

Table 1. Sedimentation rates in North American lakes

Lake		Method	N	Sedimentation Rate ($g\ m^{-2} yr^{-1}$)	Reference
Arctic Lakes					
Toolik Lake	whole lake	^{210}Pb	14	27	
(Alaska)	range	^{210}Pb	14	13–106	A
Char Lake		^{210}Pb	1	260	
(N.W.T.)		horizon	50	31	B
Imitavik	range	^{210}Pb	6	62–309	
Annak	pre-sewage	^{210}Pb	1	44	
(N.W.T.)	post-sewage	^{210}Pb	1	98	C
Eight Lake		^{14}C	1	67	
Lake A		^{14}C	1	210	
Chandler Lake		^{14}C	1	263	
Death Valley Lake		^{14}C	1	653	D
(Alaska)					
Deleen Lake		^{210}Pb	1	8000	E
Selected North American Temperate Lakes					
Costello	whole lake	^{210}Pb	12	48	
Red Chalk	whole lake	^{210}Pb	17	44	
Bob Lake	whole lake	^{210}Pb	15	70	F
ELA Lakes	range	^{210}Pb	12	34–280	G
Big Moose	range	^{210}Pb	3	60–170	H

References: A. Cornwell 1985, Kipphut unpublished; B. de March 1978, Kipphut 1978; C. Hermanson 1990; D. Livingstone 1958; E. Weiss and Naidu 1986; F. Evans & Rigler, 1980,1983; G. Anderson *et al.*, 1987; H. Gubala *et al.*, 1990.

mated on the basis of well-defined sediment horizons of a known age; the single ^{210}Pb-derived sedimentation rate (Kipphut, 1978) is vastly higher. Although both Toolik Lake and Char Lake have low deposition rates, somewhat lower rates in Char Lake may result from the harsher climatic conditions found there. Hermanson's (1990) study of lakes on Flaherty Island in Hudson Bay (North West Territories) showed, for the most part, higher sedimentation rates than Char or Toolik Lakes. Insufficient numbers of cores were analyzed to calculate a whole lake deposition rate. Early studies of deposition rates in Alaskan arctic lakes were carried out by Livingstone (1958) using ^{14}C dating and showed high sedimentation rates in three out of four foothill or coastal plain lakes. Weiss & Naidu's (1986) ^{210}Pb dating of Deleen Lake resulted in exceptionally high calculated deposition rates, likely a result of their failure to account for the ^{226}Ra-supported ^{210}Pb activity and/or sediment mixing.

Low sediment deposition rates in Toolik Lake result from climatic and biotic factors. Low rates of chemical and physical weathering result from long periods during which soils are frozen ($\sim$8 months), and from low precipitation. In addition, a high percentage of ground cover by tussock tundra plants (Brown & Berg, 1980) stabilizes the soils and hinders erosion. Production and subsequent deposition of autochthonous organic matter and biogenic SiO_2 is hindered during winter months by light limitation and during summer by severe nutrient (N and P) limitation.

Solid phase chemistry

Toolik Lake surficial sediment chemistry shows a high degree of spatial variability for most measured parameters (Table 2). Organic carbon concentrations ranged from 5–9% by weight and were similar to carbon contents estimated from arctic lake loss on ignition data (Livingstone *et al.*, 1958). Biogenic SiO_2 is a major sediment com-

Table 2. Toolik Lake sediment chemistry (0–1 cm interval)

ID	(m) Depth	All values mg g^{-1} Al	Ca	Mg	K	org–C	CO_3–C	N	P	SiO_2
A	3.5	49	6.29	4.35	10.9	73.8	2.5	8.7	1.78	121.5
B	5.5	30	5.21	2.72	6.8	62.5	3.5	6.9	6.00	227.5
C	8.0	45	4.44	3.93	9.8	49.0	4.1	6.2	1.79	184.1
D	13.0	58	5.90	4.99	12.6	58.7	4.0	6.8	2.34	39.3
E	18.0	46	7.04	4.18	9.8	55.0	5.6	6.5	1.72	53.1
F	5.5	47	4.96	4.20	11.0	85.9	1.7	10.4	1.89	199.1
G	6.0	32	5.91	2.95	7.4	72.2	5.1	7.5	2.26	119.4
H	10.0	61	3.74	5.15	13.4	63.9	1.7	7.1	2.24	111.0
I	7.5	52	5.16	4.37	11.1	62.0	3.2	7.8	1.90	142.7
J	7.0	47	4.88	4.08	10.7	55.1	3.8	6.6	1.88	114.3
K	8.0	61	5.64	4.91	12.8	90.6	2.7	7.2	2.63	70.7
L	6.0	55	4.82	4.41	11.6	89.9	2.6	9.8	2.59	97.3
Mean		49	5.33	4.19	10.7	68.2	3.4	7.6	2.42	123.3
S.D.		10	0.88	0.74	2.0	14.2	1.2	1.3	1.17	57.5

ID	mg g^{-1} Mn	Fe	Ba	Zn	µg g^{-1} Ni	Co	Cu	Cr
A	3.8	51	0.32	0.27	72	20	32	60
B	11.5	183	0.85	0.16	61	26	27	40
C	48.0	69	0.68	0.23	82	26	35	42
D	23.0	97	0.44	0.29	112	36	50	54
E	129.0	79	2.60	0.30	126	29	49	56
F	0.5	24	0.18	0.13	42	7	27	50
G	38.0	142	1.33	0.16	64	52	23	50
H	0.5	48	0.22	0.19	45	12	40	48
I	13.8	67	0.45	0.24	92	24	40	42
J	27.0	95	0.72	0.28	116	40	41	48
K	1.5	67	0.24	0.19	48	20	42	71
L	1.3	76	0.22	0.25	55	18	42	26
Mean	24.8	83	0.69	0.22	76	26	37	49
S.D.	36.4	43	0.69	0.06	29	12	9	11

ponent in shallow water sediments, presumably reflecting the presence of benthic diatoms (Cornwell & Banahan, 1992). Surficial Mn and Fe concentrations were quite high in many instances, with higher concentrations routinely found at depths below the sediment surface. In core E, high Mn concentrations were found in the form of concretions consisting of cemented Mn oxides. Concentrations of both trace and major metals in Toolik Lake sediments are similar to those found in 'oxidate crusts' and ferromanganese concretions from other unpolluted lakes (Gorham & Swaine, 1965; Calvert & Price, 1977; Sozanski & Cronan, 1979; Forstner & Wittman, 1979). The Mn and Fe enrichments in Toolik Lake sediments occur within accumulating sediment; thus, observed trace metal distributions result from diagenetic processes, rather than differences in metal inputs to different sites.

The importance of authigenic and biogenic phases to the composition of Toolik Lake surficial sediment is illustrated in Fig. 3. If it is as-

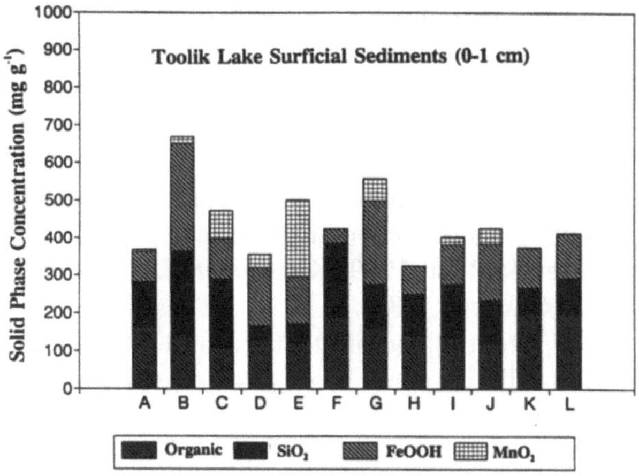

Fig. 3. The concentration of potential authigenic, biogenic and endogenic minerals assuming that all sediment Fe is FeOOH and that all sediment Mn is MnO_2. The SiO_2 is biogenic silica and the organic matter data is calculated from organic carbon data using an empirically determined ash-free dry weight to organic C ratio ($\sim 2.0 \pm 0.1$). Data for this figure are derived from Table 2 and overlying water depths may be found there.

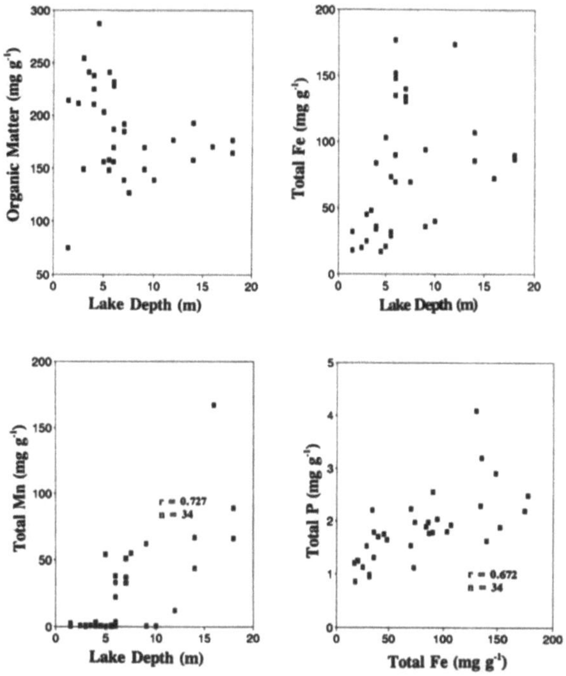

Fig. 4. Relationships between surficial sediment parameters for 34 samples from all Toolik Lake basins. The highest concentrations of organic matter occur in shallower waters within the euphotic zone. Total Fe is not correlated with water depth and total Mn shows no enrichment at depths less than ~ 5 m and variable enrichments at greater depths. Total Fe and total P are significantly correlated ($p < 0.01$).

sumed that all sediment Mn and Fe are oxides (MnO_2 and FeOOH respectively) and that total organic matter is twice the organic carbon values, we can account for much of the total sediment. The relative importance of these phase changes with sediment depth at a single site; in some sediment horizons, authigenic Mn and Fe oxides comprise over half of the sediment mass.

Toolik Lake sediment composition changes with water depth (Fig. 4). Organic matter concentrations, estimated as ash-free dry weight, are more variable in shallow-water sediments (75–287 mg g^{-1}) than in deeper sediments (139–193 mg g^{-1} at depths ≥ 10 m). The highest organic matter concentrations (≥ 200 mg g^{-1}) are found in sediments at water depths ≤ 5 m. This could result either from the presence of benthic algae (Yeakel, 1977) or from increased deposition of stream and shoreline-derived organic matter. Total Fe does not show a consistent distribution pattern in the lake, though shallow water sediments tend to have concentrations ≤ 50 mg g^{-1}. Total Mn concentrations are highest at greater water depths and low in shallow water sedimentary environments. Pore water pH and Eh data suggest that reducing conditions occur closer to the sediment-water interface in shallow-water sediments (Cornwell, 1983), with the reductive loss of Mn oxides a possible result. This distribution pattern is opposite to that found in temperate lakes such as Oneida (Dean *et al.*, 1981); there reducing conditions within the hypolimnion result in Mn dissolution and transport to the oxic epilimnion where Mn precipitation results in Mn concretion formation. Water column Mn measurements from Toolik Lake showed no evidence for dissolved Mn flux from the sediments (Johnston & Kipphut, 1988).

The positive correlation of P with Fe (Fig. 4; $P < 0.01$) observed in Toolik Lake has been noted in a number of lake sediment studies (e.g. Williams *et al.*, 1971) and arises from the adsorption of inorganic P on Fe oxides. While the authigenic formation of iron phosphates such vivianite

($Fe_3(PO_4)_2.8H_2O$) occurs in many lakes including Toolik, adsorption appears to be the main mode of Fe-P association in surficial sediments (Cornwell, 1987a). Similarly, phosphorus adsorption to surficial sediments in Barrow, Alaska ponds is related to sedimentary iron oxide minerals and has been hypothesized to be a factor limiting organic productivity (Prentki *et al.*, 1980).

Vertical profiles of organic matter, Fe and Mn show the same high degree of heterogeneity found in surficial sediments (Fig. 5). In all Toolik Lake cores, organic matter decreases with increasing sediment depth, usually by a factor of ~2 (Whalen & Cornwell, 1985). In many cores (e.g. cores 10 and 12), subsurface Fe enrichments are found. Such enrichment is not as obvious in the more organic rich shallow water sediments such as core 13, which has extremely low Fe and Mn concentrations in comparison to the other cores. Manganese is enriched in near-surface sediments in all cores. Iron concentration maxima, where present, generally occur in the more reducing conditions immediately below the Mn concentration maxima. This separation of Mn- and Fe-enrichment is consistent with the order of Mn and Fe reduction predicted by thermodynamic calculations (Krauskopf, 1957; Stumm & Morgan, 1981). Such enrichment of Mn and/or Fe has been noted in other North American arctic lakes (Coakley & Rust, 1969; Klassen *et al.*, 1975; Gilbert & Church, 1983).

The potential mobility of Fe and Mn within Toolik sediments has been studied using oxalate extraction procedures (Fig. 6), which were performed to determine the readily extractable, poorly crystalline Mn and Fe oxide mineral content of sediments (Williams *et al.*, 1971; Prentki *et al.*, 1980; Cornwell, 1983). Total and oxalate extractable Mn are virtually identical in all samples, suggesting that most Mn found in surficial sediments might be labile (ie. reactive) with respect to diagenetic processes. This is confirmed by the relatively low concentrations of Mn found deep within cores, suggesting that Mn is reduced and transported out of reducing horizons. Total and oxalate-extractable Fe profiles show that most of the total Fe in Toolik Lake sediments is poorly crystalline. The large ranges of total Fe

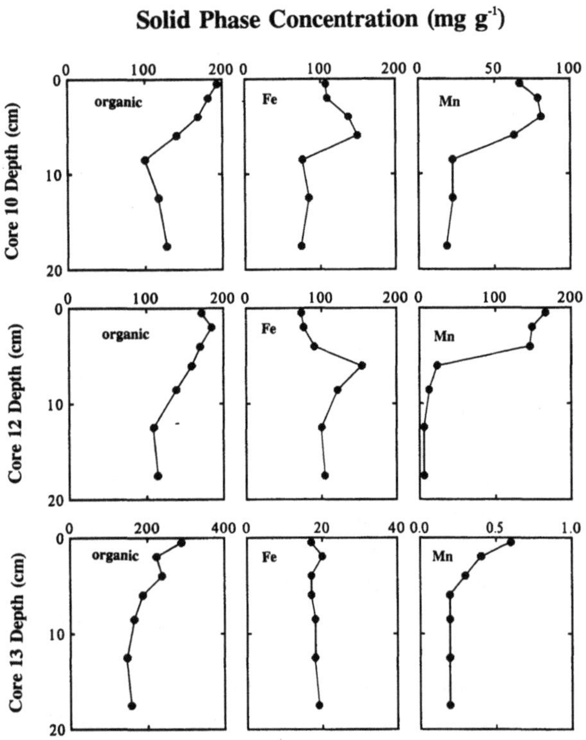

Fig. 5. Vertical profiles of organic matter, total Fe and total Mn from three Toolik Lake cores.

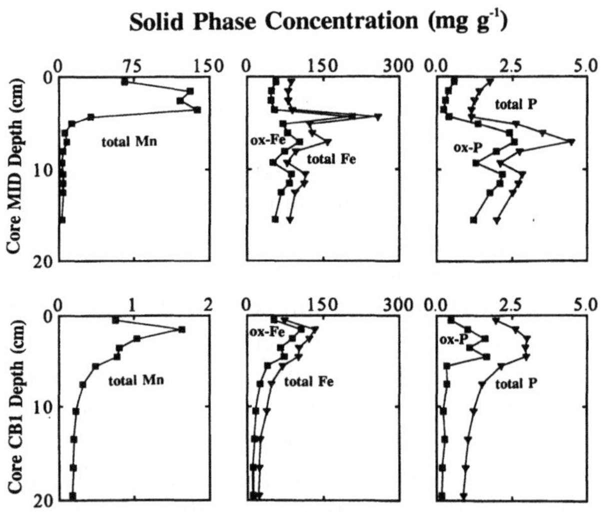

Fig. 6. Vertical profiles of total Mn, and oxalate Fe and P from sites CB1 and MID, situated in 5.5 and 18 m of water, respectively. Oxalate Mn concentrations (not shown) are virtually identical to total Mn concentrations. Note the scale change for total Mn.

concentrations found within both core MID and CB1 result from changes in the oxalate Fe fraction; the residual Fe left after oxalate extraction is relatively constant (36 ± 11 and 23 ± 8 mg g^{-1} for MID and CB1 respectively). The change in this residual, presumably non-reactive Fe pool accounts for less than 22% of the variation within a given total Fe profile. The residual fraction likely consists of Fe occurring as highly crystalline iron oxide minerals or as Fe-containing silicate minerals.

Down-core variability in total P exceeds 2 mg g^{-1} and is driven by changes in the oxalate-extractable component rather than the relatively constant residual P concentrations (0.95 ± 0.33 and 1.25 ± 0.39 mg g^{-1} for MID and CB1, respectively). Oxalate P has been considered to be inorganic P associated with high surface area iron oxide minerals (Williams *et al.*, 1971). Oxalate data from Toolik Lake cores clearly show that it is the oxalate P which is the reactive component of the sediment.

Sulfate is present at quite low concentrations in Toolik Lake (12 μmol l^{-1}; Kling *et al.*, 1991), and thus sulfate reduction is not expected to be an important metabolic process within the sediments. That at least some sulfate reduction is occurring is suggested by the total reduced S concentration profile at site E (Fig. 7). However, sulfate reduction and the production of iron sulfide minerals is clearly of little quantitative significance as it could account for <0.2% of total Fe as FeS.

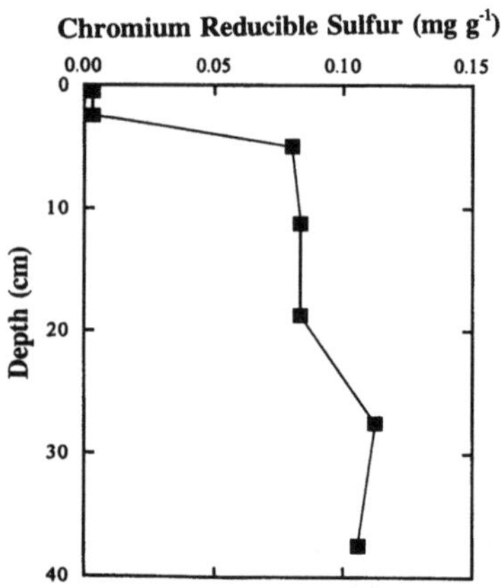

Fig. 7. Total reduced sulfur concentrations versus depth from site E in Toolik Lake.

Porewater chemistry

Measurements of chemical concentrations within pore waters provide insight into sediment zonation of redox processes and rates of solute fluxes within sediments (Froelich *et al.*, 1979; Tessenow & Baynes, 1975; Carignan & Flett, 1981). Detailed pore water data have not been collected previously for arctic lakes. A major goal of pore water chemical studies in Toolik Lake is to identify the processes resulting in Mn, Fe and P enrichment in the sediments.

Dissolved inorganic carbon (DIC) and ammonium are present at relatively low concentration in Toolik Lake pore waters (Fig. 8), suggesting low rates of organic matter decomposition. Cores E and K had surficial pore water NH_4^+ concentrations of 2.2 and 4.2 μmol l^{-1} respectively, and surficial concentrations ranged up to 18.8 μmol l^{-1} in other cores. Average surficial NH_4^+ concentrations were ($x \pm SD$, $n = 11$) 6.8 ± 5.8 μmol l^{-1}, much higher than typical water column values (<0.5 μmol l^{-1}; Whalen & Cornwell, 1985). Despite the similarity in DIC

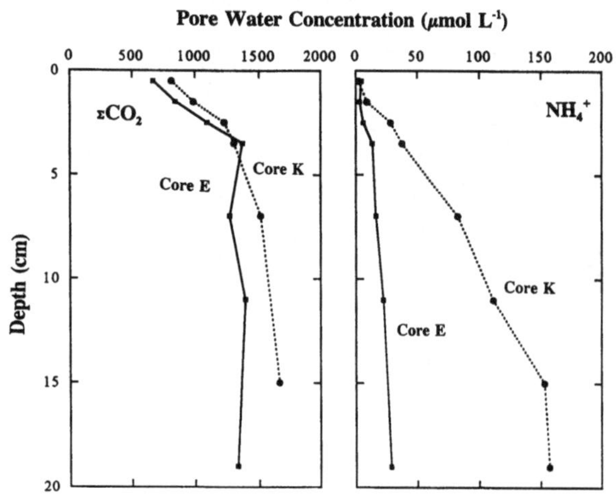

Fig. 8. Pore water dissolved inorganic carbon and ammonium profiles from site E in the Toolik Lake main basin and from near site K near the lake's secondary inlet. The data were collected in August 1981.

concentrations and gradients in cores E and K, site K appears to generate much higher amounts of pore water NH_4^+ than does E. The sedimentary flux of NH_4^+ may account for up to 20% of phytoplankton nitrogen demand within Toolik Lake (Whalen *et al.*, 1988).

Pore water profiles of Fe and inorganic P generally show mid-core maxima (Fig. 9). Higher concentrations of pore water Fe occur below solid phase Fe-enrichments, suggesting that as Fe(III) oxides are buried into more reducing horizons they are reduced to aqueous Fe(II) (Tessenow and Baynes, 1975; Gubala, 1990). Calculations show that the resulting upward flux of Fe is similar to the burial flux of Fe within the Fe-enriched zone (Cornwell, 1983; 1987a). A continuous sequence of 1) Fe oxide burial, 2) Fe(III) reduction, 3) upward flux of Fe(II) in pore water and 4) precipitation of Fe oxides in oxidizing sediment horizons maintains the Fe enrichment zones within the sediment. Pore water inorganic P shows the same general pattern as Fe, with a large increase in pore water inorganic P within the sediment. Upward fluxes of inorganic P help maintain P enrichments within sediments. Solid phase P profiles however, show no large P maxima within the sediment. This most likely results from the formation of authigenic vivianite at greater depths within the sediment.

Manganese enrichments have been studied in a variety of sedimentary environments and include deep sea (Froelich *et al.*, 1979; Burdige & Gieskes, 1983) and lacustrine deposits (Strakhov, 1966; Mothersill & Shegelski, 1973; Calvert & Price, 1977; Farmer & Lovell, 1984). In deep water sediments from Toolik Lake, high concentrations of dissolved Mn are found immediately below the solid phase Mn enrichment (Fig. 10). In a manner analagous to that described for Fe, Mn enrichment results from a sequence of Mn oxide burial, reductive dissolution, upward diffusion of aqueous Mn(II) and reprecipitation of Mn closer to the sediment-water interface. In this manner, Mn enrichments are maintained near the sediment-water interface despite the constant input of new sedimentary material (Robbins & Callender, 1975; Cornwell, 1983).

Chemical reactions involving Fe and Mn can have an impact on the cycling of other elements. Iron cycling in lake sediments has a strong influence on P cycling, particularly via adsorption/desorption of inorganic phosphate on Fe oxide minerals (Baccini, 1985; Krom & Berner, 1981). In Toolik Lake, Ba cycling follows that of Mn,

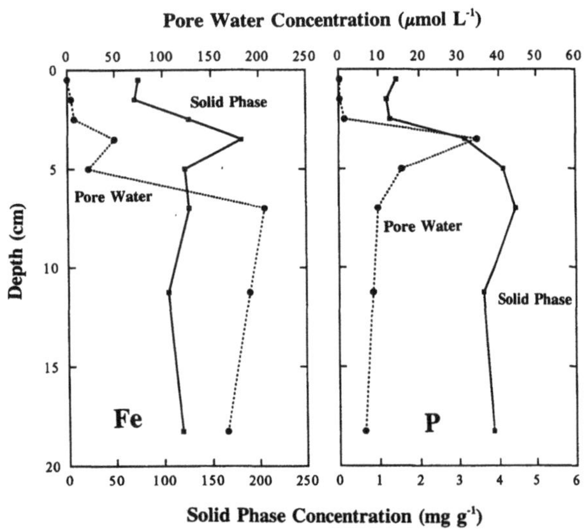

Fig. 9. Iron and phosphorus pore water and solid phase (total dissolution) profiles from site E in the Toolik Lake main basin. This core was collected in May 1980.

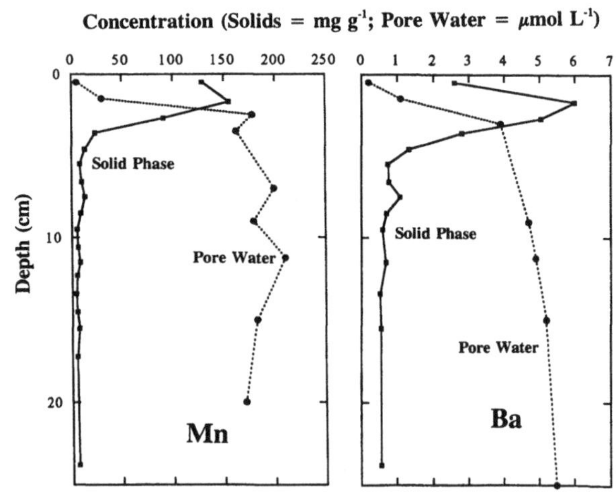

Fig. 10. Manganese and Ba pore water and solid phase profiles from site E in the Toolik Lake main basin. The Mn and Ba pore water profiles are from separate cores and the solid phase profile is from a third core. All cores were collected within 1 m of each other in May 1980.

with the formation of Ba-rich sediment zones in Mn-rich sediment horizons (Fig. 10). As with Mn, strong pore water Ba gradients result in an upward flux of Ba that approximately balances the burial flux of adsorbed or coprecipitated solid phase Ba (Cornwell, 1983). Barium may be incorporated in Mn oxide minerals to form the mineral psilomelane ($BaMn_8O_{16}(OH)_4$; Calvert & Price, 1977). Pure psilomelane could account for up to 29% of sediment Mn in some sediment samples. While direct pore water evidence is lacking for the concurrent mobility of other trace metals, solid phase profiles strongly indicate that Co, ^{226}Ra and Ni are also mobile within Toolik Lake sediments (Cornwell, 1986, 1987b).

Sediment-water chemical exchange

On the basis of nutrient mass balances for Toolik Lake, Whalen & Cornwell (1985) concluded that fluxes of carbon, nitrogen and phosphorus from the sediments to the overlying water must be small components of the nutrient cycles. Studies of chemical transport processes in arctic lakes are rare, limiting inter-site comparisons. Within Alaska, nutrient cycling in Toolik Lake contrasts markedly with the situation in shallow ponds along the arctic coast where sediments strongly influence overall P and N cycles (Prentki *et al.*, 1980).

Kipphut (1988) has presented a brief overview of the likely role of the metal-rich sediments within Toolik Lake in regulating chemical fluxes with the overlying waters. The relatively oxidizing nature of the surficial sediments appears to strongly adsorb phosphate and possibly ammonium ions. Despite the relatively high organic matter content of Toolik Lake sediments, concentrations of ammonium and phosphate within the pore waters of surficial sediments are relatively low (Fig. 8). Thus very low fluxes of ammonium and phosphate from the sediments are expected.

This supposition has been confirmed by direct measurements of sediment-water chemical transport. Benthic chambers have been used to measure chemical fluxes of several species between sediments and overlying waters within Toolik Lake. A summary of the results for nitrogen, phosphorus and dissolved oxygen are discussed here. We have not been able to measure a flux of phosphorus between sediments and the overlying water in any of our attempts to do so. We believe that this negative result clearly confirms that sediment recycling of phosphorus is a very small source of P to the Toolik Lake water column. Ammonium fluxes were also found to be low, on the order of 0.01–0.1 mmol $m^{-2} d^{-1}$. Although very small in comparison with fluxes observed in temperate aquatic environments, the sediment ammonium flux is large enough to be of significance to both planktonic and benthic algal nitrogen requirements. In the most detailed consideration of the nitrogen cycle of Toolik Lake to date, Whalen *et al.* (1988) estimated that ammonium fluxes from sediments supply 10–19% of the annual nitrogen requirements of the planktonic primary producers in this oligotrophic lake. These sediment-water N flux rates are also considerably lower than the average summer N deposition as measured by sediment traps ($\sim$0.4 mmol N $m^{-2} d^{-1}$; Whalen & Cornwell, 1985). Klingensmith & Alexander (1983) estimated the Toolik Lake sediment denitrification rate to be $\sim$0.4 mmol N $m^{-2} d^{-1}$, clearly a rate at odds with the nitrogen deposition data and most likely a result of assuming that denitrification occurs over the upper 12 cm of sediment. Under-ice ammonium release from the sediments, followed by nitrification, is clearly the explanation for elevated nitrate concentrations observed in Toolik Lake at the end of the winter season. Nitrate concentrations decrease rapidly during the period of ice-out, and Whalen & Alexander (1986) have speculated that the nitrate is utilized by the benthic algae. Rates of primary production by the benthic algae in Toolik Lake are not well characterized, but they are believed to be at least as large as rates of phytoplanktonic primary production (Yeakel, 1977).

Measurements of dissolved oxygen fluxes into the sediments are especially valuable because they provide estimates of rates of overall remineralization of organic matter within the sediments. In

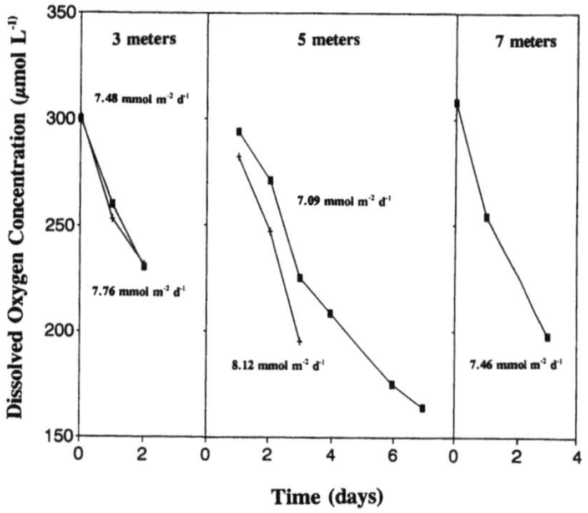

Fig. 11. Oxygen consumption by sediments measured by benthic chambers. Oxygen fluxes are computed from the rate of change of O_2 concentrations, and the dimensions of the chambers. Results are for net O_2 flux, which measures both respiration and production of O_2 by benthic algae.

Camp Bay, five measurements indicated an average net oxygen flux (as O_2) into the sediments of approximately 7.6 mmol m^{-2} d^{-1} (Fig. 11). These are extremely low rates of benthic oxygen consumption for a lake environment and they support the supposition that overall rates of sediment diagenesis are low. Annual oxygen fluxes into the sediments could be somewhat larger as our chamber measurements incorporate any oxygen production which results from the presence of benthic algae. Alternatively, annual O_2 fluxes could be smaller as our measurements do not encompass the long, cold ice-covered period. However, our oxygen consumption rates are consistent with the under-ice O_2 concentration data of Whalen & Cornwell (1985) that shows significant oxygen depletion does not occur in Toolik Lake, despite the long period of ice cover. Finally, very low rates of oxygen consumption appear to be necessary for the development of the well defined Fe and Mn enrichments which occur in Toolik Lake sediments.

Conclusions

'We are interested in arctic lakes first because they represent an extreme type, and the study of extremes often exposes or suggests general behavior patterns of our system that we would not discover by studying the more normal types. Second, we go to the arctic, as we do to hot springs and salt lakes, in search of simplicity'

Rigler (1978), cited above, has clearly pointed out a strong rationale for the study of the ecology of arctic lacustrine systems. The same arguments apply to the study of sediment processes in Toolik Lake. Although the supply of oxidants (O_2, NO_3^-, Mn(IV), Fe(III), SO_4^{2-}) clearly has a major influence on the nature of decomposition in aquatic sediments, it may be argued that the rate of supply of reactive organic matter to the sediment-water interface is also a major determinant controlling the depth zonation and overall rates of metabolic processes within low temperature sedimentary environments. In the deep sea, organic carbon deposition rates are very low (~20–150 mmol m^{-2} yr^{-1}); Bender & Heggie, 1984) while those in coastal sediments can be quite high (4000–6000 mmol m^{-2} yr^{-1}; Jorgensen, 1977). Consequently, in coastal marine sediments sulfate reduction commences within a few centimeters of the sediment-water interface, while in the deep sea, oxygen reduction may be the primary metabolic pathway in the top 5–10 cm or more of sediment (Reeburgh, 1983). Pore water studies in deep sea sediments (e.g. Froelich *et al.*, 1979) have the advantage that successive metabolic processes are spatially distinct.

The carbon depositional flux estimated from sediment traps in Toolik Lake is estimated to be ~230 mmol m^{-2} yr^{-1} (Whalen & Cornwell, 1985), with perhaps an equivalent amount of carbon input from benthic algae (Whalen *et al.*, 1988). While these fluxes are much higher than found in the deep sea, they are still an order of magnitude less than eutrophic coastal sediments, and processes such as Mn and Fe reduction occur in readily identifiable zones within Toolik Lake sediment. It is the low rate of both allocthonous and autochthonous organic matter input, low temperatures, low rates of sediment accretion and modest inputs of Mn and Fe that lead to conditions supporting the formation of Mn- and Fe-enriched sediments. In addition, the lack of significant production of sulfide simplifies the study of Fe and Mn cycling by limiting a potentially important process (formation of sulfide minerals) which would enhance burial fluxes. Instead, the oxidized conditions found at the sediment-water interface within Toolik Lake result in high retention of upwardly-diffusing Mn and Fe, and the formation of Fe- and Mn-enriched sediments. Iron and Mn geochemistry consequently exerts a strong control on the flux of phosphate and metals both within the pore waters and through the sediment-water interface.

Rigler's argument for studying extreme and simple types of systems clearly is applicable to Toolik Lake sediment studies. While Toolik Lake sediment may be an imperfect analogue to deep sea sediments, in many ways it represents an extreme type for lacustrine sediments. Ultimately, sediment biogeochemistry in Toolik Lake is shaped by the physical and biological processes within the watershed which result in strong limitation of organic production both in the watershed and in Toolik Lake.

Acknowledgements

We would like to thank S. Whalen, S. Banahan, D. McCorkle, C. Johnston and A. Doyle for assistance with sediment collection and benthic chamber deployments. We thank J. Stribling for editorial assistance and S. Whalen and R. T. Prentki for helpful comments. This work was supported by NSF grant DPP77-23475 (R. J. Barsdate), NSF grant DPP8320544 (G. W. Kipphut) and a State of Alaska marine fellowship (J. Cornwell).

References

Anderson, R. G., S. L. Schiff & R. H. Hesslein, 1987. Determining sediment accumulation and mixing rates using ^{210}Pb, ^{137}Cs and other tracers: problems due to post-depositional mobility or coring artifacts. Can. J. Fish. aquat. Sci. 44: 231–250.

Baccinni, P., 1985. Phosphate interactions at the sediment-water interface. In W. Stumm (ed.) Chemical Processes in Lakes. John Wiley & Sons, New York. 435 p.

Bender, M. L. & D. T. Heggie, 1984. Fate of organic carbon reaching the sea floor: a status report. Geochim. Cosmochim. Acta 48: 977–986.

Bray, J. T., O. P. Bricker & B. N. Troup, 1974. Phosphate in interstitial waters of anoxic sediments: oxidation effects during sampling procedure. Science 180: 1362–1363.

Brown, J. & R. L. Berg, 1980. Environmental engineering and ecological baseline investigations along the Yukon River-Prudhoe Bay Haul Road. CRREL Report 80-19.

Burdige, D. J. & J. M. Gieskes, 1983. A pore water/solid phase diagenetic model for manganese in marine sediments. Am. J. Sci. 283: 29–47.

Calvert, S. E. & N. B. Price, 1977. Shallow water, continental margin and lacustrine nodules: distributions and geochemistry. In G. P. Glasby (ed.), Marine Manganese Deposits, Elsevier, Amsterdam.

Canfield, D. E., R. Raiswell, J. T. Westrich, C. M. Reaves & R. A. Berner, 1986. The use of chromium reduction in the analysis of reduced inorganic sulfur in sediments and shales. Chem. Geol. 54: 149–155.

Carignan, R. & R. J. Flett, 1981. Post depositional mobility of phosphorus in lake sediments. Limnol. Oceanogr. 26: 361–366.

Chanton, J. C., C. S. Martens & G. W. Kipphut, 1983. Lead-210 sediment geochronology in a changing coastal environment. Geochim. Cosmochim. Acta 47: 1791–1804.

Coakley, J. P. & B. R. Rust, 1968. Sedimentation in an arctic lake. J. Sediment. Petrol. 38: 1290–1300.

Cornwell, J. C., 1983. The geochemistry of manganese, iron and phosphorus in an arctic lake. Ph.D. Dissertation, University of Alaska, Fairbanks. 238 pp.

Cornwell, J. C., 1985. Sediment accumulation rates in an Alaskan arctic lake using a modified ^{210}Pb technique. Can. J. Fish. aquat. Sci. 42: 809–814.

Cornwell, J. C., 1986. Diagenetic trace metal profiles in arctic lake sediments. Envir. Sci. Technol. 20: 299–302.

Cornwell, J. C., 1987a. Phosphorus cycling in arctic lake sediments: adsorption and authigenic minerals. Arch. Hydrobiol. 109: 161–179.

Cornwell, J. C., 1987b. Migration of metals in sediment pore waters: problems for the interpretation of historical deposition rates. In Proceedings of the 6th International Conference on Heavy Metals in the Environment, Vol. 2, p. 233–235, New Orleans.

Cornwell, J. C. & S. Banahan, 1992. A silicon budget for an Alaskan arctic lake. Hydrobiologia 240: 37–44.

Cornwell, J. C. & J. W. Morse, 1987. The characterization of iron sulfide minerals in anoxic marine sediments. Mar. Chem. 22: 193–206.

Dean, W. E., W. S. Moore & K. H. Nealson, 1981. Manganese cycles and the origin of manganese nodules, Oneida Lake, New York, USA. Chem. Geol. 34: 53–64.

de March, L., 1978. Permanent sedimentation of nitrogen, phosphorus and organic carbon in a high arctic lake. J. Fish. Res. Bd Can. 35: 1089–1094.

Evans, R. D. & F. H. Rigler, 1980. Measurement of whole

lake sediment accumulation using Pb-210 retention. Can. J. Fish. aquat. Sci. 37: 817–822.
Evans, R. D. & F. H. Rigler, 1983. A test of Pb-210 dating for the measurement of whole lake soft sediment accumulation. Can. J. Fish. aquat. Sci. 40: 506–515.
Farmer, J. G. & M. A. Lovell, 1984. Massive diagenetic enhancement of manganese in Loch Lomond sediments. Envir. Technol. Lett. 5: 257–262.
Forstner, U. & G. T. W. Wittman, 1979. Metal Pollution in the Aquatic Environment. Springer Verlag, Berlin.
Froelich, P. N., G. P. Klinkhammer, M. L. Bender, N. A. Luedtke, G. R. Heath, D. Cullen, P. Dauphin, D. Hammond, B. Hartmann & V. Maynard, 1979. Early diagenesis of organic matter in pelagic sediments of the eastern equatorial Atlantic: suboxic diagenesis. Geochim. Cosmochim. Acta 43: 1075–1090.
Gilbert, R. & M. Church, 1983. Contemporary sedimentary environments of Baffin Island, N.W.T., Canada: Reconnaissance of lakes on Cumberland Peninsula. Arct. Alp. Res. 15: 321–332.
Gorham, E. & D. J. Swaine, 1965. The influence of oxidizing and reducing conditions upon the distribution of some elements in lake sediments. Limnol. Oceanogr. 10: 268–279.
Gubala, C. P., 1990. Effects of iron cycling on ^{210}Pb dating of sediments in an Adirondack Lake, USA. Can. J. Fish. aquat. Sci. 47: 1821–1829.
Hamilton, T. D. & S. C. Porter, 1975. Itkillik glaciation in the Brooks Range, northern Alaska. Quat. Res. 5: 471–497.
Hermanson, M. H., 1990. ^{210}Pb and ^{137}Cs chronology from small, shallow arctic lakes. Geochim. Cosmochim. Acta. 54: 1443–1452.
Hobbie, J. E., 1973. Arctic limnology, a review. In M. E. Britton (ed.), Alaskan arctic tundra. Arctic Institute of North America Technical Paper 25.
Hobbie, J. E., T. L. Corliss & B. J. Peterson, 1983. Seasonal patterns of bacterial abundance in an arctic lake. Arct. Alp. Res. 15: 253–259.
Johnston, C. G. & G. W. Kipphut, 1988. Microbially mediated Mn(II) oxidation in an oligotrophic arctic lake. Appl. Envir. Microbiol. 54: 1440–1445.
Jorgensen, B. B., 1977. The sulfur cycle of a coastal marine sediment (Limfjorden, Denmark). Limnol. Oceanogr. 22: 814–832.
Klassen, R. A., I. Nichol & W. W. Shilts, 1975. Lake geochemistry in the Kaminak Lake area, District of Keewatin, N.W.T. Verh. int. Ver. Limnol. 19: 340–348.
Klingensmith, K. M. & V. Alexander, 1983. Sediment nitrification, denitrification and nitrous oxide production in a deep arctic lake. Appl. envir. Microbiol. 46: 1084–1092.
Kipphut, G. W., 1978. An investigation of sedimentary processes in lakes. Ph.D. Thesis, Columbia University, New York. 180 p.
Kipphut, G. W., 1988. Sediments and organic carbon cycling in an arctic lake. In E. T. Degens, S. Kempe & A. S. Naidu (eds), Transport of Carbon and Minerals in Major World Rivers, Lakes, and Estuaries. Mitt. Geol.-Palaont. Inst. Univ. Hamburg, SCOPE/UNEP, 66: 129–135.
Kling, G. W., W. J. O'Brien, M. C. Miller & A. E. Hershey, 1992. The biogeochemistry and zoogeography of lakes and rivers in arctic Alaska. Hydrobiologia 240: 1–14.
Krauskopf, K. B., 1957. Separation of manganese from iron in sedimentary processes. Geochim. Cosmochim. Acta 12: 61–68.
Krom, M. D. & R. A. Berner, 1981. The diagenesis of phosphorus in a nearshore marine sediment. Geochim. Cosmochim. Acta 45: 207–216.
Livingstone, D. A., K. Bryan Jr. & R. G. Leahy, 1958. Effects of an arctic environment on the origin and development of freshwater lakes. Limnol. Oceanogr. 3: 192–214.
Medlin, J. H., N. H. Suhr & J. B. Bodkin, 1969. Atomic absorption analysis of silicates employing $LiBO_2$. At. Abs. Newsl. 8: 25–29.
Miller, M. C., G. R. Hater, P. Spatt, P. Westlake & D. Yeakel, 1986. Primary production and its control in Toolik Lake, Alaska. Arch. Hydrobiol. Suppl. 74: 97–131.
Morse, J. W. & J. C. Cornwell, 1987. Analysis and distribution of iron sulfide minerals in recent anoxic marine sediments. Mar. Chem. 22: 55–69.
Mothersill, J. S. & R. J. Shegelski, 1973. The formation of iron and manganese-rich layers in the sediments of Thunder Bay, Lake Superior. Can. J. Earth Sci. 10: 571–576.
Parsons, T. R., Y. Maita & C. M. Lalli, 1984. A Manual of Chemical and Biological Methods for Seawater Analysis. Pergamon Press, New York. 173 pp.
Prentki, R. T., M. C. Miller, R. J. Barsdate, V. Alexander, J. Kelley & P. Coyne, 1980. Chemistry. In J. Hobbie (ed.), Limnology of Tundra Ponds. Dowden, Hutchinson and Ross, Stroudsburg, Pa.: 76–178.
Raad, A. T., R. Protz & R. L. Thomas, 1969. Determination of Na-dithionate and NH_4-oxalate extractable Fe, Al and Mn in soils by atomic absorption spectrophotometry. Can. J. Soil Sci. 49: 89–94.
Reeburgh, W. S., 1967. An improved interstitial water sampler. Limnol. Oceanogr. 12: 163–165.
Reeburgh, W. S., 1983. Rates of biogeochemical processes in anoxic sediments. Ann. Rev. Earth Planet Sci. 11: 269–298.
Rigler, F. H., 1978. Limnology in the high arctic: a case study of Char Lake. Verh. int. Ver. Limnol. 20: 127–140.
Robbins, J. A., 1978. Geochemical and geophysical applications of radioactive lead. In J. O. Nriagu (ed.) The Biogeochemistry of Lead in the Environment. Elsevier, Amsterdam.
Robbins, J. A. & E. Callender, 1975. Diagenesis of manganese in Lake Michigan sediments. Am. J. Sci. 275: 512–533.
Schindler, D. W., H. E. Welch, J. Kalff, G. J. Brunskill, H. Kling & N. Kritsch, 1974. Physical and chemical limnology of Char Lake, Cornwallis Island. J. Fish. Res. Bd. Can. 31: 587–607.
Sozanski, A. G. & D. S. Cronan, 1979. Ferromanganese con-

cretions in Shebandowan Lakes, Ontario. Can. J. Earth Sci. 16: 126–140.

Strakhov, N. M., 1966. Types of manganese accumulation in present-day basins: their significance in understanding manganese mineralization. Int. Geol. Rev. 8: 1172–1196.

Stainton, M. P., 1973. A syringe gas stripping procedure for gas chromatographic determination of dissolved inorganic and organic carbon in freshwater and carbonates in sediments. J. Fish. Res. Bd Can. 30: 1441–1445.

Stumm, W. & J. J. Morgan, 1981. Aquatic Chemistry. An Introduction Emphasizing Chemical Equilibria in Natural Waters, 2nd edn. Wiley Interscience, New York.

Tessenow, U. & Y. Baynes, 1975. Redox-dependent accumulation of Fe and Mn in a littoral sediment supporting *Isoetes lacustris* L. Naturwissenschaften 62: 342–343.

Weiss, H. V. & A. S. Naidu, 1986. ^{210}Pb flux in an arctic coastal region. Arctic 39: 59–64.

Welch, H. E. & J. A. Legault, 1986. Precipitation chemistry and chemical limnology of fertilized lakes at Saqvaqjuac, N.W.T. Can. J. Fish. aquat. Aci. 43: 1104–1134.

Welch, H. E. & M. A. Bergmann, 1985. Winter respiration rates of lakes at Saqvaqjuac, N.W.T. Can. J. Fish. aquat. Sci. 42: 521–528.

Whalen, S. C. & V. Alexander, 1986. Seasonal inorganic carbon and nitrogen transport by phytoplankton in an arctic lake. Can. J. Fish. aquat. Sci. 43: 1177–1186.

Whalen, S. C. & J. C. Cornwell, 1985. Nitrogen, phosphorus and organic carbon cycling in an arctic lake. Can. J. Fish. aquat. Sci. 42: 797–808.

Whalen, S. C., J. C. Cornwell & V. Alexander, 1988. Comparison of chemical and biological N budgets in an arctic lake: implications for phytoplankton production. In E. T. Degens, S. Kempe & A. S. Naidu (eds), Transport of Carbon and Minerals in Major World Rivers, Lakes, and Estuaries. Mitt. Geol. Palaont. Inst. Univ. Hamburg, SCOPE/UNEP, 66: 99–115.

Williams, J. D. H., J. K. Syers, S. S. Shukla, R. F. Harris & D. E. Armstrong, 1971. Levels of inorganic and total phosphorus in lake sediments as related to other sediment parameters. Envir. Sci. Technol. 5: 1113–1120.

Yeakel, J. D., 1977. Production and control of epilithic periphyton in an arctic lake. M.S. Thesis, Univ. of Cincinnati. 53 pp.

Hydrobiologia **240**: 61–69, 1992.
W.J. O'Brien (ed.), Toolik Lake – Ecology of an Aquatic Ecosystem in Arctic Alaska.

Nitrogen and phosphorus concentrations and export for the upper Kuparuk River on the North Slope of Alaska in 1980

Bruce J. Peterson,[1] Teresa Corliss,[1,2] Keith Kriet[3,4] & John E. Hobbie[1]
[1] *The Ecosystems Center, Marine Biological Laboratory, Woods Hole, MA 02543, USA;* [2] *Present address: 10 Park Place, Durham, NC 27712, USA;* [3] *Yale University, School of Forestry, New Haven, CT 06211, USA;* [4] *Present address: 552 Savannah Highway No. 5, Charleston, SC 29407, USA*

Abstract

Nitrogen and phosphorus concentrations were measured from May to August 1980 in the upper Kuparuk River, a tundra stream on the North Slope of Alaska. Mean values for nitrogen were 10.8 $\mu g\ N\ l^{-1}$ for ammonium, 21.4 $\mu g\ N\ l^{-1}$ for nitrate plus nitrite and 248 $\mu g\ N\ l^{-1}$ for dissolved organic nitrogen. Mean values for phosphorus were 8.1 $\mu g\ P\ l^{-1}$ for total dissolved phosphorus and 4.7 $\mu g\ P\ l^{-1}$ for fine particulate phosphorus. Nitrate concentrations were inversely correlated with flow whereas particulate phosphorus concentrations increased during high flows. Export of nitrogen and phosphorus from the watershed during 1980 was estimated to be 4.69, 3.25 and 91 $kg\ km^{-2}\ yr^{-1}$ for NO_3-N, NH_4-N and DON-N, respectively, and 2.86 and 3.03 $kg\ km^{-2}\ yr^{-1}$ for TDP-P and PP-P. Both the relative concentrations of N and P and the relative amounts exported suggest that phosphorus is in short supply but both nutrients are present in low concentrations comparable to those found previously in tundra ponds at Point Barrow, Alaska.

Introduction

Nitrogen and phosphorus are two key nutrients controlling primary production in arctic Alaskan tundra vegetation and in tundra lakes and streams (Ulrich & Gersper, 1978; Chapin *et al.*, 1980; Shaver & Chapin, 1980; Prentki *et al.*, 1980; Peterson *et al.*, 1983). For this reason, the study of nitrogen and phosphorus in tundra watersheds is important for both the basic objective of understanding the function of tundra ecosystems as well as for the applied objective of managing these systems wisely. Streams are key links between the terrestrial systems and downstream lakes or oceans, as well as being biologically active systems in their own right. Thus a study of the flows of nitrogen and phosphorus in a stream provides relevant information for the terrestrial donor systems upstream, the stream ecosystem itself, and the recipient aquatic system downstream. This paper describes the nitrogen and phosphorus concentrations in stream water and the nutrient export from a 4th-order tundra river watershed in the foothills of the Brooks Range on the North Slope of Alaska.

There are three different types of streams on the North Slope of Alaska: mountain streams, spring streams and tundra streams (Craig & McCart, 1975). The mountain streams drain snowfields and glaciers in the Brooks Range and are consequently turbid with glacial rock-flour and have low productivity. Spring-fed streams flow all year with relatively constant discharge and high productivity. Tundra streams drain the foothills and coastal plains, are not fed by glaciers or springs, and have intermediate levels of productivity. The present investigation focused on a tundra stream, the upper Kuparuk River, which crosses the Prudhoe Bay pipeline and the Dalton Highway about 25 km north of its origin in the foothills of the Brooks Range. The stream flows from south to north and upstream of the pipeline crossing the upper Kuparuk watershed is undisturbed by man.

A tundra stream has important advantages for this investigation of nutrient flux. Unlike mountain or spring streams, the water and nutrient dynamics of a tundra watershed reflect predominantly recent atmospheric inputs and biogeochemical processes of the biota and soils within the watershed. Since the entire watershed is underlain by continuous permafrost, the total nutrient output can be sampled as streamflow. Also, since the stream is frozen for most of the year, the annual nutrient export can be estimated from samples collected during the 4–5 month open water season.

Site description

The drainage basin included in this report covers the area from the headwaters of the Kuparuk River to its intersection with the Dalton Highway. The center of this headwater basin is approximately 68° 35′ N, 149° 20′ W. The river to this point has a main channel length of 25 km and drains an area of 143 km^2 which is mostly tussock tundra. A more complete description of the site is given in Kriet *et al.* (1992).

Methods

On 51 days from May 22 to August 21, 1980, the Kuparuk River discharge was estimated from Gurley current meter flow measurements or by stage height readings. The current meter measurements were taken every 0.5 m across the stream at midwater depth. At depths of >40 cm measurements were taken at two levels. Discharge was computed by multiplying current velocity times cross section area. From 1978, 1979 and 1980 data a regression ($r^2 = 0.98$, $n = 36$) of discharge versus stage allowed estimation of additional flow from stream stage (Kriet *et al.*, this volume).

Water samples for nutrient analyses were collected on 34 days from ice-out in mid-May to late August and all high flow events were sampled because they represented a large fraction of the total water, nutrient and particulate export. Nutrient samples were collected upstream of the Dalton Highway crossing by submerging a 4-liter polyethylene bottle in the river. At this location, the stream is shallow and turbulent and the fine particulates and dissolved nutrients are well mixed in the water column. No effort was made to sample particulates moving as bedload.

Water samples were processed at the Toolik research station within two hours of collection. Water was filtered through rinsed and weighed Whatman GF/C filters. The filters were stored and shipped frozen in glassine envelopes to the Marine Biological Laboratory in Woods Hole where the filters were dried at 50°–60 °C and weighed to determine total suspended sediment. Finally, the filters were analyzed for total particulate phosphorus (PP) by the acid digestion method of Stainton *et al.* (1974) with subsequent colorimetry for phosphate as described below. The filtrate to be analyzed for dissolved nutrient concentration was collected in acid-washed glass jars and stored frozen. At the end of the field season, the samples were shipped on dry ice to Woods Hole, Massachusetts.

Samples for ammonium, nitrate plus nitrite and phosphate were run on the Technicon Autoanalyzer system according to methods described in Technicon Industrial Method No. 154-71 W/B (1978), Industrial Method No. 158-71 (1977) and Industrial Method No. 155-71 W (1973) with modifications described by Glibert & Loder (1977). Samples for total dissolved nitrogen and total dissolved phosphorus were also run on the Autoanalyzer after a digestion by alkaline persulfate (D'Elia *et al.*, 1977) which oxidizes fixed nitrogen to nitrate and dissolved phosphorus forms to phosphate. The dissolved reactive phosphate (DRP) levels were consistently so low ($<6\ \mu g\ P\ l^{-1}$) that they could not be measured with the Technicon system. Because of this and because phosphorus at these low levels may have changed form during sample storage, DRP analyses are not reported here. Total dissolved phosphorus (TDP) analyses were performed using the potassium persulfate digestion procedure of Menzel & Corwin (1965) and the absorption of the samples at 882 nm was determined on a Beckman dual beam spectrophotometer with 5 cm cells.

The relationships between water flow and nutrient concentration were examined by regression of both linear and logarithmically transformed data. The watershed's nutrient output from 22 May to 21 August 1980 was calculated by linear interpolation of both discharge and nutrient concentrations to assign values for each day. Flow was multiplied by nutrient concentration and these daily export values were summed.

Spring sampling was begun May 22 when the stream was just beginning to flow in a small channel cut by snow melt runoff in the bottom-fast ice. Flow during the end of August, September and early October was not sampled but was estimated from the USGS continuous discharge record at the mouth of the Kuparuk River (190 km farther north) by a procedure outlined in Kriet *et al.* (1992). We estimate that 24% of the annual discharge was missed in 1980 by stopping our measurements on August 21 and it was assumed about 24% of the nutrient output was also missed. Total annual nutrient output from the watershed was estimated by adding this additional 24%. In further studies, an effort should be made to sample late season runoff for nutrients.

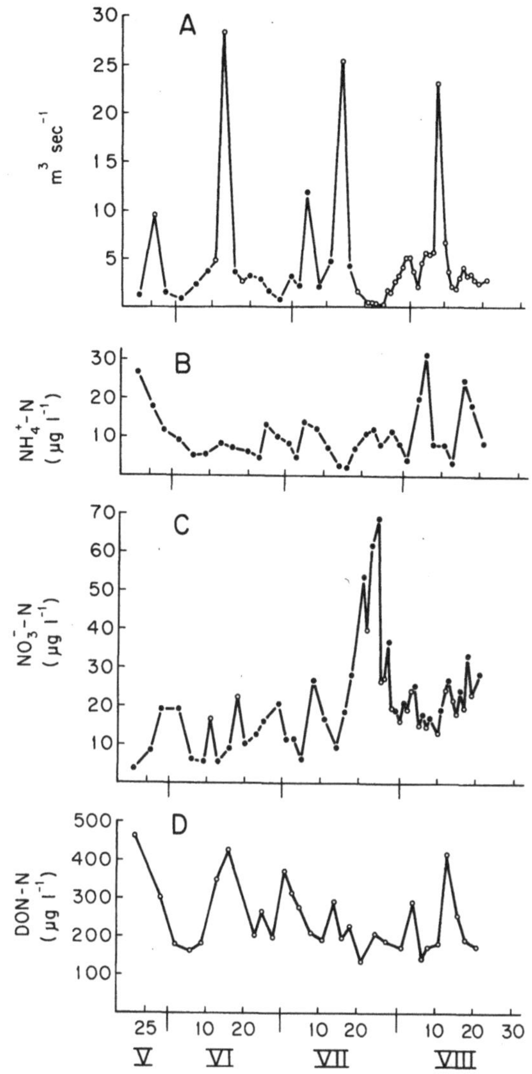

Fig. 1. Discharge and dissolved nitrogen levels in the upper Kuparuk River from May to August 1980. Panel A is flow rate in m^3 sec^{-1}. Open circles are estimates from the regression of flow on stage height. Panel B is ammonia nitrogen (NH_4^+) in μg N l^{-1}. Panel C is nitrate nitrogen (NO_3^--N) in μg N l^{-1}. Open circles are nitrate levels estimated from the regression of NO_3 concentration on flow. Panel D is dissolved organic nitrogen (DON-N) in μg N l^{-1}.

Results

The pattern of discharge is shown for comparison with the nitrogen and phosphorus concentration data in Figs 1 and 2; data are summarized in Table 1. Ammonium levels were low throughout the season and there was no correlation between discharge and ammonium concentration. The pattern was one of a decline in ammonium level during the first two weeks of the season followed by very low levels during June and July with two small peaks in concentration during August. Nitrate concentrations followed a completely different pattern with gradually increasing concentration throughout the season. Furthermore, there was a highly significant inverse correlation between nitrate concentration and discharge that could be improved by regressing nitrate on $\log_{10}$

Table 1. The range of nutrient concentrations measured in the upper Kuparuk River during May–August, 1980. N = number of sampling dates.

Parameter	N	Min	Max	Mean
NO_3-N ($\mu g\ l^{-1}$)	34	3.8	69.1	21.4
NH_4-N ($\mu g\ l^{-1}$)	34	2.3	31.7	10.8
DON-N ($\mu g\ l^{-1}$)	30	135.0	463.0	248.0
TDP-P ($\mu g\ l^{-1}$)	33	1.5	26.6	8.1
PP-P ($\mu g\ l^{-1}$)	31	0.7	24.8	4.7

flow (Table 2). This regression equation was used to estimate nitrate levels (open circles in Fig. 1C) on days when flow but not nitrate was measured. Because of the seasonal trend in the nitrate levels, the regression tended to overestimate NO_3 levels early in the season and underestimate late in the season but as can be seen from Figure 1C the bias is not large and it would not change the overall picture much if estimated values were omitted. The predominant form of dissolved nitrogen in the river water throughout the season is in the dissolved organic form. The levels of DON ranged from 135 to 463 $\mu g\ N\ l^{-1}$ and there appears to be a gradual but erratic decline in DON concentration during the summer. The linear regression of DON on flow was not significant.

The pattern of change in concentration of total dissolved phosphorus (TDP) was similar to that of ammonium (compare Figure 1B with Fig. 2B). A decline in concentration during the early part of the season was followed by several peaks later in the season. There was no correlation of TDP concentration with discharge. There was, however, a relatively weak but significant (95% level) positive correlation between TDP concentration and ammonium concentration: TDP ($\mu g\ l^{-1}$) = 0.305 NH_4^+ ($\mu g\ l^{-1}$) + 4.25 with $r = 0.41$ and $n = 32$. This may imply that similar processes are controlling the levels of the two dissolved nutrients.

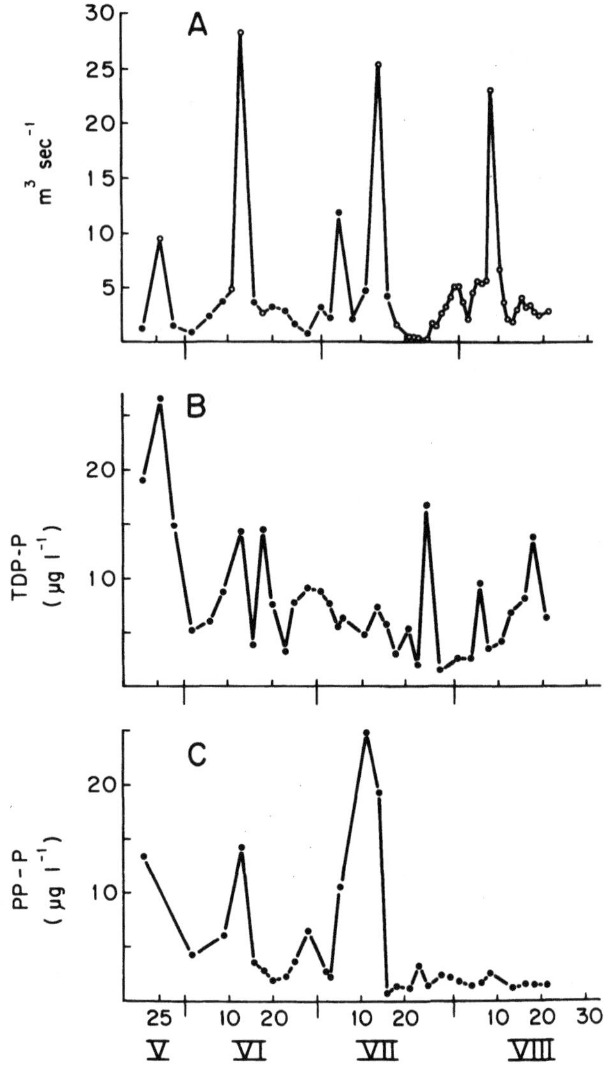

Fig. 2. Discharge and phosphorus concentrations in the upper Kuparuk River during May to August 1980. Panel A is discharge in $m^3\ sec^{-1}$. Panel B is total dissolved phosphorus (TDP-P) in $\mu g\ P\ l^{-1}$. Panel C is particulate phosphorus (PP-P) in $\mu g\ P\ l^{-1}$.

Particulate phosphorus concentrations covered the same range (0.7 to 25 $\mu g\ P\ l^{-1}$) as the TDP concentrations; however, PP was strongly and positively correlated with discharge. We also observed that increased stream flow leads to increasing suspended sediment load (Kriet *et al.*, 1992) and the percent PP of the suspended sedi-

Table 2. Statistically significant regressions relating nutrient levels to discharge for the upper Kuparuk River for May through August 1980.

1. NO_3 ($\mu g\ l^{-1}$) = -0.82 Flow ($m^3\ sec^{-1}$) + 25.6; $r = -0.37$, $n = 34$
2. NO_3 ($\mu g\ l^{-1}$) = $-21.9\ \log_{10}$ Flow ($m^3\ sec^{-1}$) + 31.5; $r = 0.66$, $n = 34$
3. PP ($\mu g\ l^{-1}$) = 0.56 Flow ($m^3\ sec^{-1}$) + 1.78; $r = 0.64$, $n = 29$

ment averaged 0.18% ± 0.15% (1 sd) for 31 determinations. Two of the particulate phosphorus observations appeared anomalous. The May 22 value was very high relative to the suspended sediment concentration, giving a PP value equivalent to 0.8% of the sediment. Unfortunately, we have no other PP values during the brief spring period when TDP levels were high. The second anomalous point is the August 8 sample when very high flow was associated with only a very minor peak in PP. The suspended sediments were also relatively low for the August 8 sample but during our three year study (Kriet *et al.*, 1992) this was the only time we observed high discharge without a marked increase in fine suspended sediment concentration. Including these two outlying points in the regression of PP on flow lowers the linear correlation coefficient from 0.64 to 0.48 (Table 2).

The mean TDN:TDP ratio during the summer was 49:1 and the ratio seldom fell below 20:1 (Fig. 3). There was no appreciable seasonal trend in the ratio and also no significant correlation with discharge.

The export of the measured forms of nitrogen and phosphorus was dominated by the flow regime (Figs 4 and 5). None of the nutrients exhibited a constant daily export. However, the pattern of nitrate export was affected by the trend of increasing nitrate levels during the season (Fig. 1). The range of daily export value is extreme for substances such as PP which increase in concentration as flow increases. Table 3 gives estimates of the nutrient output from the upper Kuparuk for 22 May through 8 August; the annual yields are based on these estimates plus the extrapolated values for fall discharge until freeze up (an additional 24%).

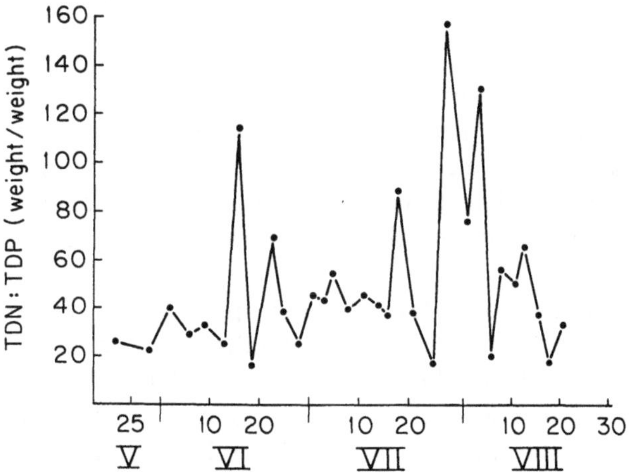

Fig. 3. The weight ratio of total dissolved nitrogen to total dissolved phosphorus in the upper Kuparuk from May to August 1980.

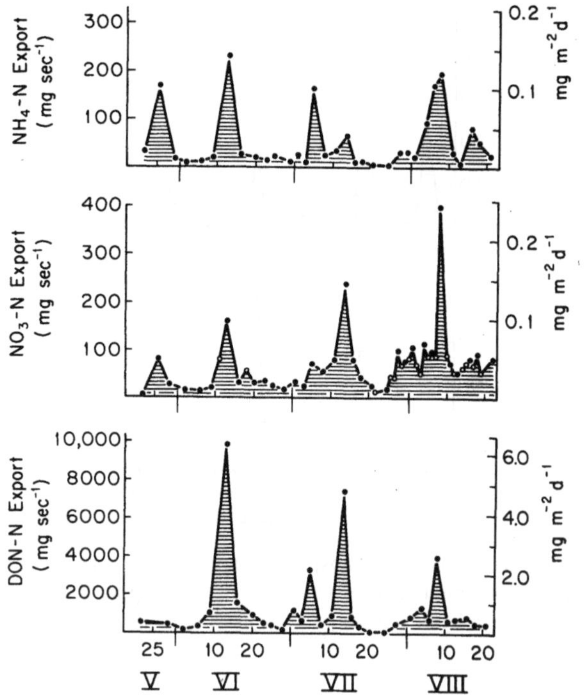

Fig. 4. Nitrogen output from the upper Kuparuk River watershed from May to August 1980. Panel A is NH_4^+ export. Panel B is NO_3^- export. Panel C is DON export.

Table 3. Nutrient output from the upper Kuparuk watershed during the interval 22 May through 21 August 1980 and the extrapolated 1980 annual output estimates.

Parameter	NO_3^--N	N_4^+-N	DON-N	TDP-N	PP-P
Specific yield (kg km^{-2} interval^{-1})	3.8	2.6	73.0	2.3	2.4
Extrapolated annual Specific yield (kg km^{-2} yr^{-1})	4.7	3.2	91.0	2.9	3.0

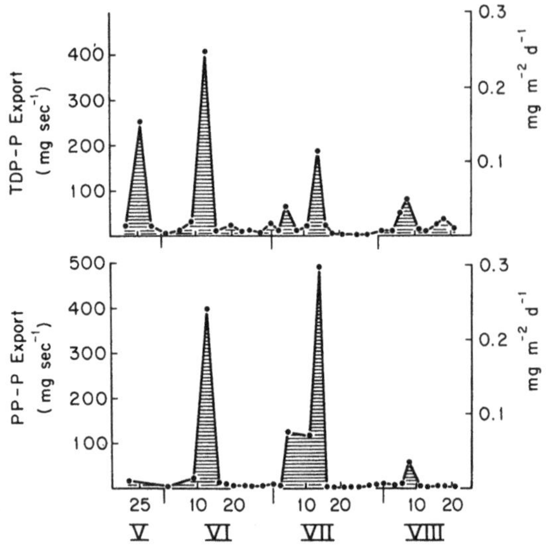

Fig. 5. Total dissolved phosphorus export (Panel A) and particulate phosphorus export (Panel B) from the upper Kuparuk River watershed from May to August 1980.

The nutrient export estimates were computed from daily export values by interpolating when necessary both flow and nutrient concentrations. Such a procedure is based on the supposition that even for nutrients with no obvious correlation with flow or season, water chemistry on a given day is more likely to be similar to the next day's value than to the annual mean. Nonetheless, such a detailed procedure had little effect on estimates for ammonium, DON and TDP. Simply multiplying the total water yield of 38.4×10^6 m^3 by the mean chemical concentration gave estimates that were within 15% of the daily summation values. Only in the cases of NO_3 and PP was a careful matching of flow and concentration sequences important. Nitrate export would be overestimated by about 50% by multiplication of total water output by mean concentration whereas particulate phosphorus or sediment output would be underestimated by 50%. Even these differences, however, can scarcely be considered critical when there may be a factor of two or more in the year-to-year variations in water and sediment yield. The 1980 measured sediment yield was 5-fold more than in 1978 because of the high frequency of high-flow events in 1980 (Kriet *et al.*, 1992).

Discussion

The nutrient concentrations determined in this study are similar to previous analyses of the upper Kuparuk River at the same study site. During 1978 a small number of ammonium and nitrate samples from the Kuparuk were run in the lab of Vera Alexander at the University of Alaska (unpublished data). The mean for seven samples for ammonium was 12 μg N l^{-1} which compares with our mean of 11 μg N l^{-1}. For nitrate, eight samples from 1978 gave a mean of 29 μg N l^{-1} as compared to our value of 21 μg N l^{-1}. For TDP, 16 determination were made during summer 1978 by J. Cornwell at the University of Alaska (unpublished data). The mean of those values was 7.6 μg P l^{-1} as compared to our 8.1 μg P l^{-1}. Studies in subsequent years from 1983 to 1990 have found very similar values for ammonium, consistently low values for dissolved reactive P and quite variable values for nitrate (Miller *et al.*, unpublished). The nutrient concentrations observed in 1980 appear to be reasonably representative for the upper Kuparuk River.

The dissolved inorganic nitrogen levels in the Kuparuk River are low. The values obtained for nitrate plus nitrite are very similar to the levels found in other studies of arctic waters (Table 4). The ammonium levels are lowest in the upper Kuparuk River although the ranges of concentration overlap considerably with other investigations. Dissolved organic nitrogen was by far the predominant form of dissolved nitrogen found in all the studies (Table 4). The levels of various nutrients at the mouth of the Kuparuk River seem to exceed the concentrations observed in its upper reaches. The USGS data are based on once a month sampling during the open water season, May to October.

Phosphorus levels in the upper Kuparuk are also extremely low. The total dissolved phosphorus levels for the Kuparuk are similar to the levels found in shallow tundra ponds at Point Barrow, the inlet streams to Char Lake and at the mouth of the Kuparuk River (Table 4). Particulate phosphorus in the Kuparuk River was usually present in lower concentration than total dis-

Table 4. Concentrations of nutrients in various arctic ecosystems given as mean and ranges (in parentheses).

Location	No_3-N + NO_2-N	NH_4-N	DON-N	TDP-P	Reference
Upper Kuparuk River	22 (4–69)	11 (2–32)	248 (135–463)	8 (1–27)	Present
Lower Kuparuk River[a]	33 (0–90)	32 (10–80)	513 (280–920)	19 (0–60)	USGS, 1979, 1980
Arctic Lakes	15[b] (15–20)	N.D.	N.D.	N.D.	Kalff, 1968
Snow Meltwater & Runoff at Barrow	14 (0–76)	64 (17–200)	N.D.	N.D.	Barsdate & Alexander, 1975
Barrow Ponds[c]	21	33	847	13	Prentki *et al.* 1980
Inlet Streams, at Char Lake, Canada	N.D.	N.D.	N.D.	11	de March, 1975

N.D. No data
[a] Samples from the mouth of the Kuparuk River at USGS gauging station.
[b] Does not include NO_2-N.
[c] Ranges are mean monthly ranges.

solved phosphorus except during high discharge. High flows usually carried elevated levels of fine suspended sediment and particulate phosphorus, whereas the TDP levels did not increase during high discharge. The total phosphorus export was about half in dissolved form and half as particulate but at flows higher than 11 $m^3\ sec^{-1}$ most of the phosphorus in the water column was in the particulate form.

Particulate phosphorus comprised a relatively constant fraction of the total suspended sediment, 0.18%. Since suspended sediment can be very easily measured gravimetrically, this suggests that particulate phosphorus export might be estimated from suspended sediment and discharge data. Rigler (1979) found that particulate phosphorus averaged 0.20 to 0.25% of total suspended sediments in streams draining granitic moorlands in Britain. Johnson *et al.* (1976) found lower values averaging 0.12% for both forested and agricultural watersheds in New York State. In another arctic study, Brunskill *et al.* (1975) found large variations in the mean phosphorus content of suspended sediments in streams in the Mackenzie and Porcupine drainages but most of the mean values for different rivers were in the range 0.10 to 0.40%. The available evidence suggests that this relationship should be checked for individual watersheds until an understanding of the causes of the variation in percent phosphorus in suspended particulates is developed.

The weight ratio of total dissolved nitrogen to total dissolved phosphorus is consistently so high that it would appear that periphyton growth in the stream would be phosphorus limited if and when nutrients are limiting. The ratio ranged from 16 to 160:1 with the average around 50:1. Algae normally use nitrogen and phosphorus in weight ratios of 5 to 10:1 (Strickland, 1960). Bioassays in this section of the Kuparuk during July and August 1980 demonstrated that periphyton growth was stimulated by added phosphorus and by phosphorus plus nitrogen but not by nitrogen alone (Peterson *et al.*, 1983).

The export of total phosphorus from the upper Kuparuk watershed was about 5.9 $kg\ km^{-2}\ yr^{-1}$ for 1980 with about half in dissolved and half in particulate form (Table 3). This value is comparable to the mean export (4.7 $kg\ km^{-2}\ yr^{-1}$) given by Dillon & Kirchner (1975) for forested igneous watersheds of the Canadian Shield. Other watersheds containing either pasture, sedimentary rocks or substantial urbanization exported much higher amounts of phosphorus. Probably a variety of processes and mechanisms interact to minimize phosphorus losses from the upper Kuparuk and the relative importance of each can not be assessed. Some important factors that may limit phosphorus export are: 1) precipitation inputs of phosphorus to the arctic tundra are low (Prentki *et al.*, 1980); 2) the tundra, especially wet tundra, may be phosphorus poor and thus retain phosphorus efficiently (Chapin, 1978); 3) epilithic algae and mosses in the stream may take up available phosphorus from the water; 4) decomposing peat which enters the stream by channel erosion may be so poor in nutrients that phosphorus is immobilized during its decomposition; and finally, 5) total water flow through the watershed is low and thus erosive power is limited. In fact, 1980 was undoubtedly a relatively high year for particulate phosphorus export due to the summer rain events with consequently high flows. Suspended sediment exports were substantially lower in 1978 and 1979 than in 1980 (Kriet *et al.*, this volume). These factors combine to keep phosphorus concentrations in waters draining this tundra watershed low. This in turn is probably one of the primary reasons for the extreme oligotrophy of lakes of the North Slope of Alaska (Hobbie, 1973).

Acknowledgements

This research was supported by the Department of Energy contract #EY-76-S-02-2989 and by NSF Polar Programs grant DPP-77-23879. We are grateful to R. Garritt for nutrient analysis on the Technicon Autoanalyzer and J. V. K. Helfrich III for field and laboratory assistance.

References

Barsdate, R. J. & V. Alexander, 1975. The nitrogen balance of Arctic tundra: Pathway rates and environmental implications. J. Envir. Qual. 4: 111–117.

Brunskill, G. J., P. Campbell, S. E. M. Elliot, B. W. Graham, W. J. Dentry & R. Wagemann, 1975. The chemistry, mineralogy and rates of transport of sediments in the Mackenzie and Porcupine River watersheds. N. W. T. and Yukon, 1971–73. Research and Development Directorate, Freshwater Institute, Winnipeg, Manitoba. Technical Report No. 546. 69 pp.

Chapin, F. S. III., 1978. Phosphate uptake and nutrient utilization by Barrow tundra vegetation, pp. 483–505. In L. L. Tieszen (ed.) Vegetation and Production Ecology of an Alaskan Arctic Tundra. Springer-Verlag, New York, 686 pp.

Chapin, F. S. III., L. L. Tieszen, M. C. Lewis, P. C. Miller & B. H. McCown, 1980. Control of tundra plant allocation patterns and growth, In J. Brown, P. C. Miller, L. L. Tieszen & F. L. Burnwell (eds) An Arctic Ecosystem, The Coastal Tundra at Barrow, Alaska. Dowden, Hutchinson and Ross, Inc., Stroudsburg, PA, 571 pp.

Craig, P. C. & P. J. McCart, 1975. Classification of stream types in Beaufort Sea drainages between Prudhoe Bay, Alaska and the Mackenzie delta, N.W.T., Canada. Arctic and Alpine Research 7: 183–198.

D'Elia, C. F., P. A. Steudler & N. Corwin, 1977. Determination of total nitrogen in aqueous samples using persulfate digestion. Limnol. Oceanogr. 22: 760–764.

de March, L., 1975. Nutrient budgets for a high arctic lake (Char Lake, N. W. T.). Mitt. Int. Ver. Theor. Angew. Limnol. 19: 496–503.

Dillon, P. J. & W. B. Kirchner, 1975. The effects of geology and land use on the export of phosphorus from watersheds. Wat. Res. 9: 135–148.

Glibert, P. M. & T. C. Loder, 1977. Automated analysis of nutrients in seawater: A manual of techniques. Woods Hole Oceanographic Institution Technical Report No. 77-47. 46 pp.

Hobbie, J. E., 1973. Arctic limnology: A review. In: M. E. Britton (ed.), Alaskan Arctic Tundra. Technical Paper No. 25, Arctic Institute of North America.

Johnson, A. H., B. R. Bouldin, E. A. Goyette & A. M. Hedges, 1976. Phosphorus loss by stream transport from a rural watershed: Quantities, processes and sources. J. Envir. Qual. 5: 148–157.

Kalff, J., 1968. Some physical and chemical characteristics of arctic fresh waters in Alaska and Northwestern Canada. J. Res. Bd Can. 25: 2545–2587.

Kriet, K., B. J. Peterson & T. L. Corliss, 1992. Water and sediment export of the upper Kuparuk River drainage of the North Slope of Alaska. Hydrobiologia 240: 71–81.

Menzel, D. W. & N. Corwin, 1965. The measurement of total phosphorus in seawater based on the liberation of organically bound fractions by persulfate oxidation. Limnol. Oceanogr. 10: 280–282.

Peterson, B. J., J. E. Hobbie, T. L. Corliss & K. Kriet, 1983. A continuous-flow periphyton bioassay: Tests of nutrient limitation in a tundra stream. Limnol. Oceanogr. 28: 583–591.

Prentki, R. T., M. C. Miller, R. J. Barsdate, V. Alexander, J. Kelly & P. Coyne, 1980. Chemistry, pp. 76–178. In J. E. Hobbie (ed.), Limnology of Tundra Ponds, Barrow, Alaska. Dowden, Hutchinson and Ross, Inc., Stroudsburg, PA. 514 pp.

Rigler, F. H., 1979. The export of phosphorus from Dartmoor catchments: A model to explain variation in phosphorus concentrations in streamwater. J. Mar. Biol. Ass. U.K. 59: 659–687.

Shaver, G. R. & F. S. Chapin III, 1980. Response to fertilization by various plant growth forms in an Alaskan tundra: Nutrient accumulation and growth. Ecology 61: 662–675.

Stainton, M. P., M. J. Capel & F. A. J. Armstrong, 1974. The chemical analysis of fresh water. Research and Development Directorate, Freshwater Institute, Winnipeg, Manitoba, Misc. Spec. Pub. No. 25. 119 pp.

Strickland, J. D. H., 1960. Measuring the production of marine phytoplankton. Bull. Fish. Res. Bd Can. 122: 172 pp.

Technicon Industrial Systems, 1973. Orthophosphate in water and seawater. Industrial Methods No. 155-71. West Tarrytown, NY.

Technicon Industrial Systems, 1977. Nitrate and nitrite in water and seawater. Industrial Methods No. 158-71 W/A. West Tarrytown, NY.

Technicon Industrial Systems, 1978. Ammonia in water and seawater. Industrial Methods No. 154-71 W/B. West Tarrytown, NY.

Ulrich, A. & P. L. Gersper, 1978. Plant nutrient limitation of tundra plant growth. In L. L. Tieszen (ed.) Vegetation and Production Ecology of an Alaskan Arctic Tundra. Springer-Verlag, NY: 457–480.

United States Geological Survey, 1979. Water resources data for Alaska. Report AK-79-1. NTIS. Springfield, VA. 365 pp.

United States Geological Survey, 1980. Water resources data for Alaska. Report AK-80-1. NTIS. Springfield, VA. 373 pp.

Hydrobiologia **240**: 71–81, 1992.
W.J. O'Brien (ed.), Toolik Lake – Ecology of an Aquatic Ecosystem in Arctic Alaska.

Water and sediment export of the upper Kuparuk River drainage of the North Slope of Alaska

Keith Kriet,[1] Bruce J. Peterson[2] & Teresa L. Corliss[3]
[1] *Yale University, School of Forestry, New Haven, CT 06511, USA; present address: 552 Savannah Highway No. 5, Charleston, SC 29407, USA;* [2] *The Ecosystems Center, Marine Biological Laboratory, Woods Hole, MA 02543, USA;* [3] *Present address: 10 Park Place, Durham, NC 27712, USA*

Abstract

Stream discharge and fine suspended sediment load were determined for the upper Kuparuk River, a clear-water tundra stream and tributary of the main Kuparuk River of the North Slope of Alaska. From 75 observations over 3 years we found a range of flows of 0.3 to 28.3 m^3 sec^{-1} and a range of sediment loads of 0.4 to 35 mg $liter^{-1}$. Specific water yields of 15.7, 29.7 and 33.2 cm and summer specific sediment yields of 0.5, 1.1 and 3.5 metric tons km^{-2} were estimated for the period 20 May through freeze-up in 1978, 1979 and 1980, respectively. The fine suspended sediment concentrations and yields for the upper Kuparuk River were less than those in many temperate streams but similar to those reported for small rivers draining the taiga of the Mackenzie Valley lowlands in subarctic Canada.

Introduction

Rivers of the North Slope of Alaska provide important and unique opportunities for hydrological and biogeochemical studies of watersheds. Many of these watersheds are still undisturbed by man except for small anthropogenic inputs of certain elements from the atmosphere. Because the streams are frozen for 7 months of the year and because the permafrost prevents deep seepage flow, fluvial outputs of water and dissolved constituents from many drainage basins can be assessed by sampling the stream water during the 5 month open water season. The striking differences in climate, vegetation and aquatic biota between arctic and temperate rivers provide important contrasts with streams in other regions.

The objectives of this paper are to describe the water discharge and fine suspended sediment concentrations during the summer in an arctic stream. Data are presented for the period May to September 1978–1980 on water flow and fine suspended sediment concentration for the upper Kuparuk River at the Dalton Highway crossing.

This study is based on data collected as part of a baseline study of the general ecology and nutrient cycles of the Kuparuk River (Peterson *et al.*, 1985). The data were not taken for the specific purpose of developing water and sediment budgets and the investigation did not employ standard USGS techniques. However, efforts were made to assess the reliability of the methods through comparisons with standard USGS current profiling and sediment collection methods. With these cautions in mind, we publish this information in order to provide as complete a record as possible for the Kuparuk watershed which has since been designated a Long Term Ecological Research site (LTER) by the U.S. National Science Foundation.

Methods

Stream flow measurements were made with two methods: dye-dilution and Gurley current meter. Measurements during high flows were made with a Rhodamine WT dye-dilution technique. Rhodamine dye was added to the stream from a carboy by means of a battery-operated pump which delivered a constant flow of 25 ml of concentrated dye per minute. Preliminary Rhodamine mixing surveys demonstrated that at least several hundred meters of rivers including several turbulent riffles were required to assure uniform mixing across the entire width of the river.

Downstream samples were collected at three times to ensure that the dye concentration had reached an equilibrium level, and at each time 5 samples were taken from points across the entire channel to ensure that the stream water was completely labelled. When either the time series or the river width series indicated that equilibrium concentration of dye was not obtained, the data were discarded as unusable.

Gurley current meter flow measurements were taken during periods of moderate to low flows when the river could be waded. A modification of two USGS methods (the two point and six-tenths-depth methods) for measuring current velocity was used (Buchanan & Somers, 1980). Readings were taken at 0.5 m intervals on a transect ranging from 8 to 15 m in width depending on discharge. No one reading was ever more than 10% of the total discharge. At depths less than 40 cm only one reading was taken at each interval. At greater depths, readings at 10 cm below the surface and 10 cm above the bottom were averaged to estimate flow at that station. Depth was never greater than 87 cm at any station. Comparisons of this procedure with the USGS technique of measurement at 0.2 and 0.8 of total depth demonstrated that this procedure gave the same result at this site for total depths of 0.4 to 1.0 meters. Flow was computed by multiplying current velocity times cross section area for each station and summing from bank to bank. Five simultaneous comparisons of the dye-dilution and current meter techniques gave discharge estimates based on current meter readings that averaged 87% of the values obtained with the dye-dilution technique. No adjustment for difference in the two techniques was made since no cause of this difference could be identified.

Stream stage was read from a staff marked in centimeters and anchored to the culverts at the Dalton Highway crossing. From 1978, 1979 and 1980 data a non-linear regression ($r^2 = 0.98$, $n = 36$) of discharge versus stream height allowed estimation of additional flows based on stream height measurements. The least squares fit for these 3 years gave the following non-linear power function:

$$\text{Discharge (m}^3\text{/sec)} = 0.240712 + 0.023787 \times [\text{stream stage (cm)}]^{1.44}. \qquad \text{(Eq. 1)}$$

Site description and climate

The Kuparuk River rises in the foothills of the Brooks Range and flows north-northeast draining an area of 8107 km^2 (Fig. 1). This study focuses on the upper reaches of the river from its headwaters near the Atigun gorge to its intersection with the Dalton Highway. The center of this headwater basin is approximately 68° 35′ N, 149° 20′ W. This headwater basin upstream of the Dalton Highway has a main channel length of 25 km, drains an area of 143 km^2, and has an average channel slope of 0.0313 m m^{-1}. Since there are no glaciers in the watershed and little deep springwater input, the Kuparuk is a clear water tundra river following the classification of Craig & McCart (1975). The other types of rivers on the North Slope are mountain rivers and spring-fed rivers. The mountain rivers drain snowfields and glaciers in the Brooks Range and are consequently turbid with glacial rock-flour and have low productivity. Spring-fed streams have relatively constant discharge year round and high productivity. Tundra streams drain the foothills and coastal plains, are not fed by glaciers or

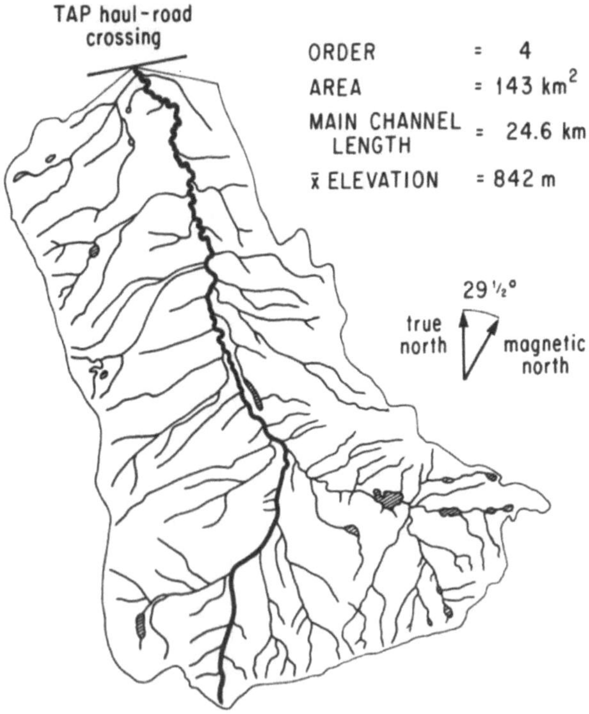

Fig. 1. The upper Kuparuk River watershed on the North Slope of Alaska.

springs, and have intermediate levels of productivity (Craig & McCart, 1975).

The upper reaches of the Kuparuk River have been influenced by glacial advance and retreat, at about 10 000 and 100 000 years ago. Soils of the basin are alluvial and glacial deposits consisting of coarse sand and gravel covered by 0.3 to 0.7 m of organic rich peat (Selkregg, 1977). Permafrost is present throughout the area and the mean thawed depth during August is about 0.4 m (K. R. Everett, personal communication).

The vegetation of the upper Kuparuk drainage consists of alpine communities at the higher elevations and moist tundra communities, predominantly *Eriophorum vaginatum*, at lower elevations closer to the river. Along the stream banks there are patches of dwarf willows and birches, some reaching 1 m in height.

At its intersection with the Dalton highway, the Kuparuk River is a 4th-stream according to the classification of Dunne & Leopold (1978). This designation is based on the extension of stream channels up all gullies and swales noticeable on a USGS topographic map of scale 1:63 630 (Fig. 1).

The drainage basin shows a palmate, dendritic channel pattern characteristic of rivers draining a uniform substrate. There is a hint of a pinnate drainage pattern in the lower portion of the basin where the Kuparuk enters a 1 km wide valley-flat of cohesive alluvial till. The length-weighted average elevation of the Kuparuk River in this basin is 842 m while its tributaries drain the foothills at elevations between 1070 m and 1370 m.

The longitudinal profile of the Kuparuk River from its source to its intersection with the Dalton highway shows the sinuosity of the main channel as it passes through the wide valley-flat described previously (Fig. 2). Sinuosity, or the ratio of channel length to down-valley distance, is a rough index of channel pattern (Leopold *et al.*, 1964). This ratio can vary from unity to four or more. Rivers of sinuosity of 1.5 or greater are said to be meandering, while rivers with sinuosity of less than 1.5 are either straight or sinuous.

Figure 3 shows the relationship between basin area and elevation. It can be seen that 50% of the drainage basin lies above 975 m. The Dalton highway crossing is at an elevation of 745 m.

The North Slope has low annual precipitation. In the foothills of the Brooks Range the average annual rainfall is around 18 cm (Selkregg, 1977). Approximately half of the precipitation falls as

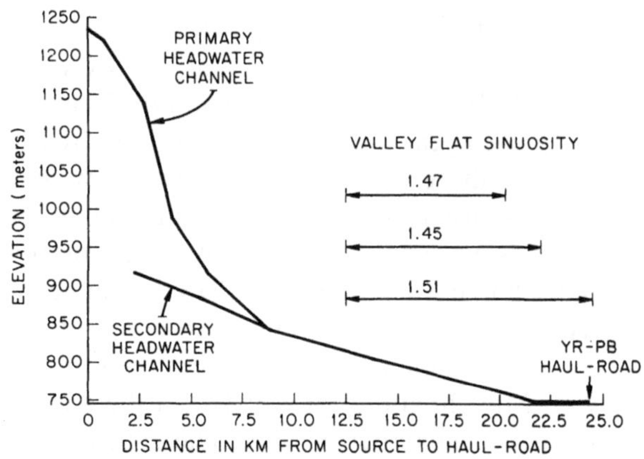

Fig. 2. A diagram of the cumulative area of the upper Kuparuk drainage basin above a given elevation.

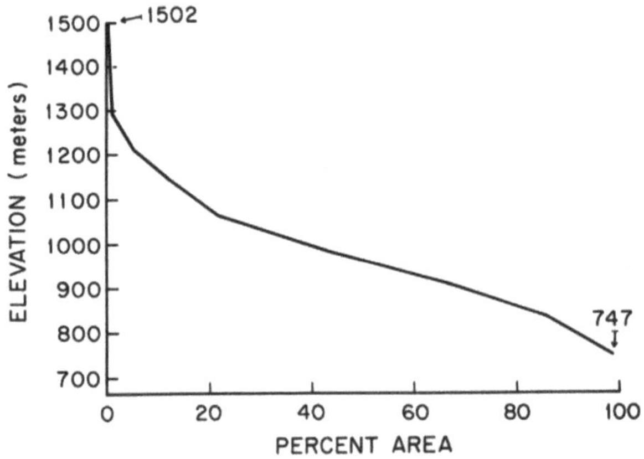

Fig. 3. The distance-elevation diagram of the primary and secondary headwater channels of the upper Kuparuk River.

snow on the upper Kuparuk River basin from September through May and the melting of this snow causes the spring flooding (Scott, 1978). The upper Kuparuk River is frozen solid with no flow from later September or October until the spring break-up floods in May.

Summer temperatures are commonly between 10 and 18 °C; winter temperatures are commonly around −30 to −40 °C (Selkregg, 1977; Scott, 1978).

Results and discussion

Water flow

The spring snowmelt usually begins in May and continues into June in the upper Kuparuk drainage (Fig. 4). Our earliest observations were on May 20, 1978, on May 20, 1979 and on May 22, 1980. In each year there was some water already flowing on these dates but flows were low and from observations and data from gauges on other watersheds we can estimate how much of the total discharge was missed. In 1978 the gauge on the Atigun tributary near Pump Station 4 of the Trans Alaska pipeline indicated first measurable flow on May 20 and a total of 1% of the annual flow for the remainder of May. The gauging station at the mouth of the Kuparuk which reflects major hydrologic events in the foothills with a lag

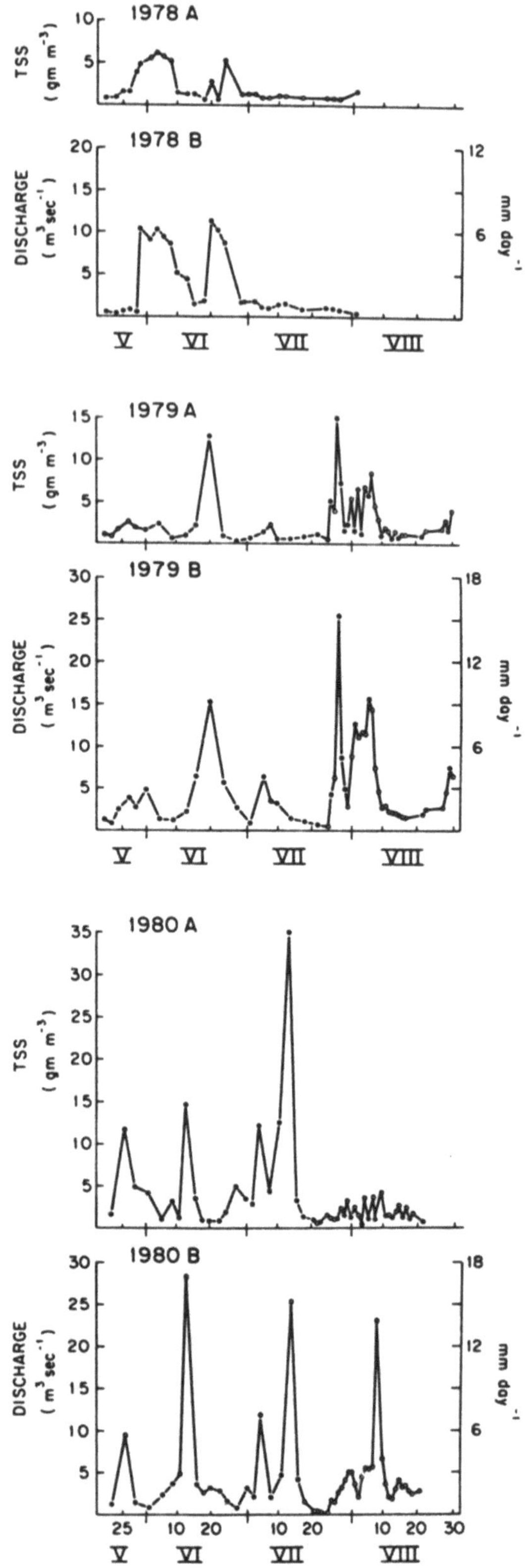

Fig. 4. Total suspended sediment concentration during the 1978, 1979 and 1980 water seasons (A panels). Open water season discharge for the upper Kuparuk River for 1978, 1979 and 1980 (B panels). (●) measured, (○) estimated.

of about 3–7 days did not indicate any major discharge until the first week in June (Fig. 6). In 1979 there is evidence to show that initial snow melting occurred well before our first sample on May 20. For example, the Atigun tributary gauge recorded low flows throughout the month of May, but the May discharge still accounted for only 1% of the annual discharge and there were no high flow events until June 7. Likewise, the gauge at the mouth of the Kuparuk recorded flow as early as May 20 but May discharge was 1.2% of the annual flow. Observations of snowbanks and ice in the Kuparuk at the Dalton Highway indicated that flow prior to 24 May had probably not exceeded 2.5 $m^3 sec^{-1}$. Discharge at the gauge at the mouth of the river peaked on 1 June. In 1980, on May 22, one of the culverts at the Dalton Highway was still blocked with ice and flow in the other was about 1 $m^3 sec^{-1}$. First discharge at the Atigun tributary was recorded on May 20 and at the Kuparuk mouth on 23 May. At both sites May discharge was 1% or less of the annual total. From this evidence we conclude that it is unlikely that more than a few percent of the upper Kuparuk's annual discharge had occurred prior to our mid-May sampling in any of these years and that our late May and early June high flow observations are consistent with the timing of the major spring runoff recorded at both the Atigun tributary and lower Kuparuk USGS gauging stations.

During 1978, spring discharge was followed by just one subsequent high discharge event during July (Fig. 4). In contrast, during 1979 and 1980, the spring discharge period was followed by several high flows associated with summer storms. North Slope rivers have been described as having nival runoff regimes where the spring break-up flood is the dominant hydrologic event (Scott, 1978). However, smaller streams of the foothills and mountains such as the upper Kuparuk do not always demonstrate this pattern of flow. The larger rivers draining to the Beaufort Sea have the drainage area to buffer local summer storms occurring in the headwater reaches and this is the case with the entire Kuparuk River with 8107 km^2 of basin area. In smaller drainages such as the upper Kuparuk River (143 km^2), local storms can result in peak discharge that may equal or exceed the snowmelt floods but are usually of short duration.

Total suspended sediments

The measured concentrations of total suspended sediment (TSS) ranged from 0.4 to 35 mg $liter^{-1}$ during the 3 years. With one exception, increased discharge resulted in increased suspended sediment load (Fig. 4). Regressions between instantaneous values of sediment concentration and flow were run for each year and for the three years combined using both the raw untransformed data and log transformation. Linear untransformed data gave the highest correlation coefficients in each case and the least squares fit for 75 observations combined over 3 years gave the following equation:

$$\text{TSS (mg liter}^{-1}) = 0.604 \text{ flow } (\text{m}^3 \text{ sec}^{-1}) + 0.30, \quad \text{with } r = 0.70. \qquad \text{(Eq. 2)}$$

The equation relating $\log_{10}$ TSS to $\log_{10}$ flow was:

$$\text{Log TSS (mg liter}^{-1}) = 0.412 \log \text{flow} + 0.839, \quad \text{with } r = 0.48. \qquad \text{(Eq. 3)}$$

The best fit power form of this relationship was $TSS = 1.213\ flow^{0.412}$. All three forms of the relationship are given because literature information appears in different forms due to both differences in the relationships in different drainages and due to statistical consideration of error terms (Finney, 1941; Miller, 1984). We chose to use the power function between flow and TSS concentration to estimate sediment concentration for days when only flow was measured.

Figure 4 shows the pattern of sediment concentration during the summer seasons of 1978, 1979 and 1980 using both measured and regression estimated points. High discharge of water was normally associated with high concentrations of total suspended sediments. However, on 8 August 1980, a high discharge of water with a low concentration of sediment was observed. This may be a result of when the sediment sample was taken, an effect called hysteresis (Gregory & Walling, 1973) or sampling error.

We have examined the sequential relationship between sediment load and discharge (the hysteresis effect) during two high flow events in 1979. During the June period of high runoff, there was a slight tendency for suspended sediment concentration during increasing discharge to exceed that during decreasing discharge. The other sequence, from a major summer storm in late July of 1979, again showed only a slight tendency for higher concentrations with rising flow. We conclude that while hysteresis may be present in the sediment versus discharge relationships during storms, it is probably not large.

The pattern of measured (•) and estimated (○) suspended sediment concentrations in late summer of 1979 and 1980 shows an oscillation where the regression gave higher values than direct observations. This suggests that there may well be a seasonal difference in sediment concentration at a given flow.

Water export

The water yield of a drainage basin refers to the amount of runoff in streams plus the amount of groundwater outflow over a period of time. However, the contributions of deep groundwater from water sources outside the drainage other than spring-fed streams (of which there are some within the Brooks Range) to water export in arctic regions are often restricted by permafrost (Hopkins *et al.*, 1955). Recent observations (Peterson, unpublished) have shown that a spring with summer discharge of 0.2 m^3 sec^{-1} flows year round. Evidence from lake level changes over winter in the upper basin and from water chemistry suggest that this spring water originates from precipitation within the basin that percolates through coarse glacial sediment and emerges about 15 km upstream of the Dalton Highway crossing. With the exception of this spring, which we believe discharges water originating from precipitation within the basin, the effects of permafrost are analogous to the effects of an impermeable soil layer at about 0.4 m. Because of this layer, runoff enters the stream channel by primarily two means. The first, overland flow, occurs when water emerges from the soil and flows to stream channels over land surfaces, e.g., as the flow of snow meltwater. The main path for runoff later in the season is seepage within the active layer of thawed ground above the permafrost table. This is a result of the high permeability of the continuous groundcover of moss, lichen and tussock-forming grasses (Zalesskiy, 1976). These water flow processes affect the response of the Kuparuk hydrograph to rainfall. If rainfall input is moderate, the response may vary depending on the antecedent moisture conditions of the tundra mat. A dry mat acts like a sponge, absorbing and holding the precipitation that falls on it. If the mat is saturated, or the amount of rainfall is large, there may be an increase in streamflow that is directly proportional to the rainfall quantity (Dingman, 1966).

A comparison of the hydrographs of the upper Kuparuk River and the entire Kuparuk River (USGS, 1978, 1979, 1980) for the years 1978, 1979 and 1980 (Fig. 5) illustrates the differences in the flow regimes between the two different size basins, and the importance of storm events in the upper basin to total discharge at the river's mouth. The runoff from summer storms in the upper Kuparuk basin was usually noticeable as only a small increase in discharge at the mouth. The time between increased discharge in the upper basin and the rise in discharge at the river's mouth varied from a low of 3 days, during the storm of 28 July 1979, to as much as 8 days during low flow periods (Fig. 5). The average discharge of 5.72 m^3 sec^{-1} in the upper Kuparuk was used to calculate an average transit time of 4 days (270 km at 0.7 m sec^{-1}). Using this transit time it was possible to estimate the percent contribution of discharge in the upper basin to discharge four days later at the river's mouth.

The above observations allowed estimation of

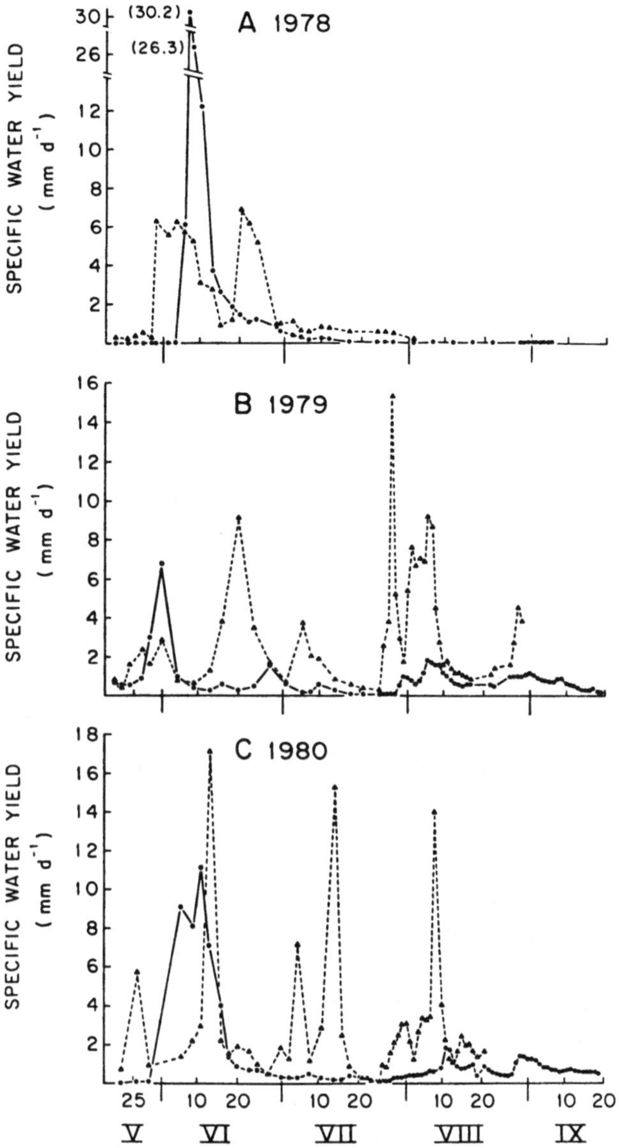

Fig. 5. A comparison of the 1979 hydrographs for the upper Kuparuk (dotted line) with the USGS record for the mouth of the Kuparuk (solid line).

the amount of flow not measured in the upper Kuparuk due to ending our measurements prior to fall freeze-up. By offsetting the flow recorded at the mouth by four days, and by assuming that the upper Kuparuk flow was 4.5% of the flow at the mouth, a rough estimation was made of daily flows during late August and September for the upper Kuparuk. Based on this calculation our sampling missed approximately 2%, 13% and 24% of the total annual discharge in 1978, 1979 and 1980, respectively. Annual water yield was calculated by determining the area under the stream hydrographs in each year and adding the estimated amounts missed (Table 1).

Our estimate of water export for the upper Kuparuk is similar to the USGS measurements for the Atigun River, a nearby mountain stream (Table 2). Both the mountain and foothill watershed have substantially greater water yields per unit area than the greater Kuparuk which drains mostly North Slope tundra, where precipitation is lower than in the foothills.

Total suspended sediment export

On days when total suspended sediment concentrations in the upper Kuparuk River were not measured, the regression $TSS = 1.213 \text{ flow}^{0.412}$ was used to calculate the predicted concentration of TSS on that day. Since high flows were normally associated with elevated sediment loads and vice versa, the range in sediment export per day was very large (Fig. 6). The extremes that we have actually measured are a low of 0.15 kg TSS km^{-2} day^{-1} on 23 July 1980 and a high of 537 kg TSS km^{-2} day^{-1} on 14 July 1980.

The annual TSS export was estimated by determining the area under the sediment yield graphs (Fig. 6) and adding an estimated amount of sediment export not measured because we terminated our sampling before freeze up. This TSS not measured at the end of each year was calculated based on the fraction of water output missed according to the gauge at the Kuparuk mouth; 2, 13 and 24% for 1978, 1979 and 1980, respectively.

We have been unable to find a detailed record of sediment output for a watershed of comparable characteristics for comparison with our estimates of suspended sediment export for the upper Kuparuk. Craig & McCart (1975) demonstrated the important differences between mountain streams, tundra streams (such as the upper Kuparuk) and spring streams on the North Slope. In addition, watershed area has a tremendous impact on sediment transport in rivers (Table 3).

Table 1. Upper Kuparuk River watershed flow and sediment load summary.

Parameter	Year 1978	1979	1980	3-year mean
Range of measured flows ($m^3\ sec^{-1}$)	0.4–11.3 (29)*	0.5–25.4 (50)	0.3–28.3 (49)	
Water yield ($m^3\ yr^{-1} \times 10^6$)				
Measured	22.1	37.6	38.4	
Extrapolated*	22.5	42.5	47.5	37.5
Annual specific water yield ($cm\ m^{-2}\ yr^{-1}$)				
Extrapolated**	15.7	29.7	33.2	26.2
Range of measured sediment ($mg\ l^{-1}$)	0.8–4.8 (13)	0.4^{-1} 5.0 (28)	0.4–35.0 (34)	3.3 (75)
Sediment yield ($mt\ yr^{-1}$)				
Measured	70	144	400	
Extrapolated**	71	163	496	
Annual specific sediment yield ($mt\ km^{-2}\ yr^{-1}$)	0.5	1.1	3.5	1.7

* Numbers in parentheses are the number of observations.

**Extrapolated water yield and extrapolated sediment yield based on fractions of water output missed were computed as described in text.

For example, the much larger Colville River drains mountainous regions and has a very large sediment load. Char Lake streams carry a large sediment load considering their small size but vegetation is very sparse in those watersheds (10% ground cover).

The streams most similar to the upper Kuparuk are the moderate to small rivers draining the taiga of the Mackenzie Valley lowlands (Brunskill *et al.*, 1975). These streams are about the same size as the Kuparuk, and in addition have similar sediment yields despite the obvious differences in vegetation and topographic relief. In fact, even the relationship between suspended sediment concentration and discharge is similar between these streams and the Kuparuk. For the Kuparuk, a linear regression of discharge versus suspended sediment concentration gave the highest correlation with sediment load, but for this comparison we have computed the regression using the logarithm of both flow and sediment concentration (Table 4). Both the slope and the intercept for the Kuparuk are within the range given for the small Mackenzie watersheds. An equation (Brunskill *et al.*, 1975) relating annual discharge (a function mainly of watershed areas) and specific sediment yield in a variety of streams in the Mackenzie and Porcupine drainages also predicts the sediment yield for the upper Kuparuk. The equation is

$$\log \mathrm{TSS}\ (\mathrm{kg\ km^{-2}\ yr^{-1}}) = -1.9259 + 0.6452 \log Q\ (\mathrm{m^3\ yr^{-1}}). \quad \text{(Eq. 4)}$$

Table 2. Comparison of specific water yields of North Slope drainage in the mountains (Atigun), foothills (Upper Kuparuk) and entire North Slope (Kuparuk).

River	Watershed area (km^2)	Specifiec water yield ($cm\ yr^{-1}$) 1978	1979	1980
Atigen[1]	84.4	30.2	32.8	36.0
Upper Kuparuk[2]	143	15.7	29.7	33.2
Kuparuk[3]	8107	14.7	11.3	17.7

[1] USGS gauge at pump station 4 – USGS data reprints.

[2] This study.

[3] USGS gauge at mouth – USGS water resources data for Alaska – 1978–1980.

Table 3. Annual water and sediment yields for arctic, sub-arctic and temperate watersheds of various sites.

Region and river	Basin area km^{-2}	Water discharge $cm\ m^{-2}\ yr^{-1}$	Specific sediment yield $mt\ km^{-2}\ yr^{-1}$	Reference
Arctic				
Kuparuk	143	25.2	1.6	1
Colville	50000	32.0	1.0	2
Char Lake				
Streams	3.5	13.7	5.5	3
Sub-Arctic				
Mackenzie Valley Small Rivers				4
Willow Lake	21600		3.7	
Martin	1980		1.6	
Harris	741		0.18	
Caribou Bar	381		11.5	
Chena, AK	5100	33.3	17.0	5
Temperate				
Hubbard Brook	0.13	76.9	2.5	6
Lake Erir	200		150.0	7
Tributaries	200		124.0	

References

1. This study
2. Arnborg *et al.* 1967
3. de March 1975 cited in 4
4. Brunskill *et al.* 1975
5. Frey *et al.* 1970 cited in 4
6. Bormann *et al.* 1969
7. Ongley 1976

Because Q (annual water discharge) for the upper Kuparuk averages $37 \times 10^6\ m^3\ yr^{-1}$, the predicted yield is 0.9 metric tons (mt) $km^{-2}\ yr^{-1}$ for the upper Kuparuk while we found 0.5, 1.1 and 3.5 mt in the 3 years of our study. This is good agreement and perhaps there is reason to think that sediment yields may be roughly predictable from discharge for certain classes of watersheds. Judging from the work of Craig & McCart (1975) this equation would probably not apply to either the mountain or spring-fed drainages of the North Slope.

Table 4. Slopes (b), interceps (a) and correlation coefficients (r) for regression of the form $SS = a\ Q^b$, relating suspended sediment concentration (SS) to discharge (Q).

River	*n*	*b*	*a*	*r*
Willow Lake[1]	11	0.60	0.52	0.64
Martins[1]	23	0.44	4.54	0.62
Harris[1]	19	0.30	1.86	0.54
Caribou Bar[1]	18	0.94	1.45	0.89
Kuparuk (this study)	75	0.52	1.21	0.48

[1] Data from Brunskill *et al.* (1975) for boreal forest watersheds on permafrost in the Mackenzie drainage.

Summary

The specific water yield for this 4th-order river draining the foothills of the Brooks Range is relatively low because of the low annual precipitation of this region. The seasonal pattern of discharge is characterized by high spring snowmelt flows in May and June followed by low flow during July, August and September. The summer

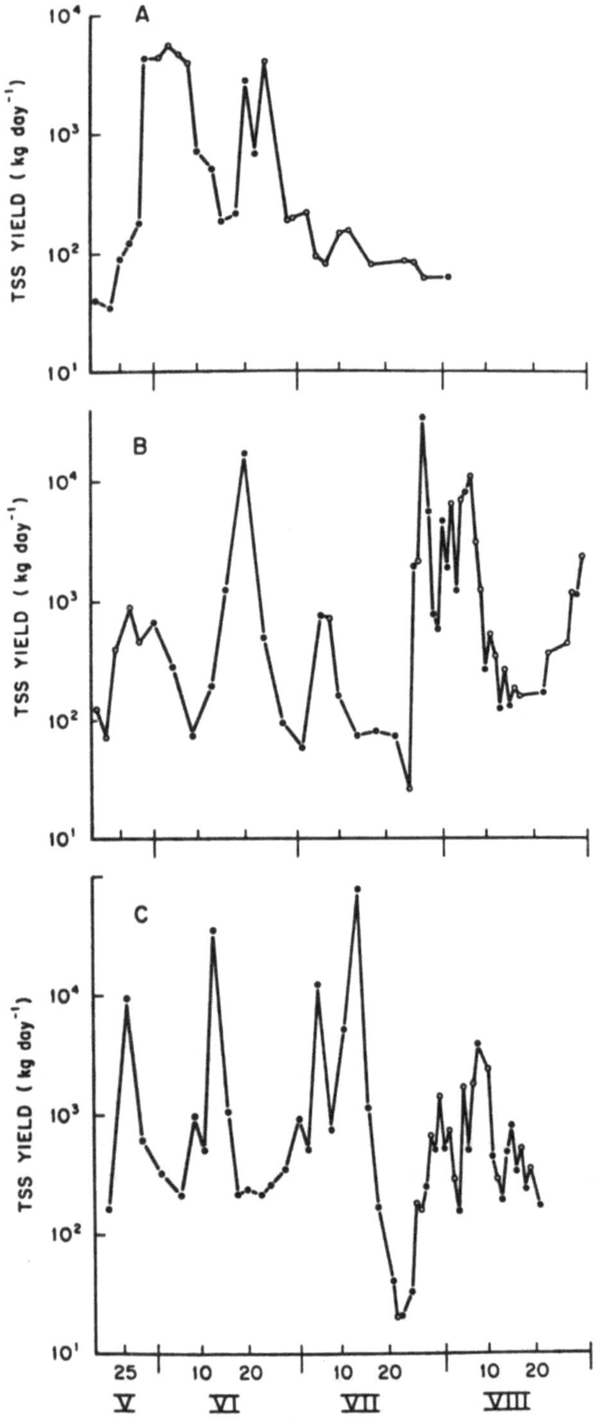

Fig. 6. Total suspended sediment yield from the upper Kuparuk River drainage basin in 1978 (A), 1979 (B) and 1980 (C).

low-flow period is often interrupted by high discharge events of short duration due to summer rainfall. The total suspended sediment concentration increases greatly when discharge increases and the annual TSS export is similar to that found in taiga watersheds of similar size in the Mackenzie drainage in Canada.

References

Arnborg, L., H. J. Walker & J. Peippo, 1967. Suspended load in the Colville River, Alaska, 1962. Geogr. Ann. 49A: 131–144.

Bormann, F. H., G. E. Likens & J. S. Eaton, 1969. Biotic regulation of particulate and solution losses from a forest ecosystem. BioScience 19: 600–610.

Brunskill, G. J., P. Campbell, S. E. M. Elliott, B. W. Graham, W. J. Dentry & R. Wagemann, 1975. The chemistry, mineralogy and rates of transport of sediments in the Mackenzie and Porcupine River watersheds, N.W.T. and Yukon, 1971–1973. Technical Report No. 546. Canada Department of the Environment, Fisheries and Marine Service, Freshwater Institute, Winnipeg, Manitoba.

Craig, P. C. & P. J. McCart, 1975. Classification of stream types in Beaufort Sea drainages between Prudhoe Bay, Alaska and the Mackenzie delta, N.W.T., Canada. Arctic and Alpine Research 7: 183–198.

de March, L., 1975. Nutrient budgets for a high arctic lake (Char Lake, N.W.T.). Mitt. Int. Ver. Theor. Angew. Limnol. 19: 496–503.

Dingman, S. L., 1966. Characteristics of summer runoff from a small watershed in central Alaska. Wat. Resour. Res. 2: 751–754.

Dunne, T. & L. B. Leopold, 1978. Water in Environmental Planning. W.H. Freeman and Co., San Francisco.

Finney, D. J., 1941. On the distribution of a variable whose logarithm is normally distributed. J. r. Statist. Soc. B7: 155–161.

Frey, P. J., E. W. Mueller & E. C. Berry, 1970. The Cheno River. The study of a subarctic stream. Federal Water Quality Administration Project No. 1810-10/70.

Gregory, K. J. & D. E. Walling, 1973. Drainage basin form and process. London, Arnold. 456 pp.

Hopkins, D. M. & T. N. V. Karstrom, 1955. Permafrost and groundwater in Alaska. United States Geological Survey, Professional Paper No. 254-F.

Leopold, L. B., M. G. Wolman & J. P. Miller, 1964. Fluvial Processes in Geomorphology. W.H. Freeman and Co., San Francisco.

Miller, D. M., 1984. Reducing transformation bias in curve fitting. Am. Statist. 38: 124–1265.

Ongley, E. D., 1976. Sediment yields and nutrient loadings from Canadian watersheds tributary to Lake Erie: An overview. J. Fish. Res. Bd Can. 33: 471–484.

Peterson, B. J., J. E. Hobbie, T. L. Corliss & K. Kriet, 1983. A continuous-flow periphyton bioassay: Tests of nutrient limitation in a tundra stream. Limnol. Oceanogr. 28: 583–591.

Peterson, B. J., J. E. Hobbie, A. Hershey, M. Lock, T. Ford, R. Vestal, M. Hullar, R. Ventullo & G. Volk, 1985. Transformation of a tundra river from heterotrophy to autotrophy by addition of phosphorus. Science 229: 1383–1386.

Scott, K. M., 1978. Effects of permafrost on stream channel behavior in arctic Alaska. United States Geological Survey, Professional Paper No. 1068. 19 pp.

Selkregg, L. L. (ed.), 1977. Alaskan Regional Profiles: Arctic Region. Arctic Environmental Information and Data Center, Anchorage, Alaska.

United States Geological Survey, 1978. Water resources data for Alaska. Report AK-78-1. NTIS. Springfield, VA. 425 pp.

United States Geological Survey, 1979. Water resources data for Alaska. Report AK-79-1. NTIS. Springfield, VA. 365 pp.

United States Geological Survey, 1980. Water resources data for Alaska. Report AK-80-1. NTIS. Springfield, VA. 373 pp.

Zalesskiy, F. V., 1976. Flash flood formation in permafrost regions. Soviet Hydrology (selected papers) 15: 95–97. Am. Geophys. Union.

Hydrobiologia **240**: 83–90, 1992.
W.J. O'Brien (ed.), Toolik Lake – Ecology of an Aquatic Ecosystem in Arctic Alaska.

Influences of slimy sculpin (*Cottus cognatus*) predation on the rocky littoral invertebrate community in an arctic lake

Benjamin E. Cuker, Michael E. McDonald & Samuel C. Mozley
Center for Marine and Environmental Studies, Hampton University, Hampton, VA 23668, USA; Department of Chemical Engineering University of Minnesota Duluth, Duluth, MN 55821, USA; Zoology Department, North Carolina State University, Raleigh, NC 27695-7617, USA

Key words: slimy sculpin, Toolik Lake, predation, community structure, benthos

Abstract

We tested the role of the slimy sculpin (*Cottus cognatus*), a benthic fish, in structuring the rocky littoral invertebrate community in Toolik Lake, Alaska. Comparisons of sculpin gut contents and prey community structure indicated that these fish forage selectively, eating proportionally more large and motile prey, and proportionally fewer small and sessile forms. Field experiments compared the effects of natural, reduced and elevated sculpin densities on benthic community structure. At natural levels of sculpin density, biomass of trichopteran larvae were reduced by more than 50%, and predatory chironomid larvae by 27%, in comparison to areas where sculpin were excluded. Tube-dwelling and small free living chironomid larvae were unaffected at normal sculpin densities. Under artificially high sculpin densities, there was some reduction of tube-dwelling chironomids, but the small free living ones remained unaffected. There appears to be a threshold length of about 3.5 mm, below which chironomid larvae are free form sculpin predation. Tube-dwelling chironomids may be longer than this threshold, but still avoid predation by having most of their body hidden in their tubes.

Introduction

Fish may structure benthic invertebrate communities by discriminate harvesting of prey. Selective consumption of large and motile prey results in communities dominated by smaller and sessile taxa (Crowder & Cooper, 1982; Gilinsky, 1984). In some structurally complex systems, fish may have little effect on the benthic community (Thorp & Bergey, 1981; Allan, 1982). However, fish predation may be more effective in regulating benthic communities in systems where there is little physical structure for prey refugia (Crowder & Cooper, 1982; Gilinsky, 1984; Anderson, 1984; Hershey, 1985).

The slimy sculpin (*Cottus cognatus*) was suggested to be an important predator of benthic invertebrates in the littoral zone of Toolik Lake (McDonald *et al.*, 1982). This was later confirmed for the soft sediments by Hershey (1985) who showed that sculpin significantly reduce numbers of small (<7 mm) chironomid larvae in macrophyte free portions of the soft-sediments of Toolik Lake. In our study we examined the influence of sculpin predation on the rocky littoral community of Toolik Lake, adjacent to that investigated by Hershey (1985). We tested the hypothesis that sculpin feed selectively in this area, and used field experiments to determine if selective feeding resulted in altered community composition.

Toolik Lake is an oligotrophic morrainal-kettle basin with an extensive rocky littoral zone (more detailed description in Cuker & Mozley, 1981). Most of the rocks in the littoral are smoothed glacial till, but some are split exposing a rougher surface. The rocks are covered by an algal mat and an associated fauna. The ectoproct *Pectinatella* and *Hydra* form a sparse canopy layer that extends 3–5 cm above the algal mat. Living atop the mat are snails (*Lymnaea*, *Gyraulus*, *Physa*, and *Valvata*), trichopteran larvae (primarily *Grensia*), copepods, cladocerans, ostracods and water mites. Insect larvae from the orders Coleoptera, Ephemeroptera, and Plecoptera are present but rare. Living within or above the mat are numerous chironomid larvae. Sculpin and to a lesser extent burbot (*Lota lota*) are found among the rocks. The rocky littoral is bounded at 3 m depth by muddy sediments which supports additional, infaunal chironomid taxa as well as part of the sculpin, gastropod, trichopteran and microcrustacean populations. The rocky littoral is illuminated for most of the period from June–August. Previous studies on the rocky littoral of Toolik lake have investigated the ecology of the *Hydra* population (Cuker & Mozley, 1981) and the influence of the grazing snail, *Lymnaea* on the structure of the epilithic community (Cuker, 1983a, 1983b).

Methods and materials

Sculpin Diet and Community Composition – To determine if sculpin feed selectively, fish diets were compared to samples of the benthic community. Sculpin for diet analysis were captured in unbaited jar traps or lift nets (McDonald, *et al.*, 1982). In the July of 1978, traps were placed systematically at 0.5, 1.5, 3.5 (mud-rock interface), and 7 m, and tended every 2–3 days. During summers of 1979 and 1980 trapping continued irregularly in the 0.5–3 m zone. Fish were preserved with visceral injections of formalin or ethanol. Prey in sculpin stomachs and intestines were identified and enumerated under compound and dissecting microscopes.

The rocky littoral benthic community composition was determined from the ambient sculpin predation treatments for the experiments described below. Since sculpin caught in the rocky littoral consumed many prey from the adjacent soft bottom, comparative information about them was obtained between 1977–1980, using Ekman grab samples sieved through 35 μm mesh nets (Cuker, 1981). These rocky littoral and soft sediment samples were used to estimate proportional composition of different faunal groups within the two communities (Cuker, 1981). Based upon this, prey items found in the sculpin guts were divided into the following groups: lithophilic, pelophilic, and eurytopic. To facilitate comparison of sculpin gut contents with the benthic communities, eurytopic prey were weighted by relative occurrence in either the rocks or the mud communities and then their fractional occurrence was added to each of these community lists.

Sculpin Enclosure and Exclosure Experiment – We tested the effects of sculpin predation on the rocky littoral community using three sculpin densities: no sculpin, ambient sculpin levels, and a 13 × sculpin enhancement. There were four replicates of each treatment. In the exclusion, benthic invertebrate biomasses were obtained from rocks kept in chambers covered with a mesh top. For the ambient treatment, the biomass of benthos on normally exposed rocks collected from near the other treatments was used. For the enhancement, sculpin were enclosed in chambers with rocks. Natural sculpin densities in the rocky littoral of Toolik Lake are about of 0.17 m^{-2} (McDonald *et al.*, 1982).

The experimental chambers were constructed from dark plastic tubs (48 × 35.5 × 17 cm deep). They were submerged at a 2 m deep portion of Lenin Shoals. From 4–8 rocks of similar size (approximately 25 × 15 × 15 cm) and appearance were collected from the site and placed in the chambers under water. All chambers were left open on the bottom for two days to allow colonization by animals to replace those that may have been dislodged during transfer of the rocks. Subsequently on 4 July, 1979, tubs were hauled to the surface for experimental manipulation. One pre-weighed sculpin (0.45–0.92 g wet mass) was added to each of the enclosures. After fitting nylon screen (500 μm mesh) over the tops of the enclosures and exclosures, they were returned to the rocky littoral. The nylon screen tops transmitted 60% of incident light (Cuker, 1983b).

On 8 August, 1979, rocks adjacent to the enclosures and exclosures were collected and placed in tubs to estimate invertebrate biomass at ambient sculpin densities. These and the two other treatments were then removed from the lake. Rocks were scrubbed with a stiff nylon brush, rinsed, and the residue collected on a 70 μm mesh net. Samples were preserved with 10% formalin. Upper rock surface area in each tub was determined by cutting the image of each group of rocks from a photograph and weighing it (Cuker, 1983a).

Processing of the Epilithic Samples – Animals >2 mm were picked from the field experiment residues using a dissecting microscope. Smaller animals were subsampled with a Stempel pipette; collecting 2–4 pooled, 2.5% fractions of the total sample. Chironomids were identified using a compound microscope. Lengths of insect larvae, microcrustaceans, and Acari were measured to the nearest 0.1 mm with an ocular micrometer. Biomass (wet mass) of the animals were computed from length (Hynes & Coleman, 1968; Vijverberg & Frank, 1976).

Chironomid Guilds – For guild assignments, species were grouped on the basis of size, habits, and feeding mode (Cuker, 1983a). Members of the subfamilies Tanypodinae and Diamesinae were classified as predators, although they may occasionally feed on algae (Armitage, 1968; Kajak *et al.*, 1968; Izvekova, 1971; Efford, 1972). These predators have elongated parapods and do not construct fixed-tubes. Most members of the subfamily Orthocladiinae in the littoral zone were grouped as free-living scrapers, feeding mainly on epilithic algae (Davies, 1974). Many orthoclads do construct tubes, but those retreats are generally poorly organized and temporary (Hershey, 1987). We did not find any orthoclads associated with tubes in this study, however this could be an artifact of our sample processing. Although *Cricotopus* nr. *sylvestris* is also an orthoclad, we segregated it into its own guild because it was larger than the other epilithic scrapers. It also seemed to be a facultative predator, as indicated by gut contents (Pankratova, 1971, and this study). Members of the subfamily Tanytarsini, such as *Paratanytarsus* sp. are classified as fixed-tube-dwellers for their habit of building a prostrate retreat on the rocks. In contrast to the orthoclads, these larvae were almost always associated with tubes in our samples. In lakes with low planktonic productivity such as Toolik, these usually filter-feeders rely on harvesting nearby benthic algae (Kajak & Warda, 1968).

Results

Selective Feeding by Sculpin – Of the 103 sculpin collected, 88 had prey within their guts. Of the 1,015 total organisms discovered in the fish guts, Chironomid larvae accounted for 73%, with cladocerans and other microcrustaceans making up another 23%. Less abundant, but important due to their large size, were trichopteran larvae (2.0%), juveniles of the snails *Lynmaea elodes* and *Valvata* sp. (1.7%) and, juvenile lake trout, *Salvelinus namaycush* (0.3%).

Sculpin captured in the shallow rocky littoral had diets of primarily lithophilic prey. Pelophilic prey dominated the diets of sculpin captured in the mud or near the mud-rock interface (Fig. 1).

Comparison of sculpin gut contents with the mud and rock prey assemblages indicated the sculpin fed selectively (Fig. 2). Among lithophilic and eurytopic chironomids, the scraper *Cricotopus* nr. *sylvestris*, predators *Procladius* sp. and *Ablabesmyia* sp. were over-represented in sculpin diets, while the fixed-tube dwelling *Paratanytarsus* sp., the predator *Arctopelopia* sp. and the guild of small scrapers were underrepresented. Similarly, the two larger 'collector-gatherers' *Heterotrissocladius* sp. and *Psectrocladius* sp., and the predators *Procladius* sp., and *Ablabesmyia* sp., and the large larvae of *Protanypus* sp. and *Monodiamesa* were the selected prey in mud habitats. Large, deep-burrowing *Stictochironomus* sp., *Arctopelo-*

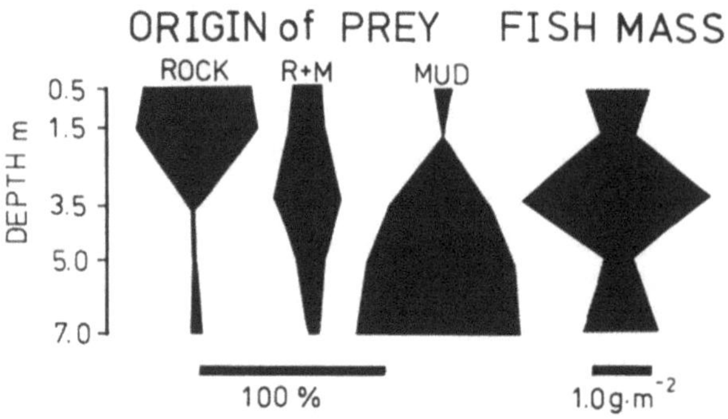

Fig. 1. Distribution of sculpin and percent composition of prey with depth. R + M refers to eurytopic forms, ROCK = lithophilic, MUD = peliophilic. Number of sculpin guts examined at each depth were 0.5 m – 13, 1.5 m – 26, 3.5 m – 44, 5.0 m – 5, 2.0 m – 21.

pia, and the small collector-gatherer *Parakiefferiella* were under-represented in diets of sculpin caught over mud. Trichopteran larvae were over-represented in the diets of mud-dwelling sculpin, but not in the diets of sculpin from the rocks. Mites were not preyed upon.

The Field Experiment – Sculpin predation at both ambient and enhanced levels yielded lower total invertebrate community biomass than the exclusion treatment; 19% less at 1 × and 64% less at 13 × (Table 1). The major taxa responded differently to sculpin predation. Ambient predation reduced trichopterans, and microcrustaceans each by more than 50%. Chironomids were less sensitive at this level, being reduced by only 15%.

Table 1. Summary of wet biomass (mg m^{-2}) for the major benthic invertebrates. Numbers in parentheses are one standerd error of the mean.

Taxon	Sculpin Density		
	0 ×	1 ×	13 ×
Chiromodiae	1883 (257)	1594 (688)	777 (204)
Trichoptera	136 (92)	53 (31)	0 (0)
Microcrustacea	108 (11)	50 (4)	57 (11)
Combined Taxa	2127	1697	834

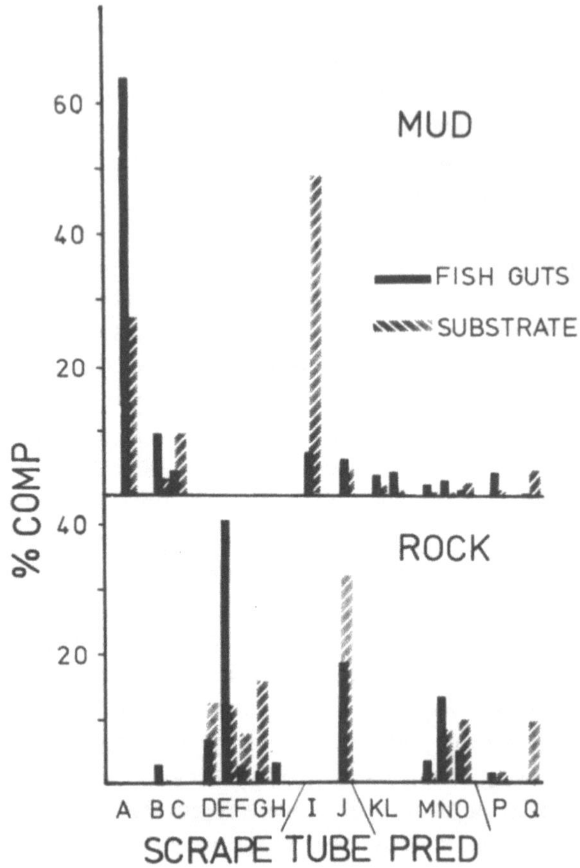

Fig. 2. Composition of major taxa form rock and mud communities versus sculpin diets. Prey are grouped in three general categories. SCRAPE = scrapers, TUBE = fixed-tube-dwelling or burrowing forms, PRED = predators. The subgroups represented are: (A) *Heterotrissocladius*, (B) *Psectrocladius*, (C) *Parakiefferiella*, (D) *Cricotopus* sp., (E) *C.* nr. *sylvestris*, (F) *Corynonerua*, (G) *Zalutschia trigonacies*, (H) *Pseudosmittia*, (I) *Stictochironomus*, (J) *Paratanytarsus* complex, (K) *Mondodiamesa*, (L) *Protanypus*, (M) *Procladius*, (N) *Ablabesmyia*, (O) *Arctopelopia*, (P) Trichoptera, (Q) Mites.

Two-tailed t-tests were significant ($P<0.05$) for comparisons between the ambient and exclusion treatments for trichopterans and microcrustaceans, but not for the chironomids. The 13 × sculpin treatment caused no further reduction in microcrustacean biomass, but did eliminate all trichopterans, and reduced chironomids by 59% relative to the enclosure.

The chironomid guilds displayed different levels of sensitivity to sculpin predation. The predator guild was the only one in which biomass was reduced at both the ambient and 13× sculpin levels. In comparison to the 0×, this guild was reduced by 27% in the 1× and 83% treatments (Fig. 3). The fixed-tube-dweller guild biomass was not effected at ambient sculpin levels, but the 59% reduction of biomass at 13× (compared to 0×) was significant ($P < 0.05$, two-sided t-test). There was an apparent trend of reduced biomass in the *Cricotopus* guild at higher sculpin levels, but little difference between the ambient and exclusion treatments. Scraper guild biomass did not differ by treatment.

Sculpin predation influenced biomass distributions within chironomid size classes. Graphical analysis of the data reveals a disproportionate loss of predator and *Cricotopus* biomass in the larger size classes due to sculpin predation, particularly at elevated sculpin densities (Fig. 4). The response of tube-dwellers was more complex, with a shift in biomass to larger size classes at the 1× and a clear reduction of larger size classes at the 13× sculpin levels.

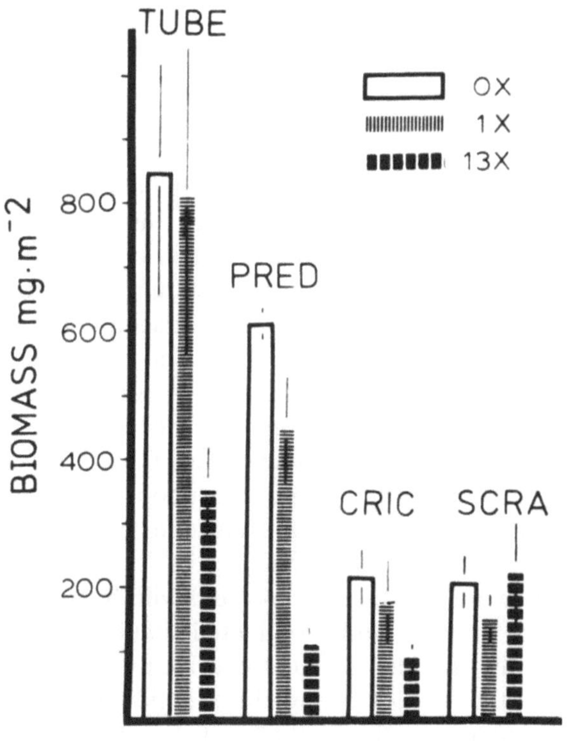

Fig. 3. Distribution of total biomass among the chironomid guilds in the field experiment. Sculpin densities were 0, 1, and 13 times normal. TUBE = fixed-tube-dwelling larvae, PRED = predators, CRIC = *Cricotopus* and SCRA = scrapers other than *Cricotopus*. Error bars are plus or minus 1SE of the mean.

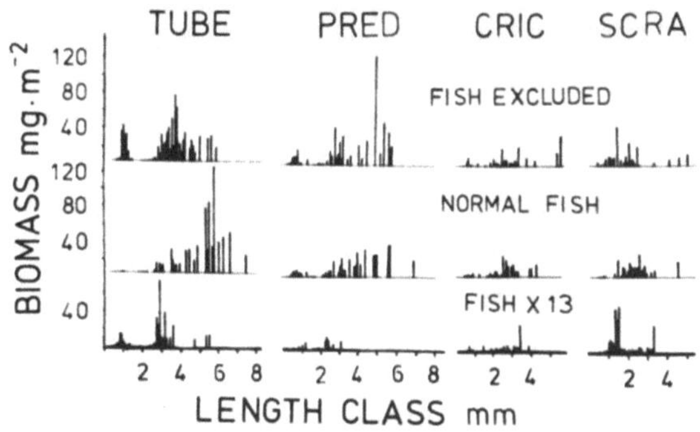

Fig. 4. Distribution of biomass by length class for the chironomid guilds. Abbreviations are as for Fig. 3.

Discussion

Sculpin influence community structure of the rocky littoral in Toolik lake by selectively cropping certain prey taxa; primarily trichopterans, microcrustaceans and the predatory guild of chironomids. This was evident from contrasting sculpin gut contents with the available prey community and, by the effects of experimental exclusion and enhancement of sculpins. In comparison to the rest of the prey community the preferred taxa are more motile, and except for the microcrustaceans are also comparatively large. This is consistent with reports from field and laboratory experiments suggesting that benthic feeding fish select large, exposed, and motile prey (Ivlev, 1961; Ware, 1983; Crowder & Cooper, 1982). That the microcrustaceans were more sensitive to predation than expected considering their small size, may be attributed to their high degree of motility exposing them to greater levels of predation.

Although the enhancement represented conditions in which sculpin densities were unrealistically high, the effects on the prey community are informative, particularly when considering the chironomids. The decline of predatory chironomids in the enhancement was consistent with over representation of this guild in sculpin guts. The predator guild tends to be larger and more active than the other guilds and as such more obvious to the visually feeding sculpin. Gilinsky (1984) also found predatory insects to be particularly sensitive to fish predation. More recently Goyke & Hershey (1992) confirmed this in a comparative study of soft sediment benthos of lakes in the area of Toolik. They found that lakes with sculpin that were uninhibited by piscivorous fish, had proportionally less biomass in the guild of predacious chironomids than was found for lakes without sculpin.

Since biomass of the scraper guild in our study was not diminished at either normal or enhancement levels of sculpin suggests that these larvae are generally safe from fish predation. This would not be expected form the inferred habits of this guild. Although we have no direct observations of behavior, taxa comprising the scraper guild seem to be among the most active of chironomid larvae in the rocky community, with *Corynoneura* sp. and others being the first to colonize cleaned substratum (Cuker, 1981). However, these scrapers are also the smallest of the chironomids. Most are less them 4 mm long when they reach the fourth instar. Their small size apparently protects all but the largest scrapers from sculpin predation. Comparing biomass distributions by length class for each of the experimental treatments does indicate some loss of the longer scraper larvae in the sculpin enhancement (Fig. 4). Others have found small non-predatory insects to increase in the presence of fish predators as an indirect effect of the fish removing predatory invertebrates (Crowder & Cooper, 1982; Gilinsky, 1984). Our study was too brief to examine this type of indirect effect.

Ordinarily the chironomid tube-dwelling guild is underrepresented in sculpin diets and little effected by sculpin predation, yet the decline of this guild in the enhancement indicates that they are potentially susceptible to sculpin predation. If prey length was the only factor in sculpin prey selection, fixed-tube-dwellers would be expected to be consumed at the same rate as the similarly proportioned larvae of the predator guild. The under-utilization of these larvae is probably due to the protection afforded by their tubes (Hershy, 1987; Hershey & Dodson, 1985). The tubes generally obscure all but the head and thorax of these larvae, making them appear shorter than their actual length of up to 7.5 mm. The resulting apparent length would be <4.0 mm, or similar to that of most larvae in the free living scraper guild, and seems to be below the normal threshold for sculpin to attack. That the protruding head and thorax are somewhat wider than free living scrapers of comparable length does not appear to render fixed-tube-dwellers any more vulnerable. Also, these larvae may retreat completely back in to tubes between foraging bouts (Hershey, 1987). As evident from biomass distribution by size-class (Fig. 4), hungry sculpin of the enrichments may lower their threshold size for attack to include fixed-tube-dwellers as short a 4–5 mm in actual length.

The effects of sculpin enhancement on length distribution of chironomid biomass also reveals the size selective nature of the predation and the role of tubes as an anti-predator device (Fig. 4). Sculpin predation was heaviest on larvae longer than 3.5 mm for all of the free-living chironomid guilds, while all larvae smaller than this are nearly immune. It also seems that living in a fixed-tube is an effective way to avoid sculpin predation under normal circumstances. Since sculpin in the enhancement apparently did feed on fixed-tube-dwellers, it is clear that the tube is not an effective suit of armor. However, that the later instars of the fixed-tube-dwellers are longer than those of the predatory guild, yet not fed upon as heavily suggests that living in a fixed-tube lessens the threat of sculpin predation.

There are also costs associated with living in a tube. In another study conducted simultaneously with this one and in the same location, it was shown that tube-dwelling chironomids are more susceptible than free-living forms to competition from the grazing snail *Lymnaea* (Cuker, 1983).

One difference in response of the fixed-tube-dwelling guild in the experiment was the presence of first instar larvae in the sculpin exclusion and enrichment, but not in the ambient treatment. This may be an artifact of the experimental chambers, both the exclusion and enrichment had a mesh cover which may have served as platform to collect egg cases. The emerging larvae could then easily slip through the mesh to colonize the rocks in the chamber.

Studying sculpin predation effects on the chironomid community of the soft sediments in Toolik lake, Hershey (1985) found no significant difference when comparing the responses of different taxa, or the functional groups of predators versus non-predators. However, she did find that smaller (≤ 7 mm) larvae were more susceptible to sculpin predation than larger ones. She attributed this to the smaller larvae living in shallow and less protected positions in the soft sediments, in contrast to larger larvae that burrow deeply. Our own comparisons of sculpin gut contents and the soft-bottom community are consistent with Hershy's observation. We also found the smaller *Heterotrissocladius* to be over represented, and the larger *Stictochironomus* to be under-represented in sculpin diets. This is despite sculpin preferring *Stictochironomus* over *Heterotrissocladius* when presented with both larvae simultaneously in aquaria experiments (Cuker, 1981). Since the experimental aquaria contained no sediments, there was no refuge for the *Stictochironomus*. Thus the situation for *Stictochironomus* in the soft sediments is similar to that for *Paratanytarsus* and other tube-dwellers on the rocks, obscurity reduces vulnerability.

Hershey's (1985) findings for the soft sediments and ours on the rocky littoral represent contrasts in the way an eurytopic benthic fish influences adjacent, but structurally different communities. In the soft sediments the influence of sculpin predation is restricted to the smaller chironomid larvae of macrophyte free areas. Larger larvae in these exposed areas, and both large and small larvae living in macrophyte beds are unaffected by sculpin. The impact of sculpin predation on the bare soft sediment community seems equally shared by all taxonomic and functional groups (Hershey, 1985). In comparison, sculpin have a more specific impact on the rocky littoral community, causing reductions of the larger and more motile taxa, with trichopteran larvae and predatory chironomids being the most susceptible.

Contrasting to the soft sediments, there is little refuge to be found on the surfaces of rocks. Perhaps sculpin will play a larger role in the benthic communities of Toolik Lake now that population size of one of their primary predators, lake trout, has recently declined (McDonald and Hershey, 1992).

Acknowledgements

This work was supported by NSF grant number DPP-7722995A to S. C. Mozley. We wish to thank T. W. Federle, J. R. Vestal, M. C. Miller, W. J. O'Brien, A. E. Hershey, J. E. Dietz, J. M. Miller, T. R. Wentworth and L. A. Real, S. S. MacCafarty, and the Toolik Lake crews of 1978–1980 for their aid and support. Reviews by A. E. Hershey and J. Janssen greatly improved the manuscript.

References

Anderson, O., 1984. Optimal foraging by largemouth bass in structured environments. Ecology 65: 851–861.

Allan, D. A., 1982. The effects of reduction in trout density on the invertebrate community of a mountain stream. Ecology 63: 1444–1455.

Armitage, P. E., 1968. Some notes on the food of the Chironomid larvae of a shallow woodland lake in South Finland. Ann. Zool. Fenn. 5: 6–13.

Crowder, L. B. & W. E. Cooper, 1982. Habitat structural complexity and the interaction between bluegills and their prey. Ecology 63: 1802–1813.

Cuker, B. E., 1981. Control of epilithic community structure in an arctic lake by vertebrate predation and invertebrate grazing. Dissertation. North Carolina State University, Raleigh North Carolina, USA.

Cuker, B. E., 1983a. Competition and coexistence among the grazing snail *Lymnaea*, Chironomidae, and microcrustaceans in an arctic epilithic lacustrine community. Ecology 64: 10–15.

Cuker, B. E., 1983b. Grazing and nutrient interactions in controlling the activity and composition of the epilithic algal community of an arctic lake. Limnol. Oceanogr. 28: 133–141.

Cuker, B. E. & S. C. Mozley, 1981. Summer population fluctuations, feeding and growth of *Hydra* in an arctic lake. Limnol. Oceanogr. 26: 697–708.

Davies, I., 1974. Selective feeding in some arctic Chironomidae. Verh. int. Ver. Limnol. 19: 3149–3154.

Efford, I. E., 1972. An interim review of the Marion Lake Project. In Kajak, Z. & A. Hillbricht-Ilkowska (eds), Productivity Problems of Freshwaters. Warszawa-Krakow: 89–109.

Gilinsky, E., 1984. The role of predation and spatial heterogeneity in determining benthic community structure. Ecology. 65: 455–468.

Goyke, A. P. & A. E. Hershey, 1992. Effects of fish predation on larval chironomid (*Diptera: Chironomidae*) communities in an arctic ecosystem. Hydrobiologia 240: 203–211.

Hershey, A. E., 1985. Effects of predatory sculpin on the chironomid communities in an arctic lake. Ecology 66: 1131–1138.

Hershey, A. E., 1987. Tubes and foraging behavior in larval Chironomidae: implications for predator avoidance. Oecologia 73: 236–241.

Hershey, A. E. & S. I. Dodson, 1985. Selective predation by a sculpin and a stonefly on two chironomids in laboratory feeding trials. Hydrobiologia 124: 269–273.

Hynes, H. B. N. & M. J. Coleman, 1968. A simple method of assessing the annual production of stream benthos. Limnol. Oceanogr. 13: 569–573.

Ivlev, V. S., 1961. Experimental ecology of the feeding of fishes. Yale Univ. Press, New Haven, 302 pp.

Izvekova, E. I., 1971. On the feeding habits of chironomid larvae. Limnologica 8: 201–202.

Kajak, Z., K. Dusoge & A. Stanczykowska, 1968. Influence of mutual interactions of organisms especially Chironomidae in natural benthic communities, on their abundance. Ann. Zool. Fenn. 5: 49–56.

Kajak, Z. & J. Warda, 1968. Feeding of benthic non-predatory Chironomidae in lakes. Ann. Zool. Fenn. 5: 57–64.

McDonald, M. E., B. E. Cuker & S. C. Mozley, 1982. Production and distribution of the slimy sculpin (*Cottus cognatus*) in an arctic lake. Env. Biol. Fishes 7: 171–176.

McDonald, M. E. & A. E. Hershey, 1992. Shifts in abundance and growth of slimy sculpin in response to changes in the predator population in an arctic Alaskan lake. Hydrobiologia 240: 219–223.

Pankratova, V. Ya., 1971. On the systematics and ecology of the subfamily Orthocladiinae (Diptera, Chironomidae). Limnologica 8: 15–17.

Thorp, J. H. & E. A. Bergey, 1981a. Field experiments on responses of a freshwater benthic macroinvertebrate community to vertebrate predators. Ecology 62: 365–375.

Vijverberg, J. & T. H. Frank, 1976. The chemical composition and energy contents of copepods and cladocerans in relation to their size. Freshwat. Biol. 6: 333–345.

Ware, D. M., 1973. Risk of epibenthic prey to predation by rainbow trout (*Salmo gairdneri*). J. Fish. Res. Bd Can. 30: 787–797.

Hydrobiologia **240**: 91–101, 1992.
W.J. O'Brien (ed.), Toolik Lake – Ecology of an Aquatic Ecosystem in Arctic Alaska.

The influence of light and nutrient addition upon the sediment chemistry of iron in an arctic lake

Susan F. Sugai & George W. Kipphut[1]
Institute of Marine Science, University of Alaska, Fairbanks, AK 99775-1080, USA; [1] Present address: Center for Reservoir Research, Murray State University, Murray, KY 42071, USA

Key words: iron, sediment, nutrients, photoreduction, arctic

Abstract

The geochemical response of sediments to increased nutrient input to an Alaskan, arctic lake was examined using direct measurements of sediment-water chemical fluxes. An unexpected increase in Fe flux occurred when sediments were exposed to high incident radiation and nutrient concentrations. Correlation between light and acid-soluble Fe concentrations suggests that photoreduction of Fe(III) oxides may occur, but nutrient addition enhanced the effect indicating that primary productivity was also important. The processes controlling the flux of Fe from sediments in this lake were complex and included the redox potential (dissolved oxygen concentration) of the water, quality of organic matter present in the sediment, light, and nutrients supplied from the sediments and/or water column. These four factors together with the possibility of direct uptake of Fe by phytoplankton and the possible release of algal reductants may contribute to Fe cycling in this lake.

Introduction

Lakes in northern Alaska are very oligotrophic and nutrient-poor resulting in low sedimentation rates. Despite the relatively high organic matter (3–5%) content of lake sediments and the short (approximately 3 month) ice-free period, oxygen-consumption rates of the sediments are extremely low. The available organic matter, primarily of terrestrial origin, tends to be refractory. The oxidizing nature of the sediments results in visible Mn- and Fe-oxide layers within the sediment that are formed by the dissolution of Mn(IV) and Fe(III) oxides under reducing conditions at depth, followed by upward diffusion of soluble Mn^{2+} and Fe^{2+} and the accumulation of these reduced metals in oxidized zones at or below the sediment-water interface (Cornwell, 1986).

As part of an interdisciplinary study examining the response of arctic freshwater ecosystems to environmental stress, we examined the geochemical response of sediments to increased nutrient input to a divided lake. Addition of N and P to these oligotrophic waters stimulated primary productivity in amounts that were nearly proportional to the amount of N and P added. Following nutrient addition, the oxygen consumption rate in sediments increased several-fold; nutrient, Fe, and Mn fluxes from the deeper sediment also increased. However, substantial fluxes of Fe and Mn from littoral sediments were also observed at times when concentrations of dissolved oxygen in the overlying water remained high. Because Fe oxyhydroxides can be important in the cycling of trace metals (Fuller & Davis, 1989; Francis & Dodge, 1990) and P (Francko, 1986; Cornwell,

1987) in fresh water systems, we designed an experiment to examine the observed 'anomalous behavior'.

Photoreduction of Fe oxides has been observed in two soft-water lakes (pH 4.1–5) in Belgium (Collienne, 1983), an acidic, humic-rich stream (pH 3.4–4.6) in the New Jersey Pinelands (Madsen *et al.*, 1986), two acidic streams (pH 3.5–4.5) with low concentrations of organic carbon in the Colorado Rocky Mountains (McKnight & Bencala, 1988; McKnight *et al.*, 1988), an acid bog lake (pH 4.6–5.1) in Ohio (Cotner & Heath, 1990) and in atmospheric water droplets collected near Zurich (Behra & Sigg, 1990). In laboratory studies of the photoreductive dissolution of colloidal Fe oxides in natural waters, Waite & Morel (1984) found that organic agents enhance the effect of light but that light-induced changes in Fe speciation occurred only at low pH. Because arctic lakes in our study area have near neutral pH values, direct photoreduction of Fe oxides seemed unlikely. However, with the possibility of continuous light during arctic summer, we felt that photoreduction of Fe oxides might contribute to the observed fluxes of Fe from littoral sediments.

The various pathways for dissolving Fe(III) oxides have been reviewed by Sulzberger *et al.* (1989). The dissolution of sedimentary Fe(III) hydroxides near the sediment-water interface in the oxic-anoxic boundary layer can occur by reductive dissolution in the presence of surface coordinating organic reductants and by Fe(II)-catalyzed dissolution. This is the likely process occurring in sediment from deeper regions of our lake. Alternatively, light-induced redox reactions might be important in the shallower regions. These reactions could utilize organic matter; for example, the reaction:

$Fe(OH)_3$ + organic solute = Fe(II) + oxidized organic solutes. In this case, Fe^{2+} concentrations correlate with light intensity. Concentrations of soluble metals could also be indirectly controlled by light through the variation in pH caused by photosynthetic activity. Adsorption-desorption reactions occurring on particle surfaces are highly sensitive to changes in pH.

To determine the important factors controlling benthic trace metal fluxes from shallow sediments, we used benthic chambers to monitor pH, dissolved oxygen, alkalinity, nutrients, Fe, and Mn over time throughout the ice-free season in 1988, and during July 1989. By varying light, nutrient, and organic matter supply to the sediment-water interface, we found that processes controlling the benthic Fe flux (and probably the flux of other associated trace metals) become far more complicated when sediments are in the euphotic zone.

Study area

Lake N-2, a 18650 m^2 lake located in an arctic tundra environment approximately 300 km south of Prudhoe Bay, Alaska, has been the site of a fertilization experiment since 1985 (see other papers in this volume). A plastic curtain extending from the surface to the sediment was emplaced in 1985 and divided the lake into east and west sections (referred to as N-2E and N-2W, respectively). The lake's main inlet is on the east side; the outlet is on the west side. Volumes for the two basins have been estimated at 29949 m^3 and 65528 m^3, for east and west sides, respectively. During the period 1985–1990, the west side of the lake received added nutrients for 6 weeks between July 1 and August 15. In 1985, 10.9 kg ammonium nitrate and 1.43 l phosphoric acid per week were added. This was equivalent to 273 mol N and 21 mol P per week. In subsequent years, the fertilization rate was 80% of the 1985 level added over the same 6 week period. The sedimentation rate determined by ^{210}Pb after the first summer's fertilization was 23 mg cm^{-2} yr^{-1} at 6 m depth in N-2E and 35 mg cm^{-2} yr^{-1} at 7 m depth in N-2W. The east side of the lake was used as a reference to assess fertilization effects.

In 1988, the lake was ice-free on 10 June; in 1989, the lake became free of ice between 15 and 20 June. The average value of total chl *a* in the epilimnion of N-2 was greater in both sides of N-2 in 1989 than in 1988. Likewise, primary productivity in N-2 was greater in 1989 when integrated production rates in late summer reached 950 mg C m^{-2} day^{-1} in N-2W and 200 mg C

m^{-2} day^{-1} in N-2E. In 1988, an earlier bloom resulted in primary production of 300 mg C m^{-2} day^{-1} in N-2W followed by a later bloom when integrated production reached 530 mg C m^{-2} day^{-1}. Peak production in the control side of the lake was approximately 130 mg C m^{-2} day^{-1} during both bloom periods.

Water column light measurements were made at 5 times in 1988 and 3 times in 1989. In N-2E (the control side of the lake), the light at 1 m was 30 to 48% of the incoming light measured at the surface in 1988 and 31 to 43% in 1989. Light at 1 m in N-2W was 26 to 52% of surface values in 1988 and 29 to 38% in 1989.

Methods

Sampling

Benthic chambers constructed from 0.25 inch thick, clear plexiglass were positioned on the bottom sediment either by diver or by lowering from the surface under calm conditions. The chambers covered 0.2 m^2 of sediment, enclosed 42 to 52 l of overlying water, and transmitted $95 \pm 1\%$ of photosynthetically active radiation (PAR). Water within the chambers was stirred continuously by a 1 rpm motor. Water samples were withdrawn through small diameter tubing leading to the surface and less than 50 ml of water was required to flush the sampling tube. Pressure was equalized by allowing bottom water to enter the chamber through a small diameter hole on the top of the chamber when samples were withdrawn. Chambers were allowed to set for approximately 1 h before initial samples were drawn. Chambers were retrieved and redeployed in a new location when dissolved oxygen concentrations dropped below about 50 μmol l^{-1}.

Light/dark experiments

The effects of light on chemical fluxes was examined in light/dark experiments conducted in July and August 1988, and in July 1989. Shallow chambers were darkened by lowering an opaque plastic sheet from above under calm conditions and then weighing down the plastic, taking care not to disturb the underlying chamber. To remove the shroud, the process was reversed. Because the chambers were deployed at water depths of 1 to 1.5 m and into very soft bottom material, we were able to perform both deployments and retrievals of chambers and shrouds with minimal disturbance of sediment within the chambers.

Nutrient enrichment experiments

In an attempt to simulate a one-time pulse of nutrients to this environment, nutrient enrichment experiments were conducted in July 1988 and July 1989. In these experiments, 30 ml of nutrient broth containing 25 mM NH_4NO_3 and 2.5 mM $KH_2PO_4.3H_2O$ were added to each chamber through the sampling tube and then flushed with 180 ml of lake water.

Analytical techniques

Dissolved oxygen (DO) was measured by Winkler titration. Fe samples were collected unfiltered and acidified during 1988 and are referred to as 'acid-soluble Fe'. In 1989, samples were collected both unfiltered and filtered through Whatman EPM 1000 glass fiber filters. Iron was analyzed using the ferrozine technique of Stookey (1970) as applied to natural waters by Gibbs (1979). Incoming radiation data were collected at the Toolik Lake weather station located approximately 2 km southeast of our study area.

Results

Early ice-free period, 1988

Figure 1 shows the incident radiation for the early ice-free period in June 1988 together with pH, DO, and acid-soluble Fe measured in chambers placed at depths of 1 and 6 m in both sides of

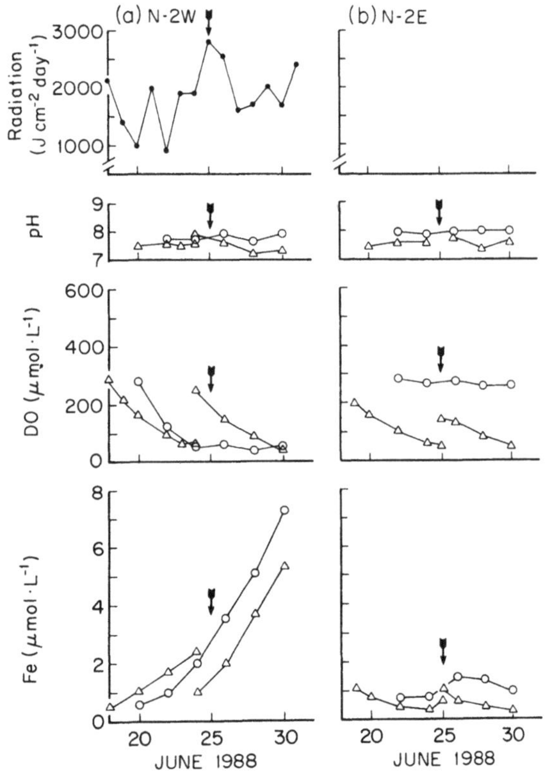

Fig. 1. (a) Changes in incoming radiation at Toolik weather station and pH, dissolved oxygen, and acid-soluble Fe measured in benthic chambers emplaced in sediments of fertilized (N-2W) side of experimental lake N-2 in June 1988. Circles (○) indicate measurements from shallow (1–1.5 m) sediments, upright triangles (△) indicate measurements taken at 6 m. Arrows (↓) indicate sampling days in which incoming radiation was relatively high. (b) pH, DO, and acid-soluble Fe measured in sediments of control (N-2E) side of lake N-2. Symbols are as in (a).

N-2. A rapid decline in DO followed deployment of benthic chambers in the shallow and deep waters of N-2W, the fertilized side of lake N-2. Coincident with this rapid decrease in DO was an increase in Fe (Fig. 1a). Measured Fe fluxes at 6 m were 67 μmol m^{-2} d^{-1} for the first deployment and 150 μmol m^{-2} d^{-1} for the second. At the shallow site, the average Fe flux for the 10 day period was 140 μmol m^{-2} d^{-1}. On the control side of the lake (Fig. 1b), the DO levels decreased rapidly in the chamber deployed at 6 m on 19 June and again on 25 June, but oxygen levels remained relatively constant in the shallow chamber. For both chamber emplacements at 6 m in N-2E, acid-soluble Fe generally decreased with time except for a slight increase between 24 and 25 June. In the shallow chamber, between 24 and 26 June, the acid-soluble Fe concentration increased from 0.68 to 1.4 μmol l^{-1} at a time when DO increased slightly and incoming daily radiation increased to 2807 J cm^{-2} day^{-1} on 25 June and 2552 J cm^{-2} day^{-1} on 26 June. Fe subsequently decreased on 28 June when both incoming radiation and DO decreased. The net Fe flux in the shallow chamber was 5.5 μmol m^{-2} day^{-1} for the 8 day deployment.

At 9 m in N-2W (not shown), strongly reducing conditions resulted in extremely high Fe fluxes. In the chamber deployed between 19–25 June 1988, an average Fe flux of 1.0 mmol m^{-2} day^{-1} resulted in response to a decrease in DO from 31 μmol l^{-1} to below detection. The flux of acid-soluble Fe was greatest during the initial 2.9 days following deployment when concentrations increased from 8.4 to 36.0 μmol l^{-1} for a flux of 2.0 mmol m^{-2} day^{-1}. Mn increased from 147 to 162 μmol l^{-1} during this same time period. Fe and Mn concentrations increased nominally in the following 3 days (0.8 μmol l^{-1} for Fe and 2 μmol l^{-1} for Mn). Following a redeployment of the chamber on 25 June, a small increase in DO (from 12 to 17 μmol l^{-1}) and Fe (from 11.6 to 18.2 μmol l^{-1}) occurred on 26 June followed by a decrease in DO to 2 μmol l^{-1} and in Fe to 13.3 μmol l^{-1} on 28 June.

Arctic summer, 1988

Beginning with 1 July 1988, incoming daily radiation increased to greater than 2500 J cm^{-2} day^{-1} on sunny days, which occurred with greater frequency (7 times) than in June (twice). The DO and Fe fluxes observed in the benthic chamber deployed on 1 July 1988 in shallow waters of N-2W (Fig. 2a) differed substantially from those observed in June. DO concentrations increased dramatically, reaching 592 μmol l^{-1} on 9 July, while Fe concentrations remained less than 0.5 μmol l^{-1}. When light was excluded from the

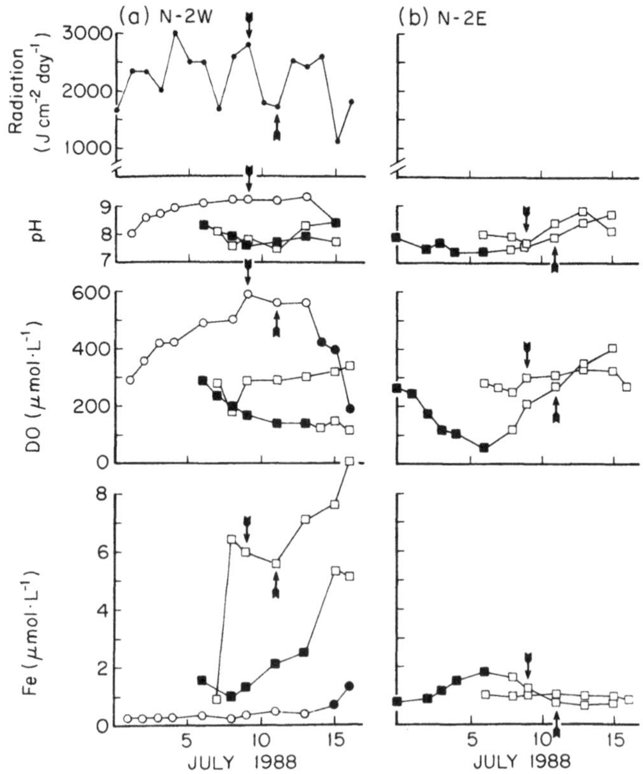

Fig. 2. (a) Changes in incoming radiation, pH, DO, and acid-soluble Fe measured in benthic chambers in N-2W during July 1988. Open symbols indicate benthic chambers exposed to natural light; filled symbols indicate dark chambers. Arrows indicate sampling dates in which incoming radiation was either high (↓) or low (↑). Circles indicate measurements from shallow (1–1.5 m) sediments, squares (□) indicate chambers in shallow sediments subjected to N- and P-enrichment on day 1 of each experiment prior to flux determinations. (b) Changes in pH, DO, and acid-soluble Fe measured in benthic chambers in N-2E.

chamber, DO decreased to 188 μmol l^{-1} while Fe increased to 1.3 μmol l^{-1}.

The influence of fertilization is shown in Fig. 2, where levels of pH, DO, and Fe *vs.* time are presented for chambers in shallow waters of both the treated and control sides of lake N-2. The scale of axes used in Figs 1 and 2 is identical to permit easier comparison.

Iron in the fertilized, light chamber in N-2W increased dramatically on 8 July (Fig. 2a) with coincident increase in light and drops in pH, DO, and soluble reactive phosphorus (SRP, not shown). Loss of SRP was more rapid in the enriched, light chamber, where concentrations dropped to 0.1 μmol l^{-1} in 49 h following deployment when compared to the enriched dark chamber, where the SRP concentration was 0.4 μmol l^{-1} after 44 h and 0.1 μmol l^{-1} after 119 h. By contrast, at the time when SRP had been reduced to 0.1 μmol l^{-1}, nitrate and ammonium were present at levels of 4.4 and 1.0 μmol l^{-1}, respectively, in the enriched, light chamber, and 0.9 and 2.0 μmol l^{-1} in the enriched, dark chamber. Fe concentrations decreased from 6.4 μmol l^{-1} on 8 July to 5.6 μmol l^{-1} on 11 July, then increased steadily to 9.1 μmol l^{-1} on 16 July when the experiment was terminated. Thus, for the 8.8 day deployment, an average Fe flux of 194 μmol m^{-2} day^{-1} was observed. Between 11 and 16 July, pH increased slightly while DO increased from 290 to 344 μmol l^{-1}. In the fertilized dark chamber, after an initial drop in Fe concentration between 6 and 8 July, the Fe concentration increased to 2.5 μmol l^{-1} on 13 July. The mean Fe flux in the dark was 26 μmol m^{-2} day^{-1}. The Fe concentration doubled between 13–15 July when the dark chamber was exposed to light resulting in a flux of 180 μmol m^{-2} day^{-1}.

The flux of Fe from the sediments enclosed in light and dark chambers on the control or east side of N-2 was related to DO concentrations (Fig. 2b) and much less influenced by nutrient addition than observed in N-2W. Loss of SRP from the enriched, light chamber in N-2E was much slower than in N-2W with 0.19 μmol l^{-1} SRP observed 121 h after deployment. Although we lacked a control chamber in N-2E during the nutrient enrichment experiment, little enhancement of Fe flux was observed in the light chamber associated with the peak radiation period or in the dark chamber when exposed to light.

Late summer, 1988

By mid-August, Fe concentrations in chambers placed in shallow waters of N-2W were relatively insensitive to changes in light conditions. Daily incident radiation during this period averaged less than 50% of that during the early ice-free period and about 40% of that during the July experiments. For the 5 day deployment, Fe concentrations were 0.44 ± 0.14 μmol l^{-1} in the light chamber and 0.45 ± 0.03 μmol l^{-1} in the dark chamber although DO decreased from 299 to 151 μmol l^{-1} in the dark chamber while DO was relatively constant at 320 ± 15 μmol l^{-1} in the light chamber.

At 6 m depth in N-2E, the DO flux in mid August (Fig. 3b) was much less than in late June (Fig. 1b) and Fe concentrations were somewhat lower. In the fertilized side of the lake, DO and Fe fluxes remained high from the chamber deployed at 9 m on 15 August. Although the initial DO concentration was much higher than during the early ice-free period, Fe and Mn fluxes of 0.95 and 4.09 mmol m^{-2} day^{-1}, respectively, were observed with concentrations reaching 21 and 96 μmol l^{-1}, or roughly 58% of the apparent 'saturation concentration' observed right after ice-out.

Summer 1989

In 1989, incident radiation during the study period averaged 1598 ± 473 J cm^{-2} day^{-1} with highest values on 9 and 11 July when 1979 and 2029 J cm^{-2} day^{-1} of radiation were recorded (Fig. 4a). This is in contrast to summer 1988 when incoming radiation reached 2961 J cm^{-2} day^{-1} and averaged 2256 ± 493 J cm^{-2} day^{-1} over the same time period. Water column light measurements taken at 1 m during the experimental periods also indicated significantly less radiation reaching the sediment in 1989. In 1988, 479 μE m^{-2} s^{-1} was measured at 1 m in N-2E

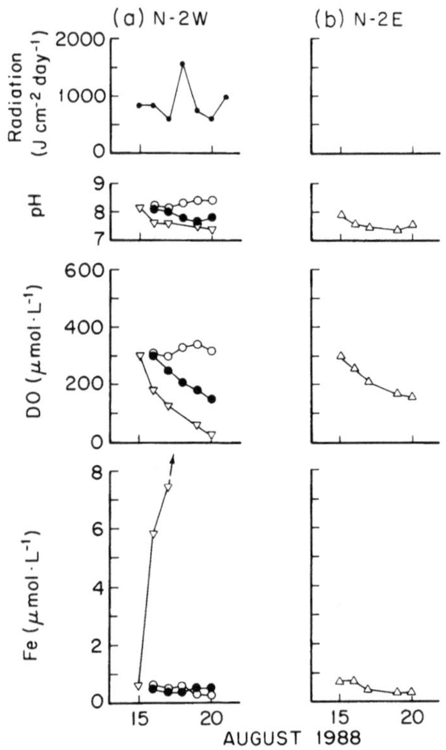

Fig. 3. (a) Changes in incoming radiation, pH, DO, and acid-soluble Fe measured in benthic chambers in N-2W in August 1988. Symbols and scale of axes are the same as in Figs. 1 and 2. Inverted triangles (▽) indicate measurements taken from 9 m depth. (b) Changes in pH, DO, and acid-soluble Fe measured in N-2E.

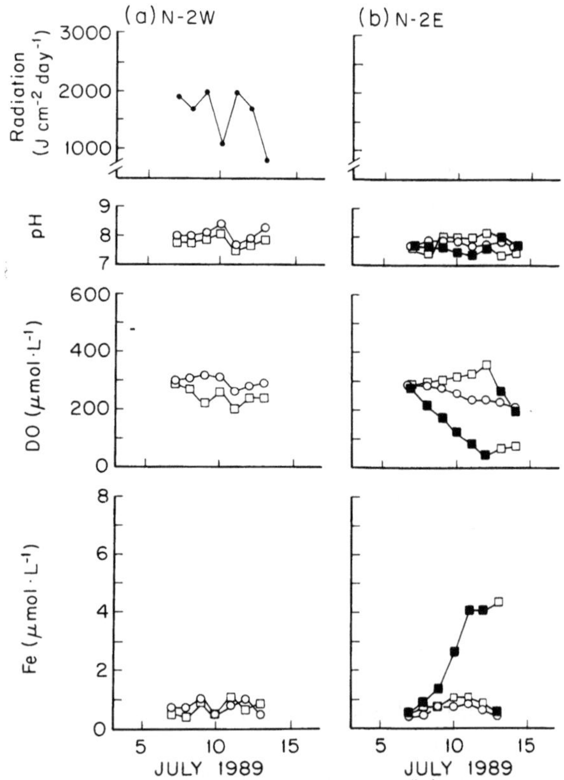

Fig. 4. (a) Changes in incoming radiation, pH, DO, and acid-soluble Fe measured in benthic chambers in N-2W in July 1989. Symbols and scale of axes are the same as in Figs. 1 and 2. (b) Changes in pH, DO, and acid-soluble Fe measured in N-2E.

on 5 and 12 July; 433 $\mu E\ m^{-2}\ s^{-1}$ was observed in N-2W on 12 July. In 1989, 113 and 103 $\mu E\ m^{-2}\ s^{-1}$ were measured at 1 m depth in N-2E on 6 and 13 July, respectively; 65 and 109 $\mu E\ m^{-2}\ s^{-1}$ were observed at 1 m in N-2W on these dates. Thus, roughly 4 times more light was reaching the shallow sediments in 1988. In N-2W, the Fe concentration in the enriched chamber had maxima on 9 and 11 July coincident with DO minima. SRP decreased from 3 μmol l^{-1} on 7 July to 0.08 μmol l^{-1} on 10 July. DO in the unfertilized chamber had a minima on 11 July but Fe increased on 12 July as well as 11 July. pH decreased significantly on 11 July for both chambers in N-2W. The Fe concentration in the nutrient-enriched chamber that was exposed to light in N-2E increased slightly on 10–11 July while DO increased steadily at a slow rate from the time the chamber was deployed on 7 July until it was darkened on 12 July. There were no observed effects on DO or pH from variations in incoming radiation and SRP remained greater than 0.1 μmol l^{-1} throughout the chamber emplacement. The DO in the unfertilized chamber decreased throughout the deployment while Fe increased slightly to 11 July and then decreased. The Fe flux from the enriched, dark chamber placed in N-2E was 170 $\mu mol\ m^{-2}\ day^{-1}$ in 1989 compared to 37 $\mu mol\ m^{-2}\ day^{-1}$ in 1988, and based on only 1 point, appeared to increase at a rate of 93 $\mu mol\ m^{-2}\ day^{-1}$ rather than decrease when exposed to light.

Discussion

Freshwater ecosystems in the arctic experience a seasonal fluctuation in incoming radiation ranging from virtually 24 h of daylight in early summer to 24 h of darkness in mid-winter. This coupled with the very short ice-free period found in arctic lakes means that the *timing* of environmental stresses will alter their effects upon biological and geochemical processes that are light dependent. Changes in the light and nutrient supply to the lake can influence lacustrine production, which in turn can alter the supply and oxygen-demand of organic matter reaching the sediment-water interface, and ultimately the magnitude of benthic fluxes of nutrients and trace metals. Although increased oxygen-consumption with resultant enhanced nutrient and trace metal fluxes was observed in the profundal sediments of the section of the lake receiving fertilizer addition, this was not strictly the case in littoral sediments.

Light/dark effects upon DO and Fe fluxes

In the absence of light (and hence, photosynthesis), the DO concentration decreased in all shallow chambers that were darkened (Figs 2, 3, and 4). DO also decreased in chambers placed at depths of 6 and 9 m, where incoming light was less than 4% of the incident light at the surface. In response to these decreases in DO, acid-soluble Fe generally increased in 6 cases, decreased in 1 case (chamber in N-2E at 6 m in June 1988), and remained relatively constant in 2 cases (shallow chambers in N-2W and N-2E, August 1988). In the majority of the experiments, under dark conditions the decomposition of organic matter reduced DO, decreased pH, and the flux of acid-soluble Fe from the sediment to the overlying water increased in response to the more reducing conditions.

The importance of nutrients, organic matter quality, and light to the sedimentary fluxes was seen in benthic chambers placed in the littoral zone in July 1988 (Fig. 2). The DO and Fe data for the light and dark chambers in the control side of the lake (N-2E, Fig. 2b) show that Fe fluxes are related to oxygen concentrations in a 'typical' arctic lake subject to a one time pulse of nutrients. The nutrient enrichment to the light chamber in N-2E had little effect on primary productivity reflected by the small changes in DO observed. The increase in DO was much greater in the chamber that after being darkened for 6 days was exposed to light. However, Fe data from N-2W (the side of the lake that had been fertilized for 6 weeks each summer since 1985) indicate that nutrients, organic matter quality, and incident light are all important in determining the magnitude of Fe fluxes. In the unfertilized chamber, extremely high pH and DO levels were observed suggesting that primary productivity was important in July 1988 (Fig. 2a). By contrast, the chamber which was placed in shallow N-2W on 7 July and enriched with a solution of NH_4NO_3 and $KH_2PO_4.H_2O$ exhibited a large drop in DO on the day following deployment (Fig. 2a). Coincident with this drop in DO, was a large increase in acid-soluble Fe (Fig. 2a). The DO in this chamber increased the following day but remained relatively constant throughout the remainder of the deployment although the Fe flux decreased and increased. The mechanism for reducing Fe during this time period was much more effective with light present than without. For the 8.8 day deployment, the average Fe flux from the enriched, light chamber was roughly 20% of the Fe flux observed in the deep sediments of N-2W shortly after ice-out when Fe and Mn concentrations within the chamber rapidly approached an apparent saturation value. The Fe concentration in the enriched, dark chamber increased dramatically when the shroud was removed although DO levels were relatively constant. From 2 July to 13 July, Fe concentrations in the control chamber were low, suggesting that although light influenced the Fe flux in the enriched chambers, the nutrient enrichment in some way changed the productivity in the chamber. Added SRP was lost much more rapidly in the enriched, light chamber than in the enriched, dark chamber in N-2W. Perhaps by adding ammonium and nitrate in equal molar concentrations allowed a different type algae to dominate over the one present in the control chamber, where nitrate was undetectable on 8 and 11 July.

Iron fluxes in the early ice-free season

Early in the ice-free season of 1988, oxygen consumption rates by littoral and deep sediment in N-2W were much greater than in N-2E although initial bottom water concentrations of DO were similar. At this time, the littoral and profundal sediments of N-2W exhibited large benthic fluxes of acid-soluble Fe. On the untreated side of the lake, we observed a much smaller rate of oxygen consumption in sediments from both shallow and deep (6 m) sites. At the deep site, the flux of acid-soluble Fe appears to be from the water column to the sediment. If the source of Fe to the water enclosed in the benthic chamber was from the lake water rather than the sediments, an observed change in Fe concentration would result from restricted exchange between the water inside and outside of the chamber. In nearby Toolik Lake, Cornwell (1983) observed that the highest concentrations of both total and dissolved Fe in inlet streams are found during spring runoff, and almost 50% of the soluble Fe is retained in the lake. Lake N-2 is much smaller than Toolik Lake, and because of the curtain dividing the lake, the pulse of soluble Fe from the inlet stream to N-2E during spring runoff could increase water column Fe concentrations to the extent seen in the initial sampling times for these chambers. Thus, if Fe measured as 'acid-soluble Fe' in the 6 m chamber is actually dissolved Fe^{2+} from the watershed, a decrease in DO would not increase observed Fe as would occur if measured Fe was actually particulate Fe-oxides. In the benthic chamber placed in shallow sediments, pH and DO concentrations remained relatively constant for 5 days while Fe concentrations increased between 24 and 26 June coincident with an increase in incident light on 25–26 June. In this case, photoreduction of Fe oxides present in the sediments may have occurred.

Iron fluxes from sediments in late summer

Results from August 1988 (Fig. 1b) reflect the decreasing daylight hours and change in sediment substrate because Fe fluxes from the dark chamber placed in N-2W remained low and variable while DO levels decreased by 50%. A much more modest decrease in DO between 13 and 14 July 1988 in the chamber that was darkened on 13 July increased Fe from 0.44 to 0.70 $\mu mol\ l^{-1}$. Lower fluxes of Fe observed in the light chamber in August 1988 may be the result of an increased rate of Fe(II) reoxidation as well as reduced photoreduction. Collienne (1983) found that two photochemical processes influenced the seasonal exchange between soluble and solid phases of Fe in soft-water, Belgium lakes. From spring to the end of summer, Fe dissolves and is reduced to soluble ferrous Fe, while organic matter is destroyed by photooxidation. At the beginning of fall, decreased light results in more organic ligands available to accelerate the rate the Fe(II) reoxidation. Alternatively, Cotner & Heath (1990) suggest that photoreduction of Fe^{3+}, coupled with slow oxidation of Fe^{2+} to Fe^{3+}, and biotic uptake of Fe^{2+} may result in decreased Fe^{3+} availability in summer.

Sediment fluxes in 1989

The fluxes of Fe from light chambers placed in the littoral zones of N-2W and N-2E were similar in July 1989. Modest increases in Fe concentrations accompanied increases in incoming radiation in enriched and control chambers in N-2W. As in July 1988, these increases in Fe concentration observed in the enriched, light chamber were accompanied by decreases in DO suggesting a physical rather than biological explanation. However, Fe fluxes from the enriched chamber in N-2W in July 1989 (Fig. 2b) were much smaller than in July 1988 (Fig. 2a). In addition to a reduction in incoming radiation in 1989, a further reduction of light intensity reaching the sediments resulted from water column turbidity caused by a water column bloom of *Peridinium* (M. Miller, pers. comm.). In N-2E, the DO levels in the enriched chamber increased although levels in the control chamber decreased slightly. By contrast, pH and DO levels in the enriched chamber in N-2W were less than those in the control chamber as was observed in July 1988. Fe fluxes from the enriched, light chamber in the control side of N-2 were comparable to those in the enriched chamber in fertilized side suggesting that although the substrate in N-2W was more readily photoreduced than the substrate in N-2E, light was limiting the Fe flux rather than organic matter or appropriate Fe oxide. However, the flux of Fe observed in the dark, enriched chamber in N-2E in 1989 was 4.6 times greater than the flux of Fe observed in the dark, enriched chamber in 1988, suggesting that perhaps a change in the quantity and/or quality of organic matter present in 1989 facilitated increased solubilization of Fe(III) oxides in the absence of light.

Lability of 'acid-soluble' Fe

In this study the behavior of 'acid-soluble' Fe was assumed to parallel that of the soluble, ionic form of Fe. This was because preliminary studies had indicated that total Fe concentrations in the lake water were very low and dominated by the dissolved phase. While this assumption may have been valid in the control side of the lake where a large portion of the Fe present in the lake water was contributed from the inlet stream, 'acid-soluble' Fe in waters of N-2W may have included a substantial (and variable) portion of particulate Fe. An attempt was made to resolve this problem in 1989 when both dissolved and total Fe were measured in a diurnal experiment conducted to measure Fe fluxes over a solar day. Unfortunately, as indicated in the above discussion, light conditions in 1989 were not favorable for photoreduction of Fe(III)-oxides and it was impossible to resolve the question as to the validity of using 'acid-soluble' rather than dissolved Fe measurements. In the diurnal experiment, dissolved Fe was 41 to 67% of the acid-soluble Fe present but there was no correlation between percent of dissolved Fe and light conditions. Further, acid-soluble Fe concentrations averaged $0.51 \pm 0.11\ \mu mol\ l^{-1}$, which were on the low end of Fe concentrations measured. Thus, the validity of using 'acid-soluble' Fe in this study remains unresolved and needs verification in future work. Evaluating the importance of the direct uptake of Fe by phytoplankton cannot be determined without making meaningful distinctions between forms of the Fe present.

Comparison with other arctic studies

In general, Fe fluxes from arctic lake sediments are among the lowest found anywhere. Cornwell (1983) used a diagenetic model to quantify the dissolution of solid phase Fe in Toolik Lake sediments and found rates of 0.03–0.8 mmol m^{-2} day^{-1}. In our study, the largest Fe fluxes were observed in the deep sediments of N-2W, where fluxes at 9 m were 1.0 mmol m^{-2} day^{-1} shortly after ice out in 1988 when bottom waters were virtually anoxic, and 0.95 mmol m^{-2} day^{-1} in late August 1988. In shallow sediments of N-2W, 0.14 mmol m^{-2} day^{-1} of Fe was observed in the early ice-free period; a maximum Fe flux of 0.19 mmol m^{-2} day^{-1} was observed when sediments were subjected to a one-time nutrient addition in July 1988. In N-2E, the control side of the lake, net Fe fluxes from shallow sediments of 5.5 μmol m^{-2} day^{-1} were observed during the early ice-free season. No further Fe fluxes were observed in either shallow or deep locations throughout the remainder of 1988 or during July 1989. This suggests that for the unfertilized portion of lake N-2, most Fe recycling occurs within the sediment and not at the sediment-water interface, in agreement with the studies of Cornwell (1983) for Toolik Lake.

To estimate the relative importance of light-mediated Fe fluxes from shallow waters of N-2, it was necessary to evaluate the potential lake area involved and the days of adequate solar irradiation. In N-2W, 20 566 m^2 of the lake's area are at depths of less than 2 m and 3436 m^2 at depths of greater than 9 m; in N-2E, 15 089 m^2 of the sediments are at depths less than 2 m (O'Brien & Kling, pers. comm.). From this study, measurable photoreduction of Fe-oxides was most favorable when incoming radiation exceeded approximately 2000 J cm^{-2} day^{-1}, which occurred on 20 days in 1988. Assuming that Fe fluxes from the deep sediments occurred at an average of 0.98 mmol m^{-2} day^{-1} (the mean of measured values in June and August 1988) for the entire 90 day ice-free period results in a Fe flux from the deep waters in N-2W of 303 mol Fe. The comparable flux for the shallower areas of N-2W using a mean flux of 0.14 mmol m^{-2} day^{-1} gives 57 mol Fe or 19% of the total Fe contributed by the deep sediments. In N-2E, where no measurable flux occurred in deep sediments, the shallow sediments have the maximum potential to contribute 1.6 mol Fe based on 20 days of favorable light conditions and an average flux of 5.5 μmol m^{-2} day^{-1}. However, based on our direct measurements during 3 periods in 1988 and one in 1989, this light-mediated reduction of Fe-oxides in naturally-occurring lake sediments requires that high incoming radiation occur when seasonally high substrate levels are available.

Summary

Previous studies have demonstrated that photoreduction of Fe oxides can result in measurable fluxes of Fe in acidic, aquatic environments. Our results suggest that in Lake N-2, Fe fluxes from sediments are possible under appropriate conditions even at relatively high pH. Important facts regulating the Fe flux include redox potential, quality of organic matter present in the sediments, and light. We were able to at least qualitatively rank these factors as they influence the Fe cycle in Lake N-2.

In the unfertilized side of the lake, N-2E, where both dissolved oxygen concentrations and pH are always quite high, and the organic matter present is more refractory, Fe fluxes from both shallow and deep water sediments were small. During mid-summer, however, the Fe fluxes were somewhat greater in the shallow water sediments, suggesting possible enhancement by photoreduction. Given the condition of continuous light during the arctic summer, we suggest that photoreduction of Fe should be considered when interpreting results of trace metal studies in arctic lakes.

In the fertilized side of the lake, N-2W, the relative roles of dissolved oxygen and organic matter are more clearly separated. By far the largest fluxes of Fe were observed from sediments in the deeper part of the lake where dissolved oxygen concentrations were small, or where overlying waters may have been anoxic. Clearly, strongly reducing conditions at the sediment-water interface result in large Fe fluxes and, when present, can dominate the other factors which have been mentioned. At shallow and intermediate depths in N-2W, where dissolved oxygen concentrations were high and light was present, Fe fluxes from sediments were smaller. They were much higher, however, than Fe fluxes at similar depths in N-2E (Figs 1 and 2). The organic matter in these sediments is much more labile than in N-2E, suggesting that the photoreduction of Fe is much more easily accomplished in the presence of labile organic matter.

Acknowledgements

We thank J. Laundre for providing the radiation data and M. Miller for water column light, chlorophyll, and primary productivity measurements. Many persons were responsible for designing and maintaining the divided lake experiment especially W. J. O'Brien, B. Moller, and M. Miller. We thank G. Gibeau for assistance in deploying and retrieving the benthic chambers, and J. Cornwell and A. Giblin for critical review. This is contribution no. 899, Institute of Marine Science, University of Alaska Fairbanks, Fairbanks, AK.

References

Behra, P. & L. Sigg, 1990. Evidence for redox cycling of iron in atmospheric water droplets. Nature 344: 419–421.

Collienne, R. H., 1983. Photoreduction of iron in the epilimnion of acidic lakes. Limnol. Oceanogr. 28: 83–100.

Cornwell, J. C., 1983. The geochemistry of manganese, iron and phosphorus in an Alaskan arctic lake. Ph.D. dissertation, Univ. of Alaska Fairbanks, 234 p.

Cornwell, J. C., 1986. Diagenetic trace-metal profiles in arctic lake sediments. Envir. Sci. Technol. 20: 299–302.

Cornwell, J. C., 1987. Phosphorus cycling in arctic lake sediment: adsorption and authigenic minerals. Arch. Hydrobiol. 109: 161–179.

Cotner, J. B. & R. T. Heath, 1990. Iron redox effects of photosensitive phosphorus release from dissolved humic materials. Limnol. Oceanogr. 35: 1175–1181.

Francis, A. J. & C. J. Dodge, 1990. Anaerobic microbial remobilization of toxic metals coprecipitated with iron oxide. Envir. Sci. Technol. 24: 373–378.

Francko, D. A., 1986. Epilimnetic phosphorus cycling: influence of humic materials and iron on coexisting major mechanisms. Can. J. Fish. aquat. Sci. 43: 302–310.

Fuller, C. C. & J. A. Davis, 1989. Influence of coupling of sorption and photosynthetic processes on trace element cycles in natural waters. Nature 340: 52–54.

Gibbs, M. M., 1979. A simple method for the rapid determination of iron in natural waters. Wat. Res. 13: 295–297.

Madsen, E. L., M. D. Morgan & R. E. Good, 1986. Simultaneous photoreduction and microbial oxidation of iron in a stream in the New Jersey Pinelands. Limnol. Oceanogr. 31: 832–838.

McKnight, D. & K. E. Bencala, 1988. Diel variations in iron chemistry in an acidic stream in the Colorado Rocky Mountains, U.S.A. Arctic Alpine Res. 20: 492–500.

McKnight, D. M., B. A. Kimball & K. E. Bencala, 1988. Iron photoreduction and oxidation in an acidic mountain stream. Science 240: 637–640.

Stookey, L. L., 1970. Ferrozine – a new spectrophotometric reagent for iron. Analyt. Chem. 42: 779–298.

Sulzberger, B., D. Suter, C. Siffert, S. Banwart & W. Stumm, 1989. Dissolution of Fe(III) (hydr)oxides in natural waters; laboratory assessment on the kinetics controlled by surface coordination. Mar. Chem. 28: 127–144.

Waite, T. D. & F. M. M. Morel, 1984. Photoreductive dissolution of colloidal iron oxides in natural waters. Envir. Sci. Technol. 18: 860–868.

Hydrobiologia **240**: 103–119, 1992.
W.J. O'Brien (ed.), Toolik Lake – Ecology of an Aquatic Ecosystem in Arctic Alaska.

Epilithic diatom community response to years of PO_4 fertilization: Kuparuk River, Alaska (68 N Lat.)

Michael C. Miller, Paulo DeOliveira[1] & George G. Gibeau
Department of Biological Sciences, University of Cincinnati Cincinnati, OH 45221, USA; [1] Zoology Department, Ohio State University, Columbus, OH 43210, USA

Key words: Epilithic algae, chlorophyll, diatom species, diversity, evenness, detrended correspondence analysis, arctic, river, nutrient bioassay tubes

Abstract

An arctic river was fertilized continuously through the ice-free season with phosphoric acid beginning in 1983. The epilithic diatom community increased in biomass in the first two years in response to the added limiting nutrient (Peterson *et al.*, 1983). The diatom community switched from one dominated by *Hannea arcus* to one dominated by species of Achnanthes and Cymbella. The immediate responses to the P-addition were decreases in both the Shannon diversity and evenness indices. By the second year, the community diversity increased downriver reaching maximal species richness (110–127 spp). In 1985–1987, the epilithic algal biomass decreased an order of magnitude with both whole-river PO_4 (1985, 1987) and $PO_4 + NH_4$ addition (1986). In the 5th summer of fertilization, the reduction in biomass was clearly caused by a numerical increase of grazing, refugia-building chironomids (Orthocladiinae, primarily) (Gibeau, 1991; Gibeau, Miller, Hershey, in prep.). We assume the algal biomass reduction in the 3rd and 4th years was similarly caused by grazers with a two year time lag in the numerical response of these monovoltine species. The evenness of the community increased in 1986 as if it might have been grazed; however the number of immigrants was reduced. The community became dominated by Eunotia, Cymbella and Achnanthes, species either fast growing or more prostrate, as the erect species of Hannea, Diatoma, and Fragillaria declined. A detrended correspondence analysis of the temporal and spatial diatom samples in species space (186 spp.) showed that the largest variation in the community was between years and less variation was associated with river fertilization.

Samples from bioassay tubes run by Peterson *et al.* (1983) in the Kuparuk River showed P and N + P limitation as found in the river in 1983–84. Like the river samples, the largest change in the diatom community occurred between 15 and 25 day samples, more than that induced by fertilization. Diatoms sampled from all treatments taken at day 25 were more similar to one another than those sampled at day 15. Diatoms colonizing glass slides used in the bioassay tubes were dominated by *Achnanthes linearis* and *Cymbella minuta*. Of the 84 species found in bioassays, 26 species were present in all river samples for 4 years. Differences in the communities discriminated by multivariate methods were cause by changes in rare species and abundance patterns of common species.

Introduction

The diatom community composition has been used extensively as a bioassay system for detecting changes in water quality (Hynes, 1960, 1970; Whitton, 1975; Patrick, 1970, 1977; Patrick *et al.*, 1968). Changes in light, current, nutrient concentration, nutrient ratios, immigration, grazing and catastrophic disturbance have all been shown to cause community change (Cattaneo, 1983; Horner & Welch, 1981; Stevenson, 1984; Lowe *et al.*, 1986; Patrick, Hoagland *et al.*, 1982). In general, nutrient enrichment leads to increased dominance, reduced species diversity, reduced evenness, loss of rare species, higher biomass (Tilman, 1985; Kilham, Kilham & Tilman, 1983; Patrick, 1970). Grazing by invertebrates, storm events, and dessciation, for example have been recognized in the intermediate disturbance hypothesis as mechanisms that maintain diversity if they act without species specificity to remove the best competitor for the limiting factor, space, light or nutrients (Connell, 1978; Kaufman, 1982; Fisher, 1987; Hoagland *et al.*, 1986). Grazing selectively removes the tall, loosely attached, filamentous, stalked forms allowing the substrate, adhering types to survive. Grazing acts to reduce biomass, increase productivity per unit biomass in the survivors, and increase or decrease the species number surviving depending upon type and severity of grazing and depth of the algal canopy (Hunter, 1980; Kehde & Wilhm, 1972; Lamberti & Resh, 1983; Mulholland *et al.*, 1983).

The Kuparuk River, has been fertilized with phosphorus every summer since 1983 (Peterson *et al.*, 1985). The Kuparuk River is a 4rd order arctic river draining the foothills of the Brooks Range (68° 33′ N. Lat.) (Peterson *et al.*, 1986). According to McCart *et al.* (1972), foothills rivers are intermediate in their stability, not subject of the abrasion of glacial-fed rivers, nor as stable as spring-fed streams which flow all year round. The Kuparuk River is frozen solid all winter (mid October to mid-late May), flushed violently by spring snow melt in mid-May and frequently by rainfall runoff spates in late July or early August. Discharge varies between <1 and $25\ m^3\ sec^{-1}$, averaging $1–4\ m^3\ sec^{-1}$ during experimental period of the enrichment studies. The river autotrophs are phosphorus-limited (Peterson *et al.*, 1983, 1985; Lock *et al.*, 1989; Gibeau & Miller, 1989). The foot-hills section of the river has 4800–5000 adult grayling, *Thyallmus arcticus*, which overwinter in a headwater lake (Peterson *et al.*, 1985). The major aquatic insects include black flies (Simulidae), a swimming mayfly, *Baetis* sp. (Ephemeroptera), a net-spinning chironomid, *Orthocladius rivulorum*, and numerous small chironomids (mostly subfamily orthocladinaea) that build tubes or refugia on the surface of rocks (Hershey *et al.*, 1988 and Hershey, personal communication). These grazers can have a significant impact on biomass of attached algae (Gibeau, 1991; Gibeau, Miller, and Hershey, in prep.).

Methods

During year 1 and 2 (1983 and 1984), phosphoric acid was delivered by a Mariot bottle to the river to increase the soluble phosphate concentration by $10\ \mu g\ P\,l^{-1}$ over ambient (Constant Concentration Addition). The delivery rate was adjusted daily for changes in discharge. From 3rd year (1985) to present the Kuparuk River was fertilized continuously with phosphorus (50% phosphoric acid) from a solar-powered, peristaltic pump set to achieve a final concentration of $10\ \mu g\ P\,l^{-1}$ at $2\ m^3\ sec^{-1}$ discharge (Constant Addition). In year 4 (1986) PO_4 and NH_4 were added at a constant rate at a second downriver station (+1.1 km below first P-dripper at $10\ \mu g$ P or $0.3\ \mu M$ P and $3\ \mu M$ N as NH_4Cl fertilizer) in order test for secondary N-limitation induced by the P-addition (Peterson *et al.*, 1985). Although the location of the dripper changed between 1983–84 and subsequent years, all distances in text and figures have been adjusted so that 0 km is the site of the P-dripper in all years and 1.1 km is the site of the second N&P dripper in 1986). This was justified for examining the algal response since it should have little memory compared to other trophic levels.

In year 1 ceramic squares ($25\ cm^2$), incubated

near the surface of river on floating plexiglass plates (July 1–Aug. 15), were harvested for diatom counts. Over the same dates in 1983, a set of floating plexiglass bioassay streams were set up by Peterson at the control station (−20 m above dripper) with duplicate treatments receiving P, N + P and controls in concentrations used in the river (Peterson *et al.*, 1983; Pringle, 1990). Diatom samples were enumerated from July 17, 1983, June 29 and July 13, 1984, Aug. 8, 1985 and Aug. 12, 1986. Between 1984–1987, the epilithon of the boulder and cobble-covered riffles were sampled by scrubbing enough flat rocks, 10–30 cm in longest dimension, to cover the bottom of a 754 cm^2 polyethylene tub. The tops, sides and bottom of the rocks were scrubbed with a nylon brush in a little stream water and made up to 1000 ml. Aliquots of this homogenate were taken for chlorophyll determination and diatom counts.

The diatom mixture, sampled for species identification was preserved in Lugol's Iodine (Edmundson, 1959). The diatoms were cleaned by boiling in concentrated nitric acid for 40 minutes in a 1 liter beaker. The cleaned diatoms were rinsed with distilled water and settled many times until the rinse water was neutral. The rinses were aspirated from 1 liter to about 100 ml between rinses. Then aliquots of the cleaned diatoms were dried on #1 coverslips, and mounted in Hyrax® mounting media. All diatoms on a slide were identified first under oil at 1000×, then strips were counted to determine relative abundance (Patrick & Reimer, 1966). The species abundance pattern was compared using diversity measures based on information theory ($\log_2$ transform) using the Shannon diversity and evenness indices (Poole, 1974).

The diatom community structure changes were analysed by detrended correspondence analysis, DCA, using a fortran program DECORANA (Gaugh, 1982; Hill, 1979; Leland & Carter, 1986) and spreadsheet programs. DCA is an iterative eigenvector ordination method based on reciprocal averaging which minimizes the influence of the first ordinate on the second ordinate. The species counts were relativized to 100 to minimize the differences between density and numbers counted. To assess the importance of rare species compared to common ones, the communities were compared by retaining all species and then eliminating those that occurred in less than 3 of the 5 samples. The percentage similarity (PS) was calculated to quantify differences in species and abundance patterns in pairs of communities (Poole, 1974).

Another aliquot (10 or 20 ml) was filtered through a Gelman A/E glass fiber filter for chlorophyll determination. After standing for 24 or 48 hours in cool darkness, the chlorophyll was measured using a Turner 111 fluorometer using a Blue T-5 bulb, a Wratten #5–60 emission filter and #64 secondary filter (Wetzel & Likens, 1979). The fluorometer was calibrated by the known specific absorbance of chlorophyll *a* (from Spinach) in 90% acetone using a Hitachi double beam spectrophotometer. The nutrients were determined weekly in fresh, filtered water using semi-automated wet chemistry methods based on the methods of Wetzel and Likens (1979): SRP using a phosphomolybdate method with ascrobate reduction (885 nm), NH_4^+ using a phenol-nitroprusside method (630 nm), and NO_3^- using Cu-chelated, Cadmium column to reduce NO_3^- to N O_2^- followed by colorimetric nitrite determination (sulfanilamide-ethylene diamine dihydrochloride) (530 nm). Conductivity, pH, alkalinity were measured routinely (using YSI Model 33 Conductivity meter, Corning pH meter and for alkalinity, a Hach cartridge titration with 0.16 N H_2SO_4 to pH 4.8 in soft waters, respectively).

Results and discussion

Whole river fertilization: species/generic changes

The addition of SRP at a nominal addition of 10 $\mu g\ P\ l^{-1}$ resulted in measured seasonal mean concentrations of SRP of 17.3, 7.6, 22.9, 10.5 $\mu g\ P\ l^{-1}$ in year 1 (1983), yr 2 (1984), yr 3 (1985), and yr 4 (1986), respectively. The highest average summer concentration below the dripper occurred in 1985, a year of low discharge (Table 1). The epilithon of the river responded within two weeks

Table 1. Variation between years of physical, chemical and biological parameters from Kuparuk River, 1983–1987.

Variation between years, Kuparuk River physical, chemical, biological differences					
parameter/yr	1983	1984	1985	1986	1987
River Discharge $m^3 sec^{-1}$	1.16	4.04	1.10	3.49	3.82
Total P added day^{-1} g P day^{-1}	1.05	2.77	1.73	3.45	1.70
Cal.added P μg P l^1	12.6	7.6	18.2	20.0	11.6
Srp found μg P l	17.3	7.6	22.9	10.5	35.9
P disappearance mg P m^2	34.6	60.5	6.3	12.4	130.5
Avg.sestonic chl. mg Chl m^2 – yr	n.a.	12.1	0.4	8.7	4.2

of beginning addition with striking increases in extracted chlorophyll in 1983 and 1984. However, there was little response to added P or added P and N + P in 1985 and 1986, respectively (Fig. 1, Table 1). Generally the algal biomass increased down river with the maximum moving down river from < 1 km in years 1 and 2 to > 2 km in year 3.

In four years 187 species of diatoms were enumerated in the Kuparuk River. Summed over four years, the most common species were *Cymbella minuta* (21% of total numbers), *Achnanthes minutissima* (16.2%), *Hannaea arcus* (13.5%) and *Cymbella minuta* (21%), and *Achnanthes linearis* (14.6%). *Diatoma tenue* var *elongatum* was present > 5% of total counts. Only another 14 species averaged greater than 1% of the total over the 4 years. An additional 135 species were present in less than 1% of the total count and their presence/absence could vary by year (Table 2). A frequency-abundance (octave) curve of the cumulative counts for 4 years × 4–8 stations is apparently lognormal as would be predicted, skewed towards the rare species (McNaughton & Wolf, 1973). The dominants varied by year, for example *Hannea arcus* (39%) and *Cymbella minuta* (27.4%) were most common in 1983, while *Tabellaria floculosa* (12%) and *Diatoma tenue* (10%) rose to maxima in 1984. By 1985 *Achnanthes minutissima* peaked (37.8%); and in 1986 *Cymbella minuta* returned (22.8%) with *Achnanthes linearis* (22.1%) as co-dominant common species summed for all river stations (Table 2).

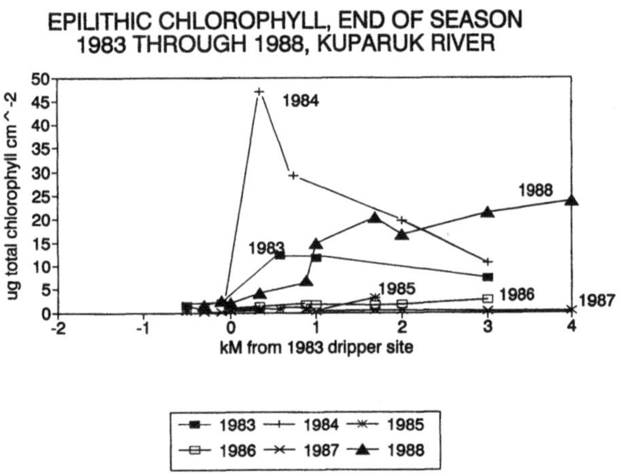

Fig. 1. Biomass changes in epilithon measured as extracted chlorophyll *a* ($\mu g\ cm^{-2}$) by river station above and below a PO_4 dripper 1983–1986.

When the species data were summed by generic abundance, the pattern became clearer, as 12 genera comprised > 95% of most samples. Over the 4 years, there were more species and/or varieties of *Eunotia* (25), *Cymbella* (23), *Navicula* (15), *Pinnularia* (15), *Fragillaria* (11) and *Achnanthes* (10). In terms of % abundance *Achnanthes* (28%

Table 2 Percentage species abundances by year (all stations) and by kilometer along fertilization gradient (all years)

Species name	Percent counts by year					Percent counts by station 83–86 Distance from dripper (Km)			
day of fertilization	15	0	13	39	43	− 0.1	0.1	1.1	2.1
year	1983	1984	1984	1985	1986	0.0	1.0	2.0	7.0
day	16jl	29jn	13jl	8ag	12ag				
Achnanthes conspicua		0.50	0.88	0.82	0.25	0.73	0.34	0.24	0.56
Achnanthes depressa		0.15	0.02	0.13	0.50	0.13	0.08	0.12	0.06
Achnanthes exigua				0.04		0.04			
Achnanthes flexella	0.06	0.30	0.27	0.35	1.25	0.62	0.13	0.24	0.11
Achnanthes linearis	0.20	8.79	8.52	8.93	22.1	10.20	4.80	6.65	5.56
Achnanthes linearis fo. curta	0.61					0.33		0.12	0.56
Achnanthes lanceolata	0.03								0.05
Achnanthes minutissima	4.10	24.6	13.8	37.8	8.02	36.41	11.46	12.35	17.43
Achnanthes sublavis va. crana			0.01			0.00			
Achnanthes sp.1	0.41	0.05				0.37			0.30
Amphora ovalis		0.06	0.03	0.04		0.05		0.01	0.06
Anoeomoeneis sp1		0.15	0.18	0.09		0.12	0.04	0.01	0.16
Caloneis sp.1				0.01		0.00			
Caloneis ventricosa var. alpina		0.01	0.01	0.22		0.04	0.05	0.02	0.15
Capatograma crucifera				0.04			0.04		
Cocconeis disculus				0.04				0.05	
Cocconeis placentula				0.13		0.08		0.06	
Cocconeis placentula v. lineata					0.25	0.04			
Cyclotella atimus				0.08		0.04			
Cyclotella bodanica		0.10	0.03	0.13		0.09	0.05	0.07	0.05
Cyclotella comta				0.04	0.25	0.04	0.04	0.01	
cyclotella glomerata				0.17				0.24	
Cyclotella kutzingiana	0.06	0.11	0.34	0.26		0.38	0.13	0.07	0.06
Cyclotella meneghiniana				0.08				0.04	
Cyclotella stelligera			0.01	0.13		0.00		0.12	0.05
Cyclotella sp.				0.01		0.00			
Cymbella bipartida			0.02			0.00		0.01	0.05
Cymbella breweria				0.08				0.04	
Cymbella arctica									0.01
Cymbella astula								0.00	
Cymbella botellus		0.34	0.24	0.43		0.45	0.25	0.12	0.05
Cymbella breweriana		0.10	0.02	0.26		0.08	0.17	0.01	0.11
Cymbella caespitosa				0.02		0.01			
Cymbella cistula	0.03	0.30				0.24	0.05	0.01	
Cymbella cuspidata		0.05	0.02	0.09		0.09	0.04	0.01	
Cymbella cymbiformis		0.30	0.22	0.17		0.13	0.13	0.24	0.12
Cymbella Ehrenbergii						0.09	0.04	0.01	0.01
Cymbella helvetica				0.04		0.04			
Cymbella microcephala				0.01		0.00		0.01	
Cymbella minuta	27.4	1.4	0.9	1.2	22.8	16.76	7.00	22.92	6.98
Cymbella minuta var. pseudogracilis	0.26	0.20	0.01	0.09	1.00	0.16	0.17	0.36	0.25
Cymbella muelleri		0.31	0.11	0.17		0.05	0.13	0.30	0.11
Cymbella naviculiformis			0.02			0.00	0.00	0.01	0.01
Cymbella parva	0.32		0.08			0.08	0.13	0.24	0.15
Cymbella proxima		0.30	0.05	0.35		0.26	0.17	0.13	0.11
Cymbella sinuata	0.26	0.26	0.73	2.25	3.01	1.47	0.85	0.85	0.86

Table 2 (Continued)

Species name	Percent counts by year					Percent counts by station 83–86 Distance from dripper (Km)			
day of fertilization	15	0	13	39	43	−0.1	0.1	1.1	2.1
year	1983	1984	1984	1985	1986	0.0	1.0	2.0	7.0
day	16jl	29jn	13jl	8ag	12ag				
Cymbella ventricosa		13.0	19.3	9.84		7.38	9.27	10.98	7.32
Cymbella vistula			0.08			0.04			
Cymbella sp.3			0.01				0.04	0.07	0.01
Cymbella sp.5			0.48					0.01	0.30
Denticula elegans		0.01	0.04	0.09		0.01	0.01	0.07	0.06
Diatoma anceps	1.89	0.40	1.85	0.52	1.25	2.04	0.80	1.49	0.96
Diatoma hiemale	0.03	0.12	0.26	0.30		0.26	0.09	0.07	0.20
Diatoma tenue var. elongatum	3.87	8.15	11.0	5.90	0.75	8.08	6.87	6.00	5.71
Diatoma vulgare				0.04				0.06	
Didymosphenia geminata	0.26	1.04	0.51	0.61	0.25	0.62	0.72	0.84	0.26
Diploneis all.							0.04		
Diploneis elliptica						0.00			
Diploneis ovalis				0.13		0.08	0.04	0.01	
Eunotia arcus			0.09	0.09		0.121	0.08	0.01	
Eunotia bidens	0.06	0.15		0.09		0.12	0.04	0.12	0.05
Eunotia bigibba			0.02			0.05			
Eunotia bidentula		0.06	0.01	0.09		0.04		0.13	
Eunotia curvata	0.03	0.16	0.01	0.22		0.04	0.17	0.12	0.11
Eunotia diodon		0.07	0.73	0.78	3.01	0.82	0.30	0.37	0.36
Eunotia flexuosa		0.01	0.02	0.17		0.08	0.05	0.01	0.05
Eunotia incisa	0.12	0.06	0.26	0.74	3.51	0.57	0.47	0.37	0.41
Eunotia maior			0.01			0.00	0.01	0.06	
Eunotia lunaris			0.01	0.04		0.00			0.05
Eunotia monodon		0.22	0.10	0.35	5.01	0.70	0.38	0.13	0.26
Eunotia paralela		0.05	0.02	0.13		0.09	0.08	0.01	0.01
Eunotia pectinalis				0.13	0.25	0.12			0.05
Eunotia pectinalis var. bidens				0.22		0.04	0.08	0.06	0.05
Eunotia pectinalis var. minor	0.03		0.01	0.04		0.04			0.05
Eunotia polydentula				0.04		0.04			
Eunotia praerupta		0.17	0;27	0.43		0.33	0.18	0.07	0.16
Eunotia praerupta var. bidens			0.02	0.22	0.50	0.17	0.04		0.10
Eunotia praerupta var. inflata		0.05	0.01	0.04		0.04		0.07	
Eunotia rostellata			0.01			0.00			
Eunotia spetentrionalis		0.25		0.13		0.12	0.13	0.06	0.05
Eunotia sudetica		0.05				0.04			
Eunotie triodon		0.07	0.03	0.22		0.18	0.09	0.02	0.01
Eunotia valida		0.06		0.04		0.04			0.06
Eunotia vanheurckii	0.06		0.05	0.95	6.52	0.38	0.85	0.90	0.31
Fragilaria brevistriata	0.06					0.08			
Fragilaria capucina var. lanceo.	2.64					2.12	0.59	0.89	0.51
Fragilaria constricta		0.01	0.01	0.04			0.05	0.01	0.01
Fragilaria construens	0.06	0.10	0.01				0.04	0.01	0.15
Fragilaria construens var. venter	0.20	0.12	0.10	0.13	0.50	0.33	0.17	0.14	0.06
Fragilaria intermedia	1.31	0.84	3.53	1.43	0;25	2.82	0.80	1.37	1.46
Fragilarie pinnata					1.00	0.08	0.08		
Fragilaria sp.1			0.41			0.21	0.01	0.07	0.05

Table 2 (Continued)

Species name	Percent counts by year					Percent counts by station 83–86 Distance from dripper (Km)			
day of fertilization	15	0	13	39	43	− 0.1	0.1	1.1	2.1
year	1983	1984	1984	1985	1986	0.0	1.0	2.0	7.0
day	16jl	29jn	13jl	8ag	12ag				
Fragilaria spA	0.03	11.2		0.09		0.04			0.10
Fragilaria virescens	0.03	0.31				0.20	0.01	0.01	0.10
Fragilaria vaucheriae		0.05			0.75	0.04		0.12	0.05
Frustulia rhomboides			0.02	0.04		0.01	0.04	0.01	
Frustulia rhomboides var; crassinervia			0.01				0.00		
Frustulia rhomboides var. saxonica			0.01	0.09		0.08		0.01	
Gomphonema angustatum	2.06	0.26	0.59	0.87	5.26	3.47	0.72	0.37	0.81
Gomphonema augur				0.04				0.06	
Gomphonema gracilis		0.15		0.26		0.16	0.08	0.12	0.05
Gomphonema mustella	0.06	0.31	0.12	0.48	0.25	0.42	0;18	0.14	0.26
Gomphonema parvulum			0.01			0.00			
Gomphonema sp1								0.01	
Gomphonema subclavatum			0.01	0.04				0.01	0.05
Gomphonema truncatum var. capitatum		0.06	0.02			0.00	0.00	0.02	0.05
Hannaea arcus	39.6	6.0	13.7	6.37	5.26	24.88	14.83	25.11	22.13
Hannaea arcus fo. recta		0.01	0.27	0.22		0.13	0.09	0.07	0.11
Hannaea arcus var. amphioxys		0.06	0.51	0.43	0.75	0.37	0.17	0.14	0.26
Hantzschia amphioxys		0.05	0.01	0.09		0.04	0.08		0.01
Melosira distans		0.26		0.30		0.13	0.13	0.24	0.11
Melosira granulata	0.12	0.15	0.06	0.61		0.42	0.09	0.30	0.21
Melosira islandica								0.01	
Meridion circulare	0.09	0.10	0.11	0.48	0.25	0.21	0.21	0.18	0.26
Meridion circulare var. construens				0.04		0.04		0.01	
Melosira islandica							0.00	0.01	
Melosira italica			0.01			0.00			
Navicula cryptocephala			0.10			0.04	0.00		0.01
Navicula cryptocephala	0.15	0.17	0.33	0.43	0.50	0.45	0.34	0.20	0.11
Navicula exigua			0.01			0.00			
Navicula lanceolata						0.00	0.00		
Navicula minima	0.06			0.04		0.08		0.06	
Navicula minuta						0.08			
Navicula minutissima	0.09			0.13		0.04	0.08	0.12	0.05
Navicula mutica	0.03		0.02	0.04		0.01	0.04	0.06	
Navicula neoventricosa			0.01			0.00			
Navicula pseudoscutiformis			0.02	0.04		0.00	0.04	0.01	
Navicula pupula v. pupula		0.05	0.01			0.00		0.06	
Navicula pupula var. rectangularis			0.01	0.04	0.25	0.04	0.04	0.01	
Navicula radiosa		0.01	0.02	0.35		0.09	0.08	0.07	0.16
Navicula sp1	0.17	0.01	0.01			0.13	0.00		0.16
Navicula sp2			0.03	0.74		0.29	0.13	0.18	0.21
Neidium bissulcatum		0.02	0.05	0.26		0.05	0.10	0.08	0.11
Neidium affini avar. undulatum			0.03			0.01		0.01	0.01
Neidium dubium			0.02			0.01			
Neidium iridis			0.02			0.00	0.00	0.01	0.01

Table 2 (Continued)

Species name	Percent counts by year					Percent counts by station 83–86 Distance from dripper (Km)			
day of fertilization	15	0	13	39	43	– 0.1	0.1	1.1	2.1
year	1983	1984	1984	1985	1986	0.0	1.0	2.0	7.0
day	16jl	29jn	13jl	8ag	12ag				
Nitzschia amphibia	0.09	0.70	2.82	1.78	0.50	1.35	0.68	0.78	1.67
Nitzschia comunis				0.04				0.06	
Nitzschia fonticola		0.54	3.13	1.30	2.76	1.22	0.55	2.08	0.66
Nitzschia frustulum							0.00		
Nitzschia gracilis	0.03							0.06	
Nitzschia linearis			0.01					0.01	
Nitzschia palea	0.26	0;44	1.53	0.26	0.25	0.41	0.63	0.71	0.35
Nitzschia romana	1.08	0.64	0.89	0.17		0.82	0.55	1.13	0.66
Nitzschia thermalis									0.01
Nizschia sp.A							0.00		
Nitzschia sp.1			0.01			0.01	0.04	0.01	
Nitzschia sp3	0.09	0.10		0.04		0.12	0.13	0.01	
Nitzchia sublinearis		0.30	0.08	0.04		0.04	0.05	0.30	0.05
Opephora montgii			0.02			0.00			0.01
Pinnularia abaujensis v. linearis			0.01			0.00			
Pinnularia abaujensis v. subundulata		0.01					0.00		0.01
Pinnularie borealis		0.01	0.01	0.04		0.01	0.05	0.01	
Pinnularia brevicostata			0.02	0.22		0.09	0.05		0.10
Pinnularia divergens			0.01					0.01	
Pinnularia gibba var. microcyla			0.01			0.00			0.05
Pinnularia lata			0.01				0.00		
Pinnularia maior			0.01			0.00			
Pinnularia mesolepta			0.03	0.13		0.04	0.00	0.01	0.11
Pinnularia microstauron			0.02	0.04		0.00	0.05	0.01	
Pinnularia molaris = Pinnularia torta				0.09		0.04			0.05
Pinnularia subcapitata			0.02				0.00		0.01
Pinnularia viridis		0.01	0.02	0.09		0.04	0.00	0.01	0.06
Pinnularia sp.3			0.02			0.01			
Pinnularia sp.2							0.00		
Pinnularia sp.1							0.01	0.01	
Stauroneis anceps fo. linearis			0.01				0.00		
Stauroneis anceps fo. gracilis		0.01	0.01	0.13		0.00	0.05		0.10
Stauroneis obtusa								0.01	
Stauroneis phoenicenteron	0.03	0.01	0.02	0.04		0.04	0.01		0.06
Stauroneis phoenicenteron fo. gracilis			0.03	0.09		0.05	0.04	0.01	0.01
Stephanodiscus sp1		0.05	0.08	0.09		0.08	0.04	0.06	
Suriella laponica			0.02				0.01	0.01	
Surirella linearis		0.01	0.02	0.09		0.01	0.05	0.07	0.01
Surirella ovata			0.01	0.04		0.00			0.05
Synedra acus	0.49					0.08			0.76
Synedra amphicepahala	0.61					0.45		0.18	0.35
Synedra cyclopum				0.04		0.04			
Synedra delicatissima	0.03			0.04		0.04		0.06	
Synedra minuscula		0.01				0.00	0.00		0.01
Synedra radians	0.03								0.05
synedra rumpens	0.03	0.35	0.49	0.09		0.17	0.29	0.12	0.16

Table 2 (Continued)

Species name	Percent counts by year					Percent counts by station 83–86 Distance from dripper (Km)			
day of fertilization	15	0	13	39	43	− 0.1	0.1	1.1	2.1
year	1983	1984	1984	1985	1986	0.0	1.0	2.0	7.0
day	16jl	29jn	13jl	8ag	12ag				
Synedra ulna	4.97	2.26	2.89	1.65	0.50	1.75	1.60	3.09	8.08
Synedra vaucheriae	2.18	0.21	1.37	0.87		2.57	0.76	1.25	0.71
Tebellaria fenestrata	0.32	0.50	0.58	0.43	0.25	0.45	0.26	0.43	0.76
Tabellaria floculosa	2.94	11.0	3.63	0.87	0.25	3.06	2.28	7.84	6.52
Tetracyclus lacustris			0.01	0.13		0.05	0.04		0.05
total count =	100	99.9	100	100	100	2453	2373	1684	1980
total species richness =	54	97	129	112	37				
avg.ind./species =	1.9	1.0	0.8	0.9	2.7				
avg.% ind./species =	1.85	1.03	0.78	0.89	2.70				

of total N) was more common than *Cymbella* (24%) > *Hannea* (21%) and *Diatoma* (8%) (Fig. 2). Averaged over the 4 years *Achnanthes* was the dominant in control reaches; *Cymbella*, *Hannea* and *Synedra* responded in fertilized reaches (Fig. 2). The down-river changes in % abundance by genus shows reduction of Achnanthes below the P dripper (Fig. 2) and an increase in *Cymbella* and *Hannea*. *Synedra* became more important far downriver (> 2 km). The inter-year variation was largely the replacement of *Hannea* in year 1 with *Achnanthes* in the following years. *Cymbella*'s were the second most abundant in all years. In the 4th year, *Eunotia*'s collectively became 3rd in % importance, replacing either *Diatoma* or *Hannea* (1984–1986) (Fig. 3).

The change in species whose percent abundance peaked in different years could have been caused by changes in that year's fertilization treatment; or by cumulative effects of fertilization during ice-free period of 4 years; by unexplained inter-annual variation; by changes in disturbance frequency and severity from high discharge and grazing pressure; or by combinations of these

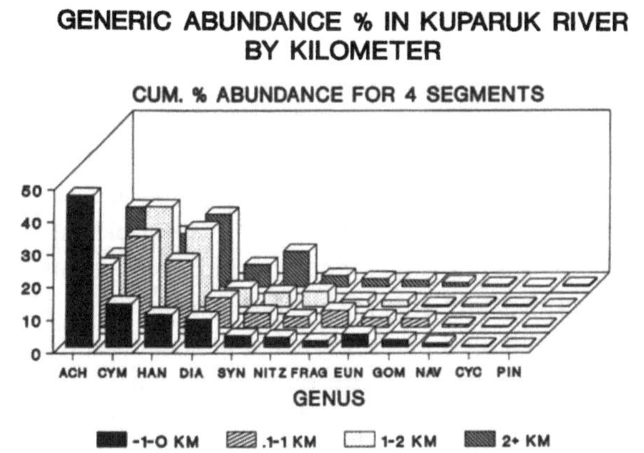

Fig. 2. Cumulative % abundance of major genera of diatoms by river kilometer above or below PO_4-dripper over 4 years (1983–1986). (Abbreviations are first three letters of the genera in order: Achnanthes, Cymbella, Hannea, Diatoma, Synedra, Nitzschia, Fragillaria, Eunotia, Gomphonema, Navicula, Cyclotella, and Pinnularia).

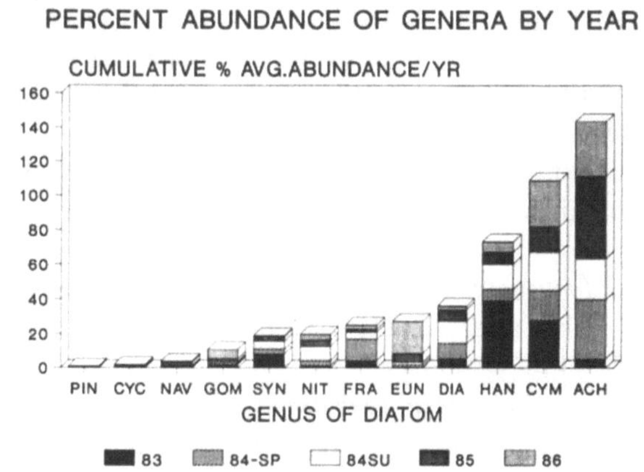

Fig. 3. Cumulative average abundance of diatom genera by year showing inter-year variation. Genus abbreviations as in Fig. 2.

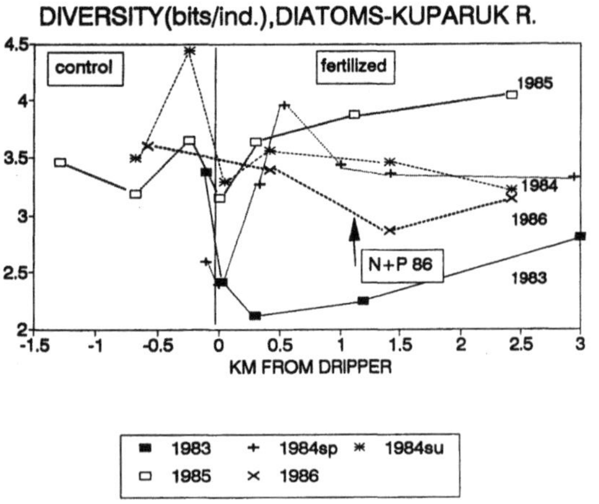

Fig. 4. Diversity (Shannon index, log 2) of diatom species enumerated from summer epilithic community in Kuparuk River, 1983–1986.

factors. The number of species enumerated per year decreased from year 2 (1984-summer) (128) > year 3 (1985) (111) > year 2 (1984) spring sampling (96) > year 1 (1983) (53). However there was no apparent pattern in the number of diatom species and the number of frustules enumerated per site (total count = 100, 250, 600–800 ind. sample varying by year), as both the highest and lowest total counts had the lowest species richness. For that reason, the sample species richness was not corrected to equivalent sized samples by any procedure, since it would have introduced more problems than it solved.

When the species were viewed downriver by kilometer units from upstream controls to below the phosphate dripper (−1–0 km, 0.01–1 km, 1.01–2 km, 2.01–6.5 km) over all years of by year (regardless of station) (Table 2), several patterns of response in the diatoms can be seen, both positive and negative. Moving downriver through the 1 km zones, the patterns of positive response were four:

a) fast sustained response – numerical increases at 0–1 km and 1–2 km compared to control and/or 2–6 km stations;

b) slow response, sustained – numerical response increasing slowly from control to 0–1, 1–2, and 2–6 km stations;

c) new colonists – arriving and growing at enriched stations only below dripper.

Of the dominant species *Cymbella minuta* and *Hannea arcus* were fast response species to the P-fertilization. Other 'fast' response diatoms included in decreasing order were: *Syndera vaucheriae*, *Nitzschia fonticola*, *Fragilaria capucina*, *Didymosphenia geminata* and *Nitzschia palea*. The dominant species with a 'slow' positive response were *Tabellaria floculosa*, *T. fenestrata* and *Syndera ulna*. No new colonists (5) were very common or consistent at all down-river stations (*A. linearis* being most common).

The negative response patterns to fertilization were

a) a fast negative response where percentage dropped at 0–1 or 1–2 km stations compared to controls or 2–6 km station (*A. conspicua*, *M. granulata*, *S. amphicephala*, *G. mustella*);

b) a slow negative response where the percentage drops at all enriched stations compared to control, usually decreasing progressively down river (*Achnanthes minutissima*, *Eunotia diodon*, *E. monodon*, *A. flexella*, *N. cryptocephala*);

c) local exclusions from the epilithon at 1–2 and/or 2–6 km stations. Two of the more common species showed no apparent response in mean% composition by river kilometer: *Achnanthes linearis* and *Diatoma tenue*.

The dominant species with the fast sustained response, *Hannaea arcus*, has a complex ecology in the Kuparuk River (Hershey *et al.*, 1988). It was clearly a colonizer of empty ceramic tiles in 1983 when its dominance was maximal (>39% of all individuals); however, this species grows as a monoculture in the silk tubes of the chironomid, *Orthocladius rivulorum* and can compose up to 12–43% of the algal biomass on the rocks at the maximal size of these chironomids before pupation in (mid-late July) (Hershey *et al.*, 1988). This chironomid feeds on the garden of *H. arcus*, where it grows better than the epilithon as a whole in response to P-addition to the river (Hershey *et al.*, 1988). Thus the proportional strength of *H. arcus* in the rock-scrubbed diatom counts is in part a function of the numerical success and mean biomass of the chironomid when the samples were taken. Thus the late August samples show this species under represented, since the *Orthocadius* have all emerged from the river (1985, 1986) (Hershey *et al.*, 1988). The mid-July samples (1983, 1984) show a higher percentage of this species because the number of 5th instar chironomids in tubes is highest then just before pupation.

Among the new colonists and the slowly-responding species, few have clear ecological preferences deduced from other studies from the temperate latitudes to help our interpretation of their response in the Kuparuk River. Those species that decreased most quickly at 0–1, 1–2 km were often characterized as being 'eutrophic' forms, *N. amphibia*, *T. fenestrata*, *M. granulata*. These could be losing in competition for space with the numerical dominant, *Cymbella minuta* or been more susceptible to grazing because of their larger size and position in the canopy. The many species (15) which had a 'slow negative' response are characterized by being acidophilic (US EPA 1974, Patrick & Reimer, 1966, 1975). The pH of the river is near neutrality; however, with fertilization and low flow there is a trend for pH to be elevated by hydrolysis of dissolved carbonate by algae (carbonic anhydrase). In the canopy of the epilithon the diel variation in pH increase may be more severe than found in the river water as a whole. Those species that disappeared from the counts were usually rare, as equilibrium (island) theory would predict (MacArthur & Wilson, 1967).

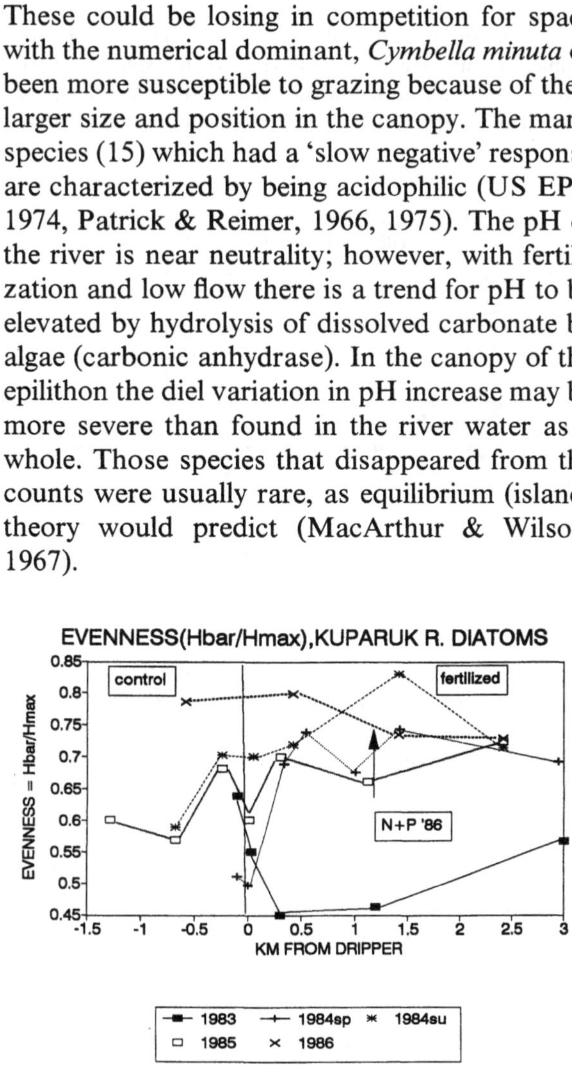

Fig. 5. Evenness index of diatom species abundances enumerated from summer epilithic community in Kuparuk River, 1983–1986.

Community changes: Biomass and diversity measures

The total amount of phosphorus added to the river varied by year, with nearly 3 × as much added in the second year (1984) when the discharge was 3 times higher than in year 1 (1983). The measured SRP concentration was lowest in that high flow year (Fig. 1, Table 1). High attached algal biomass developed during the first two years within 1 km of the dripper. Thereafter, the response of the algae to added P was reduced an order of magnitude (years 3, 4 and 5) and maximal biomass was found further down river from the dripper location. This decrease was not caused by a change in the limiting nutrient (Peterson *et al.*, 1985; Gibeau & Miller, 1989 & unpublished data). The probable reason for the decrease in epilithic algal biomass on riffle rocks in 1985 and 1986 was the delayed numeric response of several species of small (<4 mm) herbivorous chironomids (subfamily Orthocladiineae, DIPTERA). They reached densities of 30000 m^2 in 5th year (Gibeau, 1991; Hershey, unpublished; Gibeau, Miller & Hershey, in prep.). Many of these formed tubes or built refugia by accumulating the epilithic algae from the algal carpet around them. In the 4th season (1986) there was little detectable change in algal biomass in response to adding the second most probable 'limiting' nutrient, ammonia, at a second dripper at 1 km below the first P-dripper. By inference, we believe the grazers were responsible for this lack of whole river response to added P in years 3, 4, and 5. The density and effect of these grazers was significantly reduced in 1988 as the epilithic algal biomass reached 1983 levels (Fig. 1).

Thus, both the first and second years were those when P-enrichment increased attached chlorophyll significantly (Fig. 1). However, by 1985, no enhancement was found, average abundance did not appear to change down river during 1986. Fertilization should cause a simplification of the community as the concentration of limiting nutrients increases, if all other factors remain constant (Tilman, 1982). The species diversity using the information theory-based bits/individual is influenced by species richness and by the evenness of distribution of individuals among species in the community sampled. Normally a stress or change to the system causes decreases in the diversity index (Stevenson & Lowe, 1986). Diversity per individual, H_{bar}, decreased rapidly below the dripper in the first year of fertilization (at 0.34 km below the dripper) (Fig. 4). By spring of year 2, late summer of year 2 and summer of year 3, the only apparent down-river decrease in diversity occurred at the dripper site itself, 0 km station. At all upstream and downstream stations H_{bar} oscillated between 3–4.4 bits/individual.

Using the index of evenness (range 0–1), a downstream reduction was observed in the first year, but in the second and third years, the evenness increased downriver (Fig. 5). By the fourth year, no downstream trend was obvious, and the evenness was highest of all years (0.75–0.80). However, there was a small decrease at the + 1.4 km station just below the second dripper ($PO_4 + NH_4$), operational in that year. Thus only in the year of the beginning of a new whole-river enrichment was there any noticeable change in community structure based on changes in commonness.

The fertilization with H_3PO_4 caused increases in biomass of algae, which then could reduce the concentration of SRP to low levels. In the first year before the biomass had accumulated and in the forth year when grazers had reduced the biomass significantly, average SRP concentrations were the highest. In proportion to the biomass response caused by added phosphate, the river nitrate was reduced in concentration. Ammonia was always low, except in 1986 when ammonia was added 1.2 km below first dripper.

The diversity/individual and the maximal observed diversity [ln(species number)] decreased linearly ($p<0.01$ for both regressions) with increasing SRP concentrations (Fig. 6) and with increasing $NO2^{-1}+NO3^{-1}$ concentrations (Fig. 7) found in the river water on the day of sampling. This is the expectation in the Tilman model (1982) of highest diversity and highest species richness at the oligotrophic end of the trophic spectrum where many species just make it and

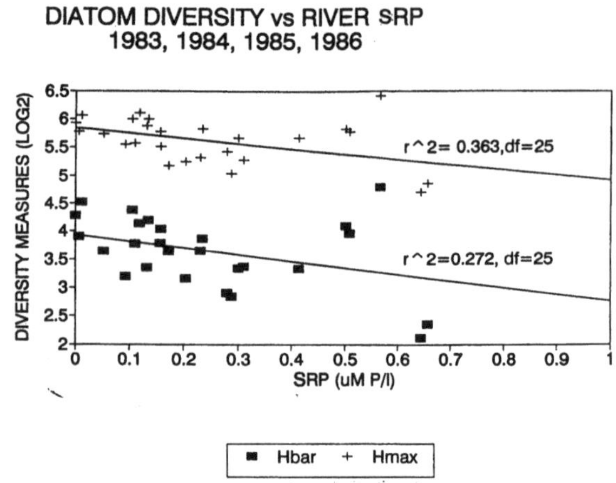

Fig. 6. Diatom diversity indices, H_{max} and H_{bar}, of epilithic samples as function of soluble reactive phosphorus, SRP, ($\mu M\ l^{-1}$) on data of diatom sampling with least squares regression and coefficient of determination for each distribution.

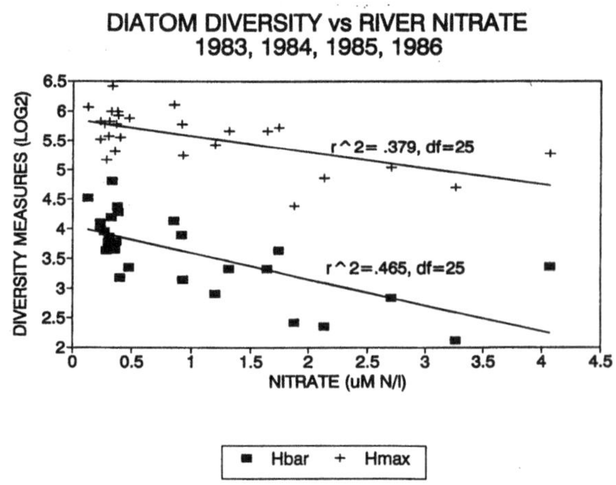

Fig. 7. Diatom diversity indices, H_{max} and H_{bar}, of epilithic samples as function of soluble (nitrate + nitrite) ($\mu M\ l^{-1}$) on date of diatom sampling with least squares regression and coefficient of determination for each distribution.

none are competitively excluded. This assumes that the instantaneous nutrient concentration is characteristic of average growing conditions of the algae, as in a steady state system. This assumption is likely true during the low flow events usually in mid-July to early-August. The low SRP and nitrate are not independent, since with the addition of phosphorus, growth of attached algal biomass, and stoichometric decrease of inorganic nitrogen are functionally linked. The highest number of species of diatoms were found in year of the highest biomass 1984, sampled prior (July 17) to the peak biomass (mid-August).

The addition of low concentrations of PO_4 changed the inorganic N:P ratio in the river making it potentially N-limited to further growth. When the river was potentially P-limited to further growth (molar N:P ratio > 10), there were relatively few diatom species (20–40/sample) and when the river was be potentially N-limited (molar N:P ratio < 10) to further growth, there are many species (20–120). The system is N-limited only when P is being added to the river.

On the other hand when substrata are limited, one species may gain a competitive edge especially with enrichment of the supply ratio of N:P or P:Si (Tilman, 1982). Such dominance may decrease diversity and evenness index values. However, a non selective disturbance, like herbivory, consuming all species in proportion to their abundance may tend to enhance species richness and evenness (Lubchenko, 1978). Samples from year 4 (1986) had the highest evenness and from year 3 (1985) the highest diversity/individual (Fig. 4 & 5). The number of species comprising more abundant than 1% of the total counts is a measure of evenness. Using that index, the community in year 4 had the greatest equality (1986 15 spp. of $> 1\%$ rel. abundance) > 1984 (13 spp.) > 1983 (12 spp.) > 1985 (10 spp.). This may be some indirect evidence from the diatom community measures, that a nonselective grazer may have been having an effect on composition by increasing evenness of distribution of individuals among species.

However, if grazing were to open new surfaces to allow colonization of new immigrant species, not previously found, that was not confirmed by the data. From mid summer of year 1 (1983); immigrants in yr 2 were 87.6% of the resident species; from spring of year 2 to summer of the same year, 26.3% were immigrants; from summer of yr 2 to the summer of yr 4 (1984–1985), 24.1% were immigrants; and from summer year 3 to summer year 4, only 8.1% were immigrants. Severe grazing may be more selective allowing only selected immigrants species to survive or the number of new species to immigrate were swamped by common species colonizing the epilithon in the enriched sections of river.

Bioassay tube analysis

In flow-through bioassay tubes used to detect nutrient limitation (P, N + P) (Peterson *et al.*, 1983) measured colonization on glass slides as chlorophyll and we measured diatom species composition on day 16 (July 16) and day 25 (July 25) (Fig. 7). Of the 84 species found searching the slides from the bioassay tubes; *Achnanthes linearis* and *Cymbella minuta* dominated, while *Hannea arcus* dominated the samples from the river. The change in species list present (Community Coefficient) between day 14 and 23 was comparable in all treatments (43.8% in controls, 35.5% in P-enriched and 42.4% in N + P-enriched). The percentage similarity (PS) using both species presence and abundance data, was 77.8 in N + P treatment between days, 56.5 in P-treatment and 26.9 in controls, respectively. In the bioassay tubes, the same species were dominant in all treatments. Using a detrended correspondence analysis of treatments (Control, + P, + N + P) with day (day 16 and 25), support this finding; ordinate distances between communities on day 16 and 25 were greater than the differences caused by treatments (Fig. 9). A change of 100 units in ordinal space in DECORANA is 1 standard deviation of change in community composition (Gaugh, 1982). The greatest change in communities in species space occurred between day 16 and day 25 in the control tubes > P-enriched > N + P-enriched. Enrichment reduced the distance between communities in successional time in these oligotrophic waters.

The diatom community on glass slides in bioassay tubes varied greatly from that in the river colonizing ceramic tiles using a DECORANA. The river and bioassay tube communities varied because of differences in rare species that may co-vary with substrata as it may affect immigration probabilities of rare species (Tuchman & Stevenson, 1980; Stevenson, 1983, 1984).

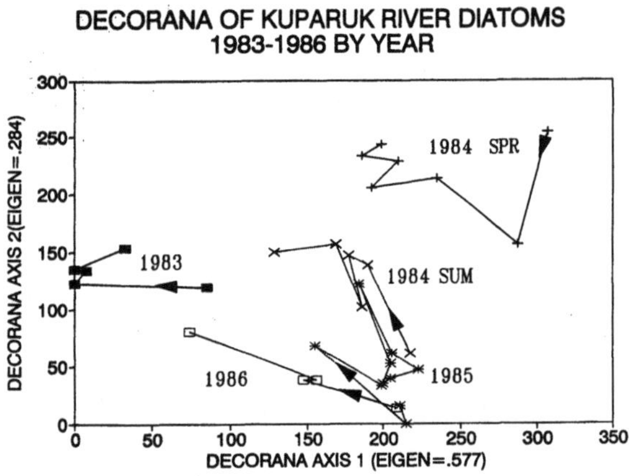

Fig. 8. Detrended Correspondence Analysis (DECORANA) of epilithic communities in diatom species space separated by year and by location along transect from control stations downriver (direction of arrow). All species were included and sample counts were normalized to 100% before analysis. Eigenvalues are shown for first and second ordinate axes.

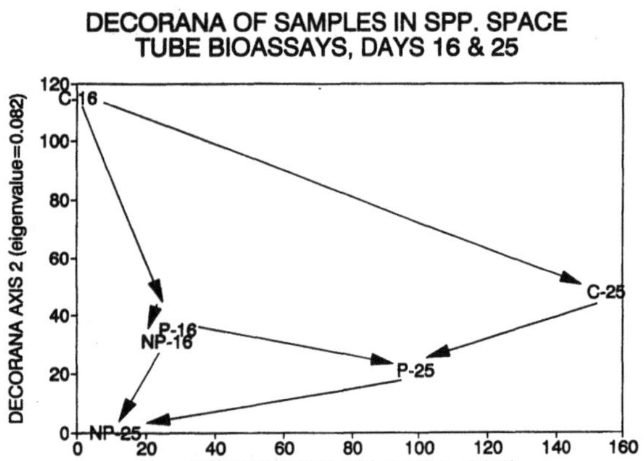

Fig. 9. DECORANA of floating tube bioassay diatom communities in species space on the first and second ordinate. Arrows connect day 16 and day 25 samples by treatment (C, P & NP) and arrows connect treatment series (C, P & NP) on day 16 and day 25 in order to emphasize the relative change in community with treatment and successional time.

Discussion

Whole river enrichment studies with a single nutrient are few and often done as bioassay systems and in artificial streams. In the Kuparuk River in 1984 the maximal autotrophic production after 21 days on artificial substrata under constant light, was 3 × greater than control stations reaching 18 μg C cm^2 hr^{-1}. In 1985, when we propose, grazer activity began to limit biomass accumulation, there was no difference upstream-downstream in autotrophic production (Hullar & Vestal, 1989). In 1985, when the biomass was 10 × lower, and SRP concentrations higher than 1984, the epilithic microbiota were not P-limited since the alkaline phosphatase activity declined by $^2/_3$ at all fertilized stations. Middle order streams in temperate latitudes are frequently light limited by a closed canopy. This could not occur in the arctic. Temperate stream flora often experience nutrient enrichment with deforestation or terrestrial habitat disturbance. Nutrient enhancement of productivity is difficult to establish (Lowe *et al.*, 1986; Webster *et al.*, 1985). Nutrient addition to the Kuparuk River should have represented a bioassay of limited nutrient since the arctic summer has much sunshine and 24 hours of light. And yet the enhancement of epilithic biomass declined after 2 years of fertilization.

We propose the transformation from a nutrient-limited, bottom-up control, to a herbivore, top-down control by grazers in the absence of an effective insectivorous fish. In 30 streams in S. Ontario, Bowlby & Roff (1986) predicted that with a nonpiscivorous fish and three trophic levels, microcommunity biomass should become larger and that with a piscivorous fish as a 4th trophic level, the microcommunity biomass should become smaller. In fact, there was no effect of ± piscivores and microcommunity biomass. Often enrichment is confounded by exposure to other toxicants which are correlated with measured decreases in diversity (De Seve & Goldstein, 1981).

The species of diatoms change with the resetting of disturbance cycles in Kentucky Streams, with *Achnanthes* being a good competitor, compared to dominants in the mature community – *Cymbella* and *Gomphnema* spp. (Stevenson, 1990). Colonists were more common in the mature rather than the disturbed community as canopy development and niches are enhanced and probability of drifting increases. Our disturbance was enrichment in which *Achnanthes* was proportionately at a disadvantage in the enriched sites (Fig. 2). Growth dominates over immigration as a source of cells following disturbance (Stevenson, 1990). We found higher immigration rates in early years of fertilization when biomass was high.

Some of the differences between years and species selection was undoubtedly caused by differences in current and substrata. In fast current, diatoms tend to dominate on rough surfaces compared to slow currents. *Achnanthes* species like faster currents, while green algae like *Ulothrix* like slow currents (Poff *et al.*, 1990). In our river, *Achnanthes* did best in 1984 and 1986, our highest flow years and in subsequent low flow years filamentous greens became important.

Intense grazing effects by snail under control conditions or with P-enrichment in stream bioassays caused increases in adherent diatoms, *Coconeis* ssp (McCormick & Stevenson, 1989). At lower intensities, *Achnanthes minutissima* and *Gomphonema* spp. dominated. Despite severe (10 ×) decreases in maximal biomass, presumably by grazers in the Kuparuk River, the year to year differences could have been induced in part by grazer intensity and achieved PO_4 concentration. P-enrichment in McCormick's streams did not change dominant diatoms at equivalent grazer intensity. Some of the lack of change seen in our whole river enrichment may be similar.

Clearly, stream ecologists are separating the effects of nutrients, current and grazer intensity and grazer type in model stream channels. Few whole river experimental applications have been attempted. The trends found are consistent with probable influence of these 3 factors on the diatom community in the Kuparuk River.

Conclusions

1. Temporal change in diatom community richness was reduced by fertilization treatments compared to oligotrophic control in floating bioassay tubes.

2. Bioassay tubes predict the community response to enrichment, but do not predict the species changes well (perhaps, depending upon substrata). Detrended correspondence analysis showed greater separation of bioassay community and river community in 1983 than between river communities on any other pair of samples. However, the direction of those changes in Decorana was similar to those in the river community (along axes 2 in Fig. 8).

3. Phosphorus-enhanced, epilithic algal growth in these cold, relatively unproductive rivers, leads to accumulation of species faster than competitive elimination.

4. Even with significant mineral nutrient addition of a whole river, the common diatom species did not change in presence and absence of a stationary down-river PO_4 treatment. Biggest change in community structure was between years.

5. Invertebrate grazing pressure caused an increase in evenness and a reduction in dominance as predicted by theory and other studies. However, there was a reduction in species in epilithon as though grazing pressure was severe. Tall canopy diatoms were replaced by more adherent *Cymbella* and *Eunotia* spp.

Acknowledgements

This work was funded by NSF grants from Division of Polar Programs to John Hobbie, Ecosystems Center, Marine Biological Laboratory, Woods Hole, MA, and by subcontracts to University of Cincinnati. This was a team project using data and ideas generated through years of work and discussion, especially with Bruce Peterson and Ann Hershey. The field work was aided by Bernard Moller, Pamela Bianconi, Robert Ries, and George Gibeau. Field logistics were contracted from the Toolik Lake Field Station was run by Institute of Arctic Biology, University of Alaska. The floating bioassay streams were set up and monitored by Bruce Peterson and assistants from MBL. The chlorophyll biomass was monitored by many workers in the program including Bruce Peterson and assistants from the MBL and by Bernard Moller.

References

Cattaneo, A., 1983. Grazing on epiphytes. Limnol. Oceanogr. 28: 124–132.

Connell, J. H., 1978. Diversity in tropical rainforests and coral reefs. Science 199: 1302–1310.

Edmondson, W. T., 1959. Freshwater Biology 2nd ed. John Wiley & Sons, N.Y. 1248 p.

Fisher, S. G., 1987. Succession in streams. In Barnes, J. R. & G. W. Minshall (eds). Stream Ecology. Plenum Press. N.Y.: 7–27.

Gaugh, H. G., Jr., 1982. Multivariate analysis in community ecology. Cambridge U. Press. Cambridge, U.K.

Gibeau, G. G. Jr. & M. C. Miller, 1989. A micro-bioassay for epilithon using nutrient-diffusing artificial substrata. Freshwat. Biol. 5: 172–176.

Hershey, A. E., A. L. Hiltner, M. A. J. Hullar, M. C. Miller, R. J. Vestal, M. A. Lock, S. Rundle & B. J. Peterson, 1988. Nutrient influence on a stream grazer: Orthocladius microcommunitites respond to nutrient input. Ecology 69: 1383–1392.

Hill, M. O., 1979. Decorana. Ecology and Systematics Program, Cornell University, Ithaca, N.Y. 52 p.

Hoagland, K. D., S. C. Roemer & J. R. Rosowski, 1982. Colonization and community structure of two periphyton assemblages, with emphasis on the diatoms (Bacillariophyceae) Am. J. Bot. 69: 188–213.

Hoagland, K. D., A. Zlotsky & C. G. Peterson, 1986. The source of algal colonizers on rock substrates in a freshwater impoundment. In L. V. Evans & K. D. Hoagland (eds), Algal Biofouling. Elsevier Science Publishers. Amsterdam: 21–39.

Horner, R. R. & E. B. Welch, 1981. Stream periphyton development in relation to current velocity and nutrients. Can. J. Fish. aquat. Sci. 38: 449–457.

Hullar, M. A. & J. R. Vestal, 1989. The effects of nutrient limitation and stream discharge on the epilithic microbial community in an oligotrophic arctic stream. Hydrobiologia 172: 19–26.

Hunter, R. D., 1980. Effects of grazing on the quantity and quality of freshwater aufwuchs. Hydrobiologia 69: 251–259.

Hynes, H. B. N., 1960. The biology of polluted waters. Liverpool Univ. Press, U.K.

Hynes, H. B. N., 1970. The ecology of running waters. Univ. Toronto Press, Canada 555 p.

Kaufman, L. H., 1982. Stream aufwuchs accumulation: disturbance frequency and stress resistance and resilience. Oecologia 52: 57–63.

Kehde, P. M. & J. L. Wilhm, 1972. The effects of grazing by snails on community structure of periphyton in laboratory streams. Am. Midl. Nat. 87: 8–24.

Lamberti, G. A. & V. H. Resh, 1983. Stream periphyton and insect herbivores: an experimental study of grazing by a caddishfly population. Ecology 64: 1124–1135.

Leland, H. V. & J. L. Carter, 1986. Use of detrended correspondence analysis in evaluating factors controlling species composition of periphyton. pp. 101–117. In Isom, B. G. (ed.) Rationale for sampling and interpretation of ecological data in the assessment of freshwater ecosystems. ASTM STP 894. Am. Soc. Testing and Materials. Philadelphia.

Lock, M. A., T. E. Ford, D. M. Fiebig, M. C. Miller, M. Hullar, M. Kaufman, J. R. Vestal, B. J. Peterson & J. E. Hobbie, 1989. A biogeochemical survey of rivers and streams in the mountains and foot-hills province of arctic Alaska. Arch. Hydrobiol. 115: 499–521.

Lowe, R. L., S. W. Golladay & J. R. Webster, 1986. Periphyton response to nutrient manipulation in streams draining clearcut and forested watersheds. J. N. Am. Benthol. Soc. 5: 221–229.

Lubchenco, J., 1978. Plant species diversity in a marine intertidal community: importance of herbivore food preference and algal competitive abilities. Am. Nat. 112: 23–39.

MacArthur, R. H. & E. O. Wilson, 1967. The theory of island biogeography. Princeton Univ. Press, Princeton, N.J.

McCormick, P. V. & R. J. Stevenson, 1989. Effects of sanil grazing on benthic algal community structure in different nutrient environments. J. N. Am. Benthol. Soc. 8: 162–172.

McNaughton, S. J. & L. L. Wolf, 1973. General Ecology. Holt Reinhart & Winston, Inc. N.Y. 710 p.

Mulholland, P. J., J. D. Newbold, J. W. Elwood & C. L. Hom, 1983. The effect of grazing intensity on phosphorus spiralling in autotrophic streams. Oecologia 58: 358–366.

Patrick, R., 1970. Benthic stream communities. Am. Scientist 58: 546–549.

Patrick, R., 1977. Ecology of freshwater diatoms and diatom communities. In D. Werner (ed.) Biology of Diatoms. Univ. California Press, Berkeley: 284–332.

Patrick, R., J. Cairns Jr. & A. Scheier, 1968. The relative sensitivity of diatoms, snails and fish to twenty common constituents of industrial wastes. Prog. Fish Culturist 30: 137–140.

Peterson, B. J., J. E. Hobbie, T. L. Corliss & D. Kriet, 1983. A continuos flow periphyton bioassay; Tests of nutrient limitation in a tundra stream. Limnol. Oceanogr. 28: 582–595.

Peterson, B. J., J. E. Hobbie, A. E. Hershey, M. A. Lock, T. E. Ford, J. R. Vestal, V. L. McKinley, M. A. J. Hullar, M. C. Miller, R. M. Ventullo & G. S. Volk, 1985. Transformation of a tundra river from heterotrophy to autotrophy by addition of phosphorus. Science 229: 1383–1386.

Peterson, B. J., J. E. Hobbie & T. L. Corliss, 1986. Carbon flow in a tundra stream ecosystem. Can. J. Fish. aquat. Sci. 43: 1259–1270.

Poff, N. L., N. J. Voelz & J. V. Ward, 1990. Algal colonization under four experimentally-controlled current regimes in a high mountain stream. J. N. Am. Benthol. Soc. 9: 303–318.

Poole, R., 1974. An introduction to quantitative ecology. McGraw-Hill. N.Y. 532 p.

Pringle, C. M., 1990. Nutrient spatial heterogeneity: effects on community structure, physiognomy, and diversity of stream algae. Ecology 71: 905–920.

Rosowski, J. R., K. D. Hoagland & J. E. Aloi, 1986. Structural morphology of diatom-dominated stream biofilm communities under the impact of soil erosion. In Evans, L. V. & K. D. Hoagland (eds), Algal Biofouling. Elsevier. Amsterdam: 247–297.

Stevenson, J. R., 1984. How currents on different sides of substrates in steams affect mechanisms of benthic algal accumulation. Int. Revue ges. Hydrobiol. 69: 241–262.

Stevenson, J. R., 1990. Benthic algal community dynamics in a stream during and after a spate. J. N. Am. Benthol. Soc. 9: 277–288.

Stevenson, J. R. & R. L. Lowe, 1986. Sampling and interpretation of algal patterns for water quality assessments. In B. G. Isom (ed.) ASTM STP 894. Am. Soc. Test. & Materials, Philadelphia: 118–149.

Tilman, D., 1982. Resource competition and community composition. Monographs in Population Ecology 17. Princeton U. Press. N. J. 296 p.

Tilman, D., S. S. Kilham & P. Kilham, 1982. Phytoplankton community ecology: the role of limiting nutrients. Ann. Revue Ecol. Syst. 13: 349–372.

Wetzel, R. G. & G. E. Likens, Limnological analyses. W. B. Saunders & Co., Ltd. N.Y.

Whitton, B. A. (ed.), 1975. River Ecology. Univ. California Press, Berkeley. 725 p.

Hydrobiologia **240**: 121–131, 1992.
W.J. O'Brien (ed.), Toolik Lake – Ecology of an Aquatic Ecosystem in Arctic Alaska.

Epilithic chlorophyll *a*, photosynthesis, and respiration in control and fertilized reaches of a tundra stream

William B. Bowden,[1,*] Bruce J. Peterson,[2] Jacques C. Finlay[1] & Jane Tucker[2]
[1] *Department of Natural Resources, University of New Hampshire, Durham, NH 03824, USA (* author for correspondence);* [2] *The Ecosystems Center, Marine Biological Laboratory, Woods Hole, MA 02543, USA*

Key words: streams, arctic, tundra, epilithon, photosynthesis, respiration

Abstract

Photosynthesis and respiration by the epilithic community on cobble in an arctic tundra stream, were estimated from oxygen production and consumption in short-term (4–12 h), light and dark, chamber incubations. Chlorophyll *a* was estimated at the end of each incubation by quantitatively removing the epilithon from the cobble.

Fertilization of the river with phosphate alone moderately increased epilithic chlorophyll *a*, photosynthesis, and respiration. Fertilization with ammonium sulfate and phosphate, together, greatly increased each of these variables. Generally, under both control and fertilized conditions, epilithic chlorophyll *a* concentrations (mg m^{-2}), photosynthesis, and respiration (mg O_2 m^{-2} h^{-1}) were higher in pools than in riffles. Under all conditions, the P/R ratio was consistent at ~1.8 to 2.0. The vigor of epilithic algae in riffles, estimated from assimilation coefficients (mg O_2 [mg Chl *a*]$^{-1}$ h^{-1}), was greater than the vigor of epilithic algae in pools. However, due to the greater accumulation of epilithic chlorophyll *a* in pools, total production (and respiration) in pools exceeded that in riffles. The epilithic community removed both ammonium and nitrate from water in chambers. Epilithic material, scoured by high discharge in response to storm events and suspended in the water column, removed ammonium and may have increased nitrate concentrations in bulk river water. However, these changes were small compared to the changes exerted by attached epilithon.

Introduction

The 'river continuum' concept introduced by Vannote *et al.* (1980), suggests that community respiration typically exceeds primary production in headwater streams within woodland areas because the overstory tree canopy shades incident light and introduces metabolizable organic matter in the form of leaf litter. Such streams have P/R (production:respiration) ratios <1.0 and were termed heterotrophic by Odum (1956). Literature in support of this model has been reviewed recently by King and Cummins (1989a).

Individual sites may diverge from this model due to local conditions (Minshall, 1978). For example, King & Cummins (1989b) found that closed-canopy reaches of a first to third order stream in Michigan were heterotrophic ($P/R < 1$). However, open reaches of the same stream were autotrophic ($P/R > 1$) due to increased light input. Peterson *et al.* (1985) found that fertilization with phosphorus significantly increased the ratio of autotrophic activity (light metabolism) to heterotrophic activity (dark metabolism) in an arctic tundra stream under similar light conditions. Thus, both light and nutrient status may significantly alter the metabolic balance of headwater streams.

Tundra streams such as the Kuparuk River studied by Peterson *et al.* (1985) are an interesting variant in the river continuum. Because they have no overstory canopy and because the photoperiod is 24 h day^{-1} for most of the growing season, light inputs are high. Unlike mountain streams, which are often highly turbid from glacial melt, tundra streams are usually much clearer, with only traces of color from dissolved organic carbon (Craig & McCart, 1975). Consequently, light readily penetrates to the stream bottom in tundra streams and they are, therefore, autotrophic. For example, Peterson *et al.* (1986) found P/R ratios that averaged about 2.5 for epilithon in unfertilized reaches of the Kuparuk River. Like closed-canopy, woodland streams, allochthonous organic inputs to tundra rivers are very high. However, in contrast to the 'river continuum' model, Peterson *et al.* (1986) found that the flux of coarse particulate organic matter (CPOM) was considerably smaller than the flux of fine particulate organic matter (FPOM). They estimated that organic loading from dissolved organic carbon and eroded peat particles was at least 100 times greater than gross photosynthesis. This material did not stimulate respiration as would leaf litter in a temperate stream, which suggested that it is largely recalcitrant carbon that resists decomposition. Thus, tundra streams have characteristics of both autotrophic and heterotrophic temperate streams.

The primary objective of the Kuparuk River research program is to better understand how nutrient input and predation control the structure and function of this tundra stream ecosystem (i.e. bottom-up versus top-down control). To address bottom-up controls on river structure and function, the Kuparuk River has been fertilized each year since 1983 with phosphate P, the single nutrient found to most limit production in this river (Peterson *et al.*, 1983) and most other tundra rivers, as well (Lock *et al.*, 1989). Nitrogen (as ammonium) was added to the river with P in 1986 and again in 1989.

Epilithic photosynthesis and respiration have been measured previously in the Kuparuk River: before fertilization in 1978 and 1980 (Peterson *et al.*, 1986) and after fertilization in 1984 and 1985 (B. J. Peterson, unpublished data). However, neither of these variables was measured during the + N fertilization experiment of 1986 and no systematic comparison exists of metabolic activity in pool versus riffle habitats under fertilized conditions. In fact, there are relatively few reports in the literature that include simultaneous measurements of riffle and pool benthic metabolism. Because pools make up half of the benthic habitat in meandering streams and have hydrodynamic and nutrient characteristics that differ substantially from riffles, it is important to include this habitat in any assessment of total stream metabolism (Bott *et al.*, 1978). This report describes results of epilithic chlorophyll *a*, photosynthesis, and respiration measurements from pools and riffles in control and fertilized reaches of the Kuparuk River during the + P and N + P fertilization experiments done in 1989.

Methods

The Kuparuk River originates from partially glaciated tundra in the northern foothills of the Brooks Range, and flows north-northeast draining an area of 8107 km^2. The study sites described in this report were located near the intersection of the Kuparuk River and both the trans-Alaska oil pipeline from Prudhoe Bay to Valdez and the Dalton highway between Fairbanks and Prudhoe Bay, Alaska (Fig. 1). The center of the headwater basin is located at approximately 68° 35′ N and 149° 20′ W. At the crossing with the Dalton Highway, the river drains about 143 km^2 of alpine and moist tundra and has a substratum composed almost entirely of rounded cobble that range in size from roughly 10 to 30 cm in their broadest dimension.

According to Craig & McCart (1975), productivity in tundra streams such as the Kuparuk River

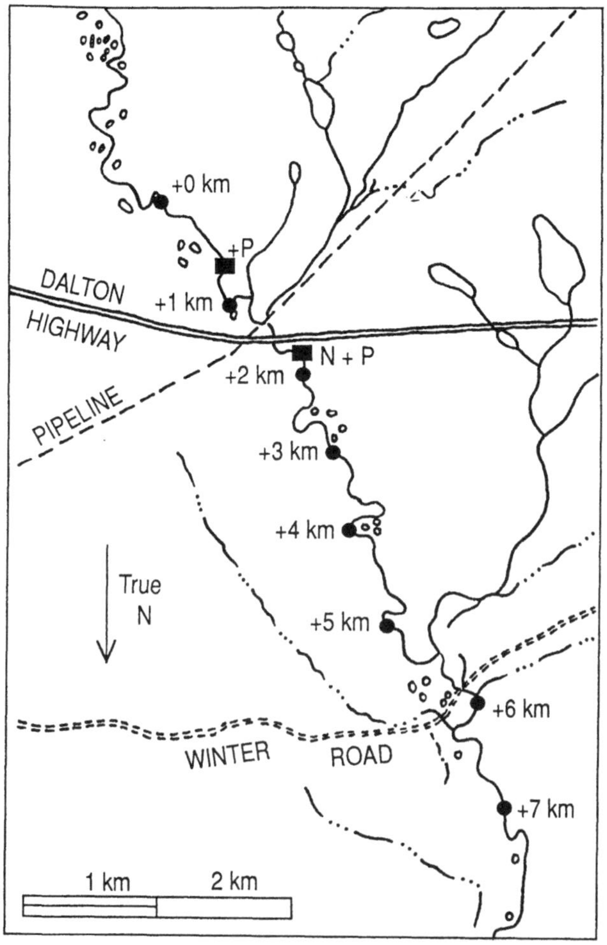

Fig. 1. Location map of the study area. The Kuparuk River watershed drains south to north (down in this figure). River stations are indicated by ● at 1 km intervals from an arbitrary benchmark. The +P and N+P dripper locations are indicated by ■ at about +0.59 km and +1.9 km, respectively.

is intermediate between turbid, glacial streams and clear, nutrient-rich, spring-fed streams. However, little information about benthic community metabolism exists for arctic rivers (Lock *et al.*, 1989). Peterson *et al.* (1986) estimated that net primary production was about 13 g C m^{-2} yr^{-1}. Under normal conditions, the Kuparuk River water is highly oligotrophic. Mean inorganic nutrient concentrations are 11 μg NH_4-N l^{-1}, 21 μg NO_3-N l^{-1}, and 8 μg TDP-P l^{-1} (total dissolved phosphorus). The mean dissolved organic nitrogen concentration (250 μg DON-N l^{-1}) is considerably greater than the mean inorganic N concentrations. The ecology and biogeochemistry of this site are described more fully by Peterson *et al.* (1985), Peterson *et al.* (1986), Lock *et al.* (1989), and Peterson *et al.* (Biological responses of fertilization of a tundra river: a four-year synthesis 1983–1986, in preparation).

Phosphorus, and in some years nitrogen, has been added to the Kuparuk River every year since 1983. The mechanisms used to deliver the nutrients to the river has varied from year to year, but in 1989 separate peristaltic pumps at two locations on the river delivered P as phosphoric acid and N as ammonium sulfate from reservoirs on the bank of the river. Locations on the river are referenced to a common benchmark that is located about 1 km upstream (south) of the road and are identified by their position in kilometers below (+) or above (−) this benchmark. In 1989 the P dripper was located at about +0.59 km; the N dripper was at about +1.9 km. Data from previous years showed that under low-flow conditions, P was rapidly removed from the water column, so a second +P dripper was located with the +N dripper at +1.9 km to insure that both N and P were at elevated levels below +1.9 km. Sample locations are identified in the same manner.

The drippers delivered P and N at constant rates that would nominally result in 10 μg P l^{-1} and 100 μg N l^{-1} at the mean summer discharge rate of 2 m^3 sec^{-1}. In 1989, flow ranged from 1 to 4 m^3 sec^{-1}, with a maximum flow of 12.6 m^3 sec^{-1} during a storm in early August. The drippers defined three zones in the river: a control (no fertilization) zone above +0.59 km, a +P zone between +0.59 and +1.9 km and an N+P zone below +1.9 km. These will be referred to as the control, +P and N+P zones, respectively.

Photosynthesis and respiration in these three zones were estimated from changes in oxygen within closed chambers during separate light and dark incubations (Bott *et al.*, 1978; Pennak & La Velle, 1979; King, 1982; Duff *et al.*, 1984). Rocks with epilithon were placed in sealed Plexiglas chambers filled with river water. Dissolved oxygen decreased in the dark due to respiration alone and normally increased in the light due to the net effect of photosynthesis and respiration. Photosynthesis was estimated as the sum of dissolved oxygen changes in the light and dark incubations.

The chambers (Fig. 2) were constructed from a 25 cm length of stock cast acrylic tubing (21.590 cm O.D. by 0.635 cm wall). A piston, constructed from a 5 cm length of stock cast acrylic tubing (20.320 cm O.D. by 0.635 cm wall), was used to isolate the interior of the chamber from the atmosphere. The piston fit snugly, but not tightly, into the chamber. Oxygen leaks into the chamber were limited to the minute area between the piston and chamber and the long diffusion path, through water, along the piston edge. A BOD port on the top of the piston accepted a stirred, BOD probe from an Orbisphere Model 2714 oxygen meter. Water was recirculated in the chamber with an external, submersible pump. Water samples were removed for analysis by displacing the piston with the sample tube open. When not in use, the sample tube was pinched closed and the BOD port was stoppered with a # 1 rubber stopper.

Rocks were collected at random locations within pools or riffles at each site. Surface rocks with a 'typical' development of epilithon, based on visual inspection of the site, were selected for study. We rejected rocks with heavy moss growth and favored rocks of a uniform size and shape, that would fit neatly in the chamber bottom. We found that 3 to 4 rocks of an average shape and size would essentially cover the chamber bottom in a single layer, at a surface density similar to that found in the river.

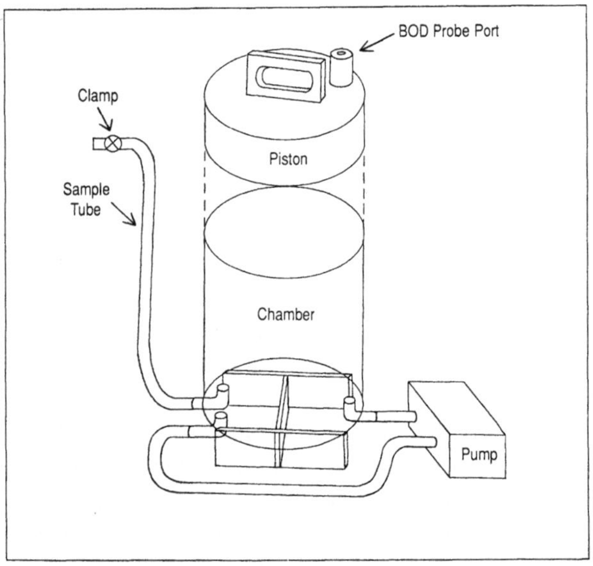

Fig. 2. Schematic of the river benthic metabolism chamber used for this study. The piston slides snugly into the chamber body to seal the contents from atmospheric contamination.

Water for the incubations was collected in carboys. Rocks were collected from the river and placed in coolers, without water, to keep them cool and moist. We found that if the rocks were kept submerged in water in the coolers, delicate epilithic material was dislodged during transport from the field to the lab. Without water, we were able to transport the rocks and epilithon, and place them in the chambers with the epilithon essentially intact, even with flocculent pool epilithon. The time from collection and transport to installation in a chamber was generally 1–2 h.

At the lab, rocks were carefully lowered into an open chamber previously filled with fresh river water. The piston was inserted into the chamber and all bubbles were removed via the open BOD port. The piston was pushed into the chamber as far as practical, to minimize the headspace volume and so maximize the change in dissolved oxygen concentration per unit time. The entire chamber and pump were submerged in a recirculating water bath to maintain constant water temperatures near those found in the Kuparuk River. Toolik Lake was used as a source of cooling water. Each chamber could be illuminated by two, fluorescent, plant and aquarium bulbs that together provided about 190 $\mu E\ m^{-2}\ sec^{-1}$ of energy at the chamber surface. The lights, water bath, and chambers were operated under a thick, rubberized tarp to enable dark incubations.

In a typical experiment, four chambers were run simultaneously. After rocks were in place and chambers were sealed and submerged, we started with dark incubations. Dissolved oxygen and chamber temperature were measured at three time points, until the dissolved oxygen had been reduced by 3 to 4 mg $O_2\ l^{-1}$ or approximately 4 h had elapsed. After the third time point, we turned on the lights and measured dissolved oxygen at two additional time points, until the dissolved oxygen returned to the initial reading or approximately 4 h had elapsed in the light (8 h from the beginning of the experiment). Typically dissolved oxygen decreased linearly in the dark and increased linearly in the light. If non-linearity was encountered (e.g. supersaturation in the light), only the first two data points in a sequence were used in the calculations.

At the end of each experiment, the epilithic material on all rocks within a single chamber was carefully scrubbed off with a stiff wire brush into a known volume of lake water (usually 2 or 5 l). The total pigment content in the lake water used for scrubbing was negligibly small. This 'scrubbate' was mixed vigorously and replicate volumes (usually 5 or 10 ml) were removed by pipette and filtered onto 25 mm, Whatman, GF/F filter papers. Each filter paper was placed in 10 ml of 90% acetone-10% water in a polyethylene, screw-cap, centrifuge tube and extracted in the dark for 24 to 48 h. Chlorophyll *a* in the acetone-water extract was estimated by fluorometry (APHA, 1985). The amount of epilithic chlorophyll *a* originally on the rocks in each chamber was calculated from replicate determinations of this concentration plus the total 'scrubbate' and sub-sampled volumes.

Epilithic chlorophyll *a*, photosynthesis, and respiration were converted to an areal basis by multiplying each variable by the area of the chamber bottom (0.0324 m^2). Essentially, this assumes that the cobble density in the chamber is similar to that in the river and that only the top layer of cobble is important for the variables measured. More complicated methods could be used to ascertain rock surface area per unit area of stream bed (e.g. Naiman, 1983; Murphy, 1984). However, this method was quick, unbiased, and is probably not in error by more than a factor of two (i.e. misrepresented the true surface area of the top layer of rocks by $^1/_2$ to 2 times) as long as the chambers were loaded with similar cobble sizes, shapes, and densities. This approach (as well as most other benthic chamber methods employed elsewhere) underestimates any respiration that may occur deeper in the hyporheic zone.

In a separate series of experiments, water samples were withdrawn from the chambers at three times during light-only incubations of pool and riffle rocks from each of the sampling stations. Concentrations of ammonium-N and nitrate-N were measured in each sample to compare N dynamics in the chambers with those observed in the river. Ammonium was analyzed by the phenol-hypochlorite method, as an indophenol dye at 640 nm (Grasshoff & Johannsen, 1972). Nitrite was measured as a diazo dye at 540 nm (Bendschneider & Robinson, 1952). Nitrite was typically at or below detection limits. Nitrate was converted to nitrite using a manual, mossy cadmium reduction method (MacKereth *et al.*, 1978; Jones, 1984; Elliot & Porter, 1971).

Results

Epilithic chlorophyll *a* in fertilized reaches of the Kuparuk river increased significantly in response to N + P additions and perhaps to + P additions alone (Fig. 3). Epilithic chlorophyll *a* concentrations were measured after oxygen incubations in late July and early August, about 3 and 5 weeks after fertilization began. In both cases, epilithic chlorophyll *a* concentrations in the control-reach riffles were < 10 mg m^{-2}. By late July, the added N + P had significantly increased epilithic chlorophyll *a* concentrations in riffles at both N + P sites (+ 2 km and + 4 km). Epilithic chlorophyll *a* concentrations in the control-reach pools were significantly higher than concentrations in the control-reach riffles. As was the case for the riffles, added N + P significantly increased epilithic chlorophyll *a* concentrations in pools within both fertilized reaches. Chamber incubations were not done in the + P reach in late July.

In early August, the largest storm of the season increased discharge in the Kuparuk River from pre-storm rates of 1 to 4 m^3 sec^{-1}, to a peak discharge of 12.6 m^3 sec^{-1}. As a consequence, epilithic chlorophyll *a* was scoured from the river bottom at all locations (Fig. 3). Nevertheless, the

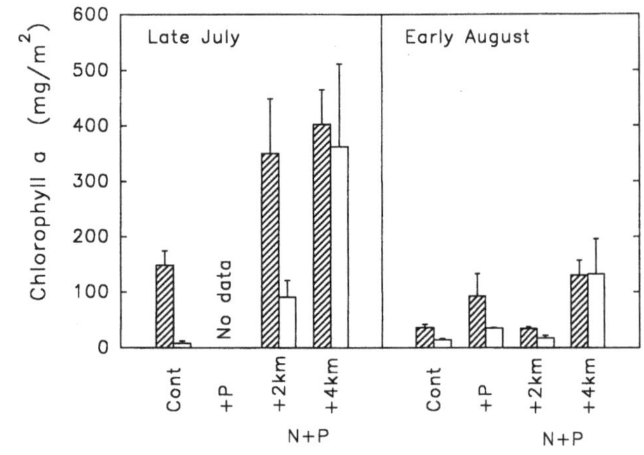

Fig. 3. Responses of epilithic chlorophyll *a* to fertilization by + P and N + P in the Kuparuk River in late July and early August, 1989. The control station was located at about + 0.35 km or about 150 m above the + P dripper. The + P station was located at about + 1.5 km or about 900 m below the + P dripper and 400 m above the N + P dripper. The two N + P stations at + 2 and + 4 km were about 100 m and 2100 m below the N + P dripper. Shaded bars are pools; light bars are riffles. The error bars indicate ± 1 S.E. about the mean ($n = 4$ in July, $n = 2$ in August).

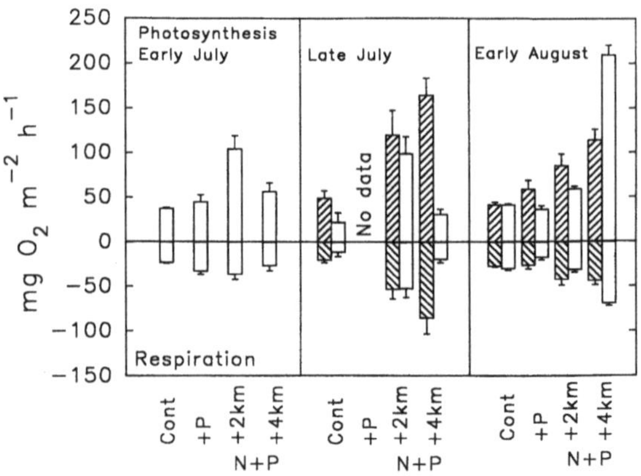

Fig. 4. Responses of epilithic photosynthesis and respiration to fertilization by + P and N + P in the Kuparuk River in early July, late July, and early August, 1989. Sample locations, shading, and error bars are the same as described in Fig. 3.

major patterns observed in late July prevailed in the early August, post-storm, samples. Specifically, epilithic chlorophyll *a* concentrations in control riffles were low. Epilithic chlorophyll *a* concentrations were not significantly greater in riffles at the + 2 km N + P station but were significantly greater at the + 4 km N + P station. In addition, for this sample period, we also measured epilithic chlorophyll *a* concentrations in the + P reach and found that they were also higher than in the control-riffle reach. As was the case in late July, epilithic chlorophyll *a* concentrations in pools were significantly greater than in riffles, except for the + 4 km N + P station where pool and riffle epilithic chlorophyll *a* concentrations were similar.

Patterns of photosynthesis and respiration were similar to those observed for epilithic chlorophyll *a* concentrations (Fig. 4). Light/dark oxygen incubations were done in early July, late July, and early August (after the storm). Only riffle rocks were used in the early July incubations. Rocks from the + P reach were not used in the late July incubations. Photosynthesis in the control reach was < 50 mg O_2 m^{-2} h^{-1} on all dates for both pools and riffles. Respiration was typically ≤ 20 mg O_2 m^{-2} h^{-1}.

Fertilization with N + P (and perhaps + P alone) significantly increased both epilithic photosynthesis and respiration in the Kuparuk River, especially late in the season (Fig. 4). In late July and early August, photosynthesis in pools was equal to or greater than (often significantly greater than) photosynthesis in riffles. This pattern held for all incubations except the + 4 km N + P station in early August, where riffle photosynthesis significantly exceeded pool photosynthesis. In all cases, patterns of epilithic respiration were well correlated with patterns of epilithic photosynthesis (Fig. 5), with a consistent P/R of ~2.5.

Rates of epilithic photosynthesis and respiration did not change dramatically after the early August storm, as was the case for epilithic chlorophyll *a* concentrations. Photosynthesis and respiration in the control reach pools and riffles changed only slightly, although significantly, in some cases. Photosynthesis and respiration were lower after the early August storm, in all but the + 4 km N + P pool incubations. In this particular case, field notes indicated an obvious abundance of epilithic chlorophyll *a* on the rocks used for this incubation, indicating that this site may not have been scoured as severely as the others.

Assimilation coefficients for epilithic algae (mg

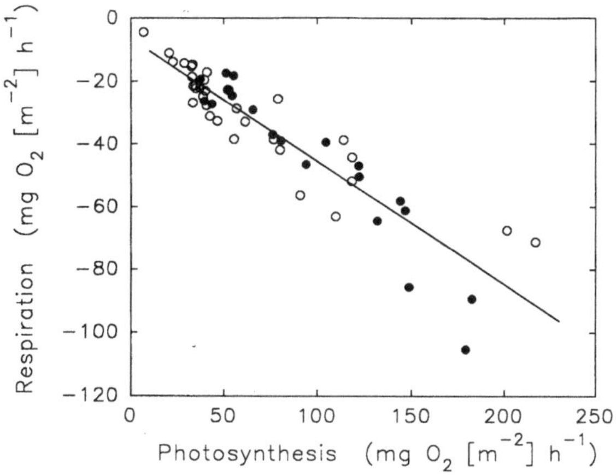

Fig. 5. Respiration as a function of photosynthesis for individual chamber incubations of cobbles from the Kuparuk River, in early July, late July, and early August, 1989. Open circles (○) are incubations with riffle rocks; closed circles (●) are incubations with pool rocks. Data from control, + P, and N + P reaches are included in this figure. The regression line is Y = − 6.5 − 0.39 × X($n = 58$, $r^2 = 0.85$).

O_2 [mg Chl *a*]$^{-1}$ h^{-1}) differed between riffles and pools (Fig. 6). However, unlike the pattern observed for epilithic chlorophyll *a*, photosynthesis, and respiration, assimilation coefficients were lower in pools than in riffles. In general, assimilation coefficients were high when chlorophyll *a* accumulation was low and were low when chlorophyll *a* accumulation was high. Thus, assimilation coefficients were usually higher in riffles than in pools and were higher after the storm in early August than before, due to the effects of scouring on chlorophyll *a* accumulation.

Ammonium and, to a lesser extent, nitrate were both removed from the headspace water in the chambers during the oxygen incubations (Fig. 7). Furthermore, rates of ammonium loss from water in the chambers increased as the concentration of ammonium initially present in the river water in-

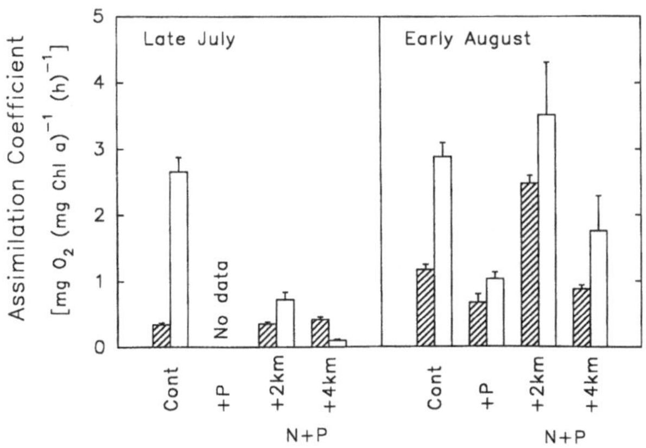

Fig. 6. Assimilation coefficient response to fertilization by + P and N + P in the Kuparuk River in late July and early August, 1989. Sample locations and shading are as described in Fig. 3. The error bars indicate ± 1 S.E. about the mean ($n = 4$ in July, $n = 2$ in August).

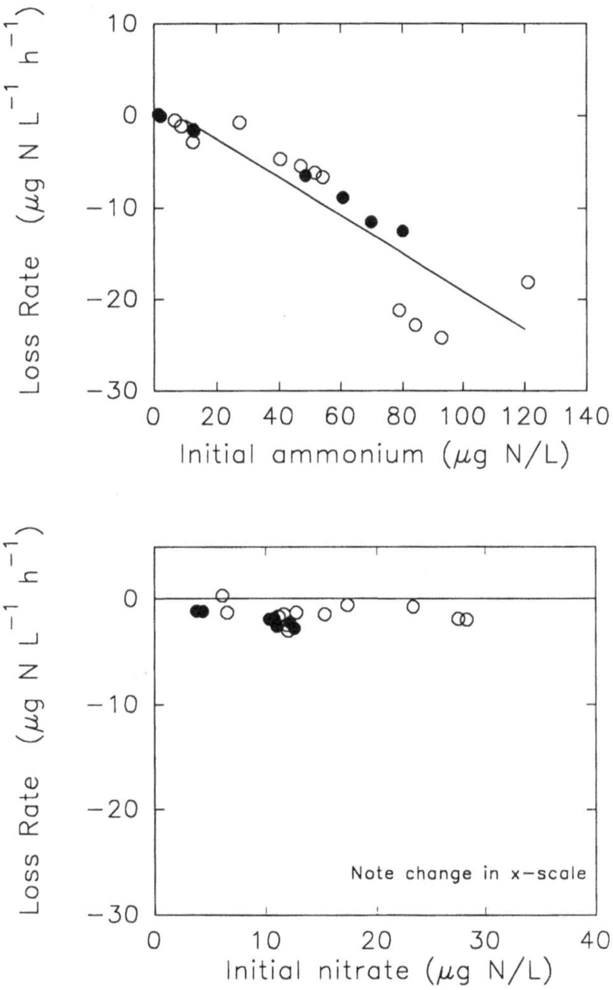

Fig. 7. Ammonium (upper panel) and nitrate (lower panel) dynamics versus initial N concentration in the incubation chambers for experiments done with pool (●) and riffle (○) rocks collected in control, + P, and N + P reaches of the Kuparuk River in late July and early August 1989. The line through the ammonium data is the best-fit linear regression. The line in the nitrate panel is the 0-slope line.

creased. This relationship was similar for both pool and riffle rocks. A similar trend for nitrate loss was not apparent.

Under normal circumstances, most of the metabolic activity in the river is located in the epilithon rather than in the water column. Control (no rock) incubations typically supported low rates of photosynthesis and respiration in the headspace water (data not shown). However, under conditions of high discharge, current scour removes epilithon and transfers it to the water column. Under these conditions, scoured epilithon may cause a transient increase in water column photosynthesis, respiration, and nutrient dynamics. Nevertheless, these changes are small compared to similar changes caused by attached epilithon (e.g. Fig. 7).

Discussion

Fertilization of the Kuparuk River in 1989 with ammonium-N plus phosphate-P increased the standing stock of epilithic chlorophyll *a* and the rates of epilithic photosynthesis and respiration. Addition of phosphate-P alone may have stimulated these variables as well, however, the effect was clearly less than when N and P were added together.

Previous responses by the epilithic community to fertilization of the Kuparuk River have been qualitatively consistent but quantitatively variable. Bioassays done in 1980 (Peterson *et al.*, 1983) showed clearly that the river was P limited. Addition of N alone (as NH_4NO_3) had no affect on periphyton growth on tiles, whereas P additions (as PO_4^{3-}) and N + P additions both stimulated periphyton growth to similar, high levels. In 1983, a series of whole-river fertilization experiments started that, at first, corroborated the bioassay results. In 1983 and 1984, P alone stimulated epilithic chlorophyll *a* growth in the Kuparuk. However, in 1985 there was no observable effect. On the premise that N might become limiting due to the addition of surplus P, both P and N were added simultaneously to the river in separate reaches in 1986. Although both nutrients significantly increased epilithic chlorophyll *a* accumulation, the increase was small. There continued to be little or no accumulation of epilithic chlorophyll *a* in response to fertilization with P alone in 1987 and 1988. In 1989, the 1986 N + P experiment was repeated, and this time epilithic chlorophyll *a* accumulated to very high levels in response to the N + P addition.

Recent experiments (Miller *et al.*, in preparation) suggest that after 1984, an increase in grazing pressure prevented epilithic chlorophyll *a* accumulation. However, under low-flow conditions, filamentous algae (which are not effectively grazed) may quickly overgrow the river bottom in all but the swiftest riffle areas. This may explain the fertilization response observed in 1989. Thus, P appears to always stimulate productivity and stimulates additional productivity given additional N. Accumulation of epilithic chlorophyll *a* is variable, however, depending on grazing pressure and scouring effects.

The photosynthesis and respiration rates measured on control riffle rocks during this experiment, are similar to rates measured in earlier years (Fig. 8). Photosynthesis measured during this and four previous seasons, ranged from about 11 to 48 mg $O_2\ m^{-2}\ h^{-1}$; respiration ranged from about 4 to 21 mg $O_2\ m^{-2}\ h^{-1}$. In two previous fertilization experiments, + P alone stimulated photosynthesis and respiration by over 100% (1984) and about 44% (1985). In the 1989 experiment, + P alone stimulated photosynthesis and respiration by only 20%. However, N + P additions almost doubled respiration rates and tripled photosynthesis rates.

P/R ratios for the epilithon were remarkably consistent at about 2.5 for all sites and treatments (Fig. 5). Groeger & Kimmel (1989) found a consistent linear relationship between respiration and photosynthesis, very similar to that found in this study. In our study, this relationship was apparently not affected by site (pool versus riffle) or treatment (control, + P, or N + P).

Direct comparisons of riffle and pool metabolism are relatively rare in the literature (Naiman & Sedell, 1980; Naiman, 1983). Such a comparison was reported for control reaches of the Ku-

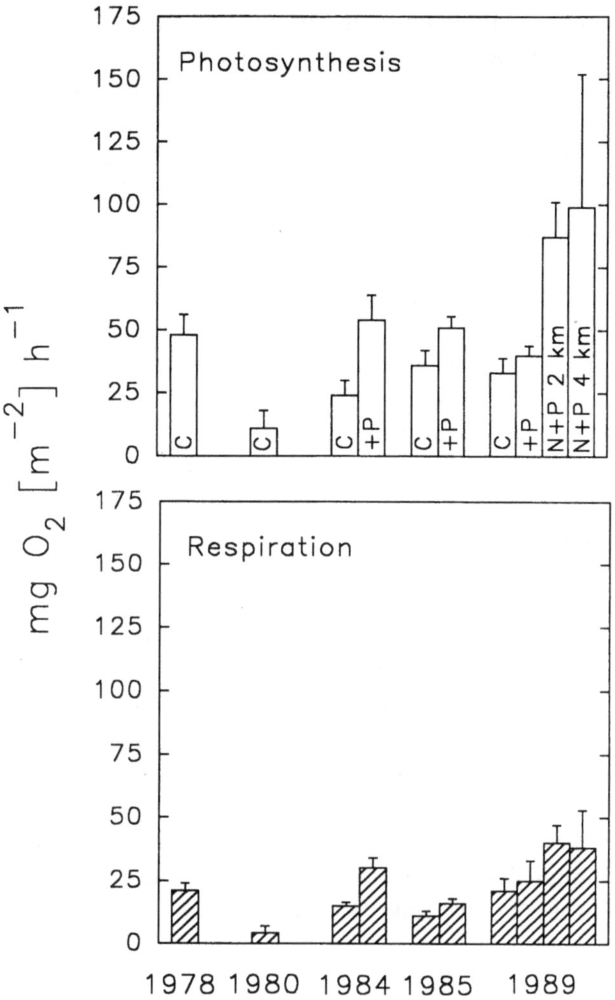

Fig. 8. Photosynthesis and respiration by riffle epilithon for four years prior to the 1989 fertilization experiment. Control (C) and + P locations for years prior to 1989 are similar to those described in Fig. 3. Error bars are ± 1 S.E. for $n = 2$ to 3 chamber incubations.

paruk River by Peterson *et al.* (1986), however, no data exist for fertilized conditions. We found that the amount of epilithic chlorophyll *a* in pools was typically greater than that found in adjacent riffles, under both control and fertilized conditions. Algal biomass is able to accumulate in the deep, relatively quiet pools, while in riffles algal biomass is quickly scoured by currents or grazed by insects and usually does not accumulate.

Largely as a consequence of the large epilithic chlorophyll *a* standing stock, autotrophic production in the pools was often greater than in the riffles, under both control and fertilized conditions. It is reasonable to expect that autotrophic production might be high in pools of the Kuparuk River. Unlike glacial-fed streams, the Kuparuk river is comparatively clear because it originates in headlands that are not currently glaciated. Consequently, although the pools are considerably deeper than riffles (up to 1 or 2 m), the water clarity and lack of overstory vegetation allow light to penetrate to the bottom of most pools.

While total production was higher in pools than in riffles, rates of photosynthesis and respiration per unit epilithic chlorophyll *a* were higher in riffles than in pools (Fig. 4). This suggests that the algae in the riffles were more vigorous than those in the pools, possibly as a result of differences in available light and the effects of physical scouring. Although light typically penetrated to the bottom of pools in the Kuparuk River, epilithic algae on rocks at the bottom of quiet pools were coated with a thick floc of detrital material. This material was removed continuously from riffle rocks but accumulated on pool rocks due to lower current velocities. Consequently, pool algae may be shaded by both their neighbors and by accumulated detrital material. In contrast, high current scour and grazing in the riffle environment may maintain riffle epilithon in a state of logarithmic growth and high vigor.

Nevertheless, productivity by pool epilithon was high and required, in turn, a supply of nutrients sufficient to support this production. High epilithic productivity in the riffle environment is probably maintained by high throughput of low-nutrient water. However, in pools, high productivity in the thick floc or biofilm that develops on the rocks may be maintained by internal recycling of accumulated nutrients (Lock *et al.*, 1984). Consequently, although the sources may differ, nutrient supply rates in pools and riffles may not differ greatly.

Changes in chamber headspace ammonium concentrations (Fig. 7) are in general agreement with observed declines of ammonium concentration in the river, downstream of the N + P dripper location. Concentrations of ammonium just below the N + P dripper were typically 50–250 μg N l^{-1}, depending on the river discharge rate and fell to values of about 20–40 μg N l^{-1} by +4 km (a reach of about 2 km). The cross-sectional area of the river is highly variable but at a typical discharge rate of 2 m^3 sec^{-1}, averages roughly 6 to 7 m^2. Thus the average velocity of water under these conditions is ~0.3 m sec^{-1}.

Using the information from Fig. 7 and this estimate of average water velocity in the river under typical conditions, it is possible to predict the rate at which ammonium ought to be removed from the water column. There were typically 5 l of headspace water in each chamber, with a chamber bottom surface area of 0.0324 m^2. Thus the y-axis unit in Fig. 7 can be converted to an approximate, areal, ammonium demand by the epilithon by multiplying by (5/0.0324), to get μg NH_4-N m^{-2} h^{-1}. Given the above values for discharge, the physical dimensions of the river, and the calculated uptake rate from above, the predicted concentration of ammonium in the river matches reasonably well with the observed concentrations (Fig. 9).

Nitrate data from the chambers do not match observations from the river as well as ammonium concentrations. Nitrate in chambers was taken up, although the uptake rate was generally quite small. Data from river surveys done in 1989 show that, as added ammonium was removed from the water column, nitrate sometimes increased, but usually not before about the +4 km mark. The chamber experiments were done with rocks collected from +2 to +4 km and may have just missed the zone where net nitrification began in 1989. Experiments with ammonium-^{15}N and nitrate-^{15}N additions have been done and will be reported separately. Preliminary information from these experiments indicates that some nitrification did occur in the chambers, even in rocks collected above the +4 km mark.

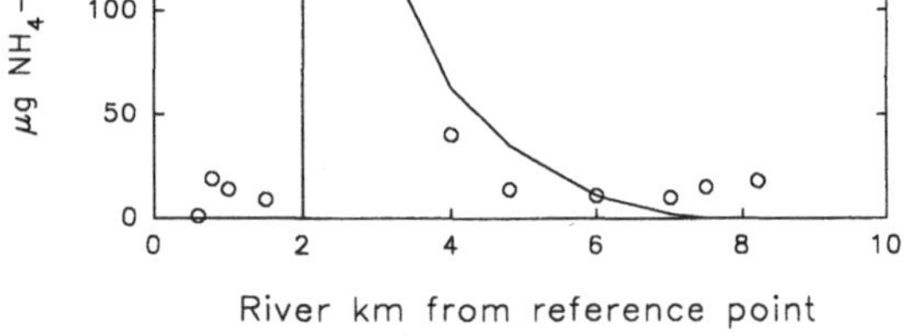

Fig. 9. Observed (O) and predicted (-) concentrations of ammonium in the Kuparuk River on 18 July 1989, as a function river km downstream from the fixed reference point. The model prediction employs chamber data to predict the behavior of observed river water concentrations. See text for details.

Acknowledgements

We thank C. Baumann and B. Moeller for assistance in the field and D. Witt and M. Ables for logistic support. M. Miller and W. Rowe did the chlorophyll *a* analyses. L. Isaacson drafted Figs 1 and 2. A. Baker, W. H. McDowell, and an anonymous reviewer provided helpful comments on a draft of this manuscript. This research was supported by grant DPP 8722015 from the National Science Foundation and is a contribution to the Toolik Lake/Kuparuk River LTER effort.

References

APHA, 1985. Standard methods for the examination of water and wastewater, 16th edition. Greenberg, A. E., R. R. Trussell & L. S. Clesceri (Eds.). American Public Health Association, Washington.

Bendschneider, K. & R. J. Robinson, 1952. A new spectrophotometric method for the determination on nitrite in sea water. J. mar. Res. 11: 87–96.

Bott, T. L., J. R. Brock, C. E. Cushing, S. V. Gregory, D. King & R. C. Petersen, 1978. A comparison of methods for measuring primary productivity and community respiration in streams. Hydrobiologia 60: 3–12.

Craig, P. C. & P. J. McCart, 1975. Classification of stream types in Beauford Sea drainages between Prudhoe Bay, Alaska and the Mackenzie Delta, N.W.T., Canada. Arctic and Alpine Res. 7: 183–198.

Duff, J. H., K. C. Stanley & R. J. Avanzino, 1984. The use of photosynthesis-respiration chambers to measure nitrogen flux in epilithic algal communities. Verh. int. Ver. Limnol. 22: 1436–1443.

Elliot, R. J. & A. G. Porter, 1971. A rapid cadmium reduction method for the determination of nitrate in bacon and curing brines. Analyst 96: 522–527.

Grasshoff, K. & J. Johannsen, 1972. A new sensitive and direct method for the automatic determination of ammonia in seawater. J. Cons. int. Explor. Mer. 34: 516–521.

Groeger, A. W. & B. L. Kimmel, 1989. Relationship between photosynthetic and respiratory carbon metabolism in freshwater phytoplankton. Hydrobiologia 173: 107–117.

Jones, M. N., 1984. Nitrate reduction by shaking with cadmium. Wat. Res. 18: 643–646.

King, D. K., 1982. Community metabolism and autotrophic-heterotrophic relationships of woodland stream riffle sections. Ph.D. Thesis. Michigan State University, East Lansing. 356 pp.

King, D. K. & K. W. Cummins, 1989a. Autotrophic-heterotrophic community metabolism relationships of a woodland stream. J. Freshwat. Ecol. 5: 205–218.

King, D. K. & K. W. Cummins, 1989b. Factors affecting autotrophic-heterotrophic relationships of a woodland stream. J. Freshwat. Ecol. 5: 219–230.

Lock, M. A., T. E. Ford, D. M. Fiebig, M. C. Miller, M. Hullar, J. R. Vestal, B. J. Peterson & J. E. Hobbie, 1989. A biogeochemical survey of rivers and streams in the mountains and foot-hills province of arctic Alaska. Arch. Hydrobiol. 115: 499–521.

Lock, M. A., R. R. Wallace, J. W. Costerton, R. M. Ventullo & S. E. Charlton, 1984. River epilithon: toward a structural-functional model. Oikos 42: 10–22.

MacKereth, F. J. H., J. Heron & T. F. Talberg, 1978. Nitrate. In: MacKereth, F. J. H., J. Heron & T. F. Talberg (Eds.), Water analysis: some revised methods for limnologists, Freshwater Biological Association, Windermere, U.K.: 72–73.

Minshall, G. W., 1978. Autotrophy in stream ecosystems. BioScience 28: 767–771.

Murphy, M. L., 1984. Primary production and grazing in freshwater and intertidal reaches of a coastal stream, southeast Alaska. Limnol. Oceanogr. 29: 805–815.

Naiman, R. J., 1983. The annual pattern and spatial distribution of aquatic oxygen metabolism in boreal forest watersheds. Ecol. Monogr. 53: 73–94.

Naiman, R. J. & J. R. Sedell, 1980. Relationships between metabolic parameters and stream order in Oregon. Can. J. Fish. aquat. Sci. 37: 834–847.

Odum, H. T., 1956. Primary production in flowing rivers. Limnol. Oceanogr. 1: 102–117.

Pennak, R. W. & J. W. Lavelle, 1979. *In situ* measurements of net primary production in a Colorado mountain stream. Hydrobiologia 66: 227–235.

Peterson, B. J., J. E. Hobbie & T. L. Corliss, 1983. A continuous flow periphyton bioassay: tests of nutrient limitation in a tundra stream. Limnol. Oceanogr. 28: 583–591.

Peterson, B. J., J. E. Hobbie & T. L. Corliss, 1986. Carbon flow in a tundra stream ecosystem. Can. J. Fish. aquat. Sci. 43: 1259–1270.

Peterson, B. J., J. E. Hobbie, A. E. Hershey, M. A. Lock, T. E. Ford, J. R. Vestal, V. L. McKinley, M. A. H. Hullar, M. C. Miller, R. M. Ventullo & G. S. Volk, 1985. Transformations of a tundra river from heterotrophy to autotrophy by addition of phosphorus. Science 229: 1383–1386.

Vannote, R. L., K. W. Cummins, G. W. Minshall, J. R. Sedell & C. E. Cushing, 1980. The river continuum concept. Can. J. Fish. aquat. Sci. 37: 130–137.

Hydrobiologia **240**: 133–141, 1992.
W.J. O'Brien (ed.), Toolik Lake – Ecology of an Aquatic Ecosystem in Arctic Alaska.

Community structure and bottom-up regulation of heterotrophic microplankton in arctic LTER lakes

Parke A. Rublee
Biology Department, University of North Carolina at Greensboro, Greensboro, NC 27412, USA

Key words: microplankton, rotifers, arctic lakes

Abstract

Microplankton community structures and abundance was assessed in lakes at the Toolik Lake LTER site in northern Alaska during the summers of 1989 and 1990. The microplankton community included oligotrich ciliates, but rotifers and zooplankton nauplii comprised greater than 90% of total estimated heterotrophic microplankton biomass. Dominant rotifer taxa included *Keratella cochlearis*, *Kellicottia longispina*, *Polyarthra vulgaris*, *Conochilus unicornis* and a *Synchaeta* sp. Microplankton biomass was lowest in highly oligotrophic Toolik Lake ($< 5\ \mu gC\ l^{-1}$ at the surface) and highest (up to $55\ \mu gC\ l^{-1}$) in the most eutrophic lakes, experimentally fertilized lakes, and fertilized limnocorrals, consistent with bottom-up regulation of microplankton abundance.

Introduction

Recognition of the potential trophic significance of the microbial loop (Azam *et al.*, 1983; Porter *et al.*, 1985) has generated considerable study of heterotrophic nanoplankton (2–20 μm) and microplankton (20–200 μm) in aquatic ecosystems (e.g. Stockner & Porter, 1988). Heterotrophic nano- and microplankton graze on bacteria (Hobbie & Helfrich, 1988; Sanders *et al.*, 1989) and small autotrophs (Stockner & Porter, 1988; Rublee & Gallegos, 1989; Gallegos, 1990). Thus, the microplankton represent a potentially important trophic link from bacterial utilization of dissolved organics and algal utilization of inorganics to zooplankton and higher order consumers.

The significance of the microbial loop to higher trophic levels remains uncertain. From an energy and carbon standpoint, numerous studies confirm bacterivory and algal consumption in freshwaters (e.g. Sherr & Sherr, 1984; Sanders *et al.*, 1989) and that microplankton can be consumed by crustacean zooplankton (e.g. Williamson & Butler, 1986). Nevertheless, many investigators suggest that the microbial loop acts as an energy sink (cf. Pomeroy & Wiebe, 1988), since insertion of new trophic levels must result in respiratory losses as energy passes along them. However, if the microbial loop acts primarily as a 'salvage pathway' by returning dissolved organic carbon (which otherwise would be lost by excretion and decomposition) back into grazing food webs, then regardless of the conversion efficiency, the microbial loop represents a net addition to resources available to the higher trophic levels. Further, the role of the microbial loop in recycling nutrients and regulating microbial populations may actually stimulate bacterial and algal production (Porter *et al.*, 1985).

Regulation of microbial populations is thought to be dependent on both bottom-up and top-down controls. Bottom-up regulation is represented by food resources which appear to set limits on the attainable biomass within aquatic systems (cf. Pace, 1984; McQueen *et al.*, 1989). Top-down regulation is represented by predators which are known to control population abundance (see e.g. Hobbie & Helfrich, 1988).

Arctic freshwater ecosystems may be ideal sites to study aspects of the microbial loop. Reduced diversity of organisms in arctic freshwaters relative to temperate systems (Hobbie, 1973) should make it easier to determine interactions and thus regulation and impact of microbial loop components. Here I report data on microplankton abundance and distribution from some lakes of the LTER site at Toolik Lake, Alaska, during the summers of 1989 and 1990, and discuss bottom-up control of heterotrophic microplankton populations.

Materials and methods

The Toolik Lake LTER site (68° N, 149° W) is located in the northern foothills of the Brooks Mountain Range of Alaska. The site includes a number of ponds and lakes which have been under study for over a decade. These lakes are naturally oligotrophic systems with varying zooplankton and fish populations. Most data presented here are from eight lakes of the Toolik LTER site (Table 1), some of which have been experimentally manipulated (nutrient or fish additions). Additional data are reported from an experimental study conducted in limnocorrals within Toolik Lake during 1989.

Water samples were collected by Van Dorn sampler at the surface and at depths of 1, 3, 5, 8, 12, 16, and 20 m or to the deepest depth of shallower lakes. Two liters of sample were then gently concentrated to 60 ml by reverse flow through a 20 μm net (Dodson & Thomas, 1964). Glutaraldehyde was added as a preservative to 1% final concentration and samples were stored refrigerated until counted. Live samples were also examined to assure that glutaraldehyde was an appropriate preservative and so that microplankton could be identified after preservation.

Samples were enumerated following the procedure of Baldock (1986). An aliquot of the concentrated samples (5–20 ml, depending on particle concentration of sample) was stained with 0.25% Rose Bengal solution, and then drawn onto an 8.0 μm pore cellulose acetate filter under gentle vacuum. Filters were mounted on slides in a 43% sucrose solution and examined under a compound microscope. The entire surface area of the filter was scanned at 100 × or 200 × and individual microplankton identified. Magnification up to 400 × was used to aid in identification as necessary. Taxonomic guides to protozoa (Jahn *et al.*, 1979; Lee *et al.*, 1985) and rotifers (Ruttner-Kolisko, 1974) were used to identify organisms. Nauplii were enumerated but not taxonomically identified. Carbon biomass was estimated based

Table 1. Characteristics of Toolik LTER study lakes.

Lake	Depth (max)	Chlor *a*[1] ($\mu g\, l^{-})^1$	Description
Toolik	25 m	1.69	Highly oligotrophic control lake.
E–1	12 m	4.65	No manipulation, abundant small grayling
N–1	12 m	5.09	Lake trout present, fertilization of lake begun in 1990
N–2	7 m	3.46(C) 12.81(F)	No fish, lake divided by plastic curtain, one side fertilized (N + P), 1985–1990.
I–6	12 m	0.26	No manipulation, in Toolik watershed.
I–8	8 m	0.31	Lake trout removed, in Toolik watershed.
S–6	6 m	1.07	Lake trout added beginning in 1986.
S–7	4 m	0.26	No manipulation, few fish.

[1] Chlorophyll values are time and depth integrated means from 2 (I–6, I–8) to 9 (Toolik, N–1, N–2) samples from these lakes. (Data from M.C. Miller, Univ. of Cincinnatti)

on measured sizes and literature values (Ruttner-Kolisko, 1977; Pauli, 1989; Putt & Stoecker, 1989).

The experimental study conducted in limnocorrals was a follow-up to previous work reported in this volume (O'Brien *et al.*, 1992) and details will be presented elsewhere (G. Kipphut & S. Whalen, in prep.). Briefly, 1 mole nitrogen, as either nitrate or ammonia, was added in a single dose to 5 m^3 limnocorrals on July 4, 1989. The corrals were sampled three times during the summer for microplankton, once immediately prior to nitrogen amendments and twice afterwards. Data from the limnocorrals is reported as the average values from 1 and 4 m samples.

Results

The microplankton observed in lakes of the Toolik LTER site included ciliated protozoans, rotifers, and crustacean zooplankton nauplii. Most ciliates were small oligotrichs (< 50 μm in largest dimension) of the genera *Halteria*, *Strombidium* and *Strobilidium*. A *Vorticella* sp. was also present and dominated in a few samples. Other protozoans were rarely seen. Four species of rotifers dominated the microplankton: *Keratella cochlearis*, *Kellicottia longispina*, *Polyarthra vulgaris* and *Conochilus unicornis*. Five additional species were occasionally observed: *Keratella quadrata*, *Ascomorpha* (*Chromogaster*) *ecaudis*, *Filinia terminalis*, *Gastropus stylifer*, and an unidentified *Synchaeta* sp.

Microplankton assemblages demonstrated similar temporal and spatial patterns of distribution among lakes, but species composition and biomass varied. Results from Toolik Lake and the control side of Lake N-2 (Figs 1, 2) serve to demonstrate these patterns. The highest protozoan abundance was usually found during late June, the beginning of the ice-free season. Initially, protozoans were more numerous than rotifers (100 to 1000 l^{-1}, compared to 10–300 rotifers l^{-1} at the surface), but rotifer biomass was usually larger than protozoan biomass because of their larger individual size. Protozoan abundance and biomass declined rapidly in all lakes after the initial peak. Secondary peaks in numbers were evident in some lakes, but protozoan biomass generally remained low (<0.5 μg C l^{-1}). An exception to this pattern was found in Toolik Lake in 1990, where a late summer peak of ciliates was found in the upper 3 m of the water column.

Rotifer abundance and biomass was generally low early in the summer, and increased to peak values during mid July to early August in all lakes. Maximum rotifer abundance (3000 individuals l^{-1}) was found at 3 m in the fertilized side of Lake N-2 on July 20, 1989. Nauplius biomass generally equalled or exceeded rotifer biomass and tended to increase during the summer. Over all lakes, protozoan abundance decreased with depth ($r = -0.25$, $p < 0.01$), $n = 497$), rotifer abundance showed no significant correlation with depth ($r = -0.07$, $p = 0.14$, $n = 497$), and nauplius abundance showed a positive correlation with depth ($r = 0.15$, $p < 0.01$, $n = 497$).

Although seasonal patterns were similar for all lakes, absolute abundance and biomass of microplankton varied by more than an order of magnitude among the lakes (Fig. 3). Oligotrophic Toolik Lake consistently had the lowest microplankton biomass, generally less than 4 μg C l^{-1}. Lakes E-1 and N-2 are more productive than Toolik and microplankton biomass was generally an order of magnitude higher, reaching values as high as 55 μg C l^{-1} (Lake E-1, 8 m, July 19, 1989).

The distribution of rotifer species varied over time and depth. For example, in Lake E-1 in 1989, the peak abundance of rotifers was initially found in the epilimnion (0–3 m) during June and early July. Increases in number of individuals at 5–11 m depths resulted in displacement of the peak abundance to the hypolimnion by mid July. Three rotifer species demonstrated significant positive correlations with depth in Lake E-1 (*Kellicottia longispina*, $r = 0.21$, $p < 0.05$, $n = 93$; *Polyarthra vulgaris*, $r = 0.49$, $p < 0.01$, $n = 93$; and *Filinia terminalis*, $r = 0.48$, $p < 0.01$, $n = 93$). This pattern varied among lakes, however. For example, in Toolik Lake, both *K. longispina* and *P. vul-*

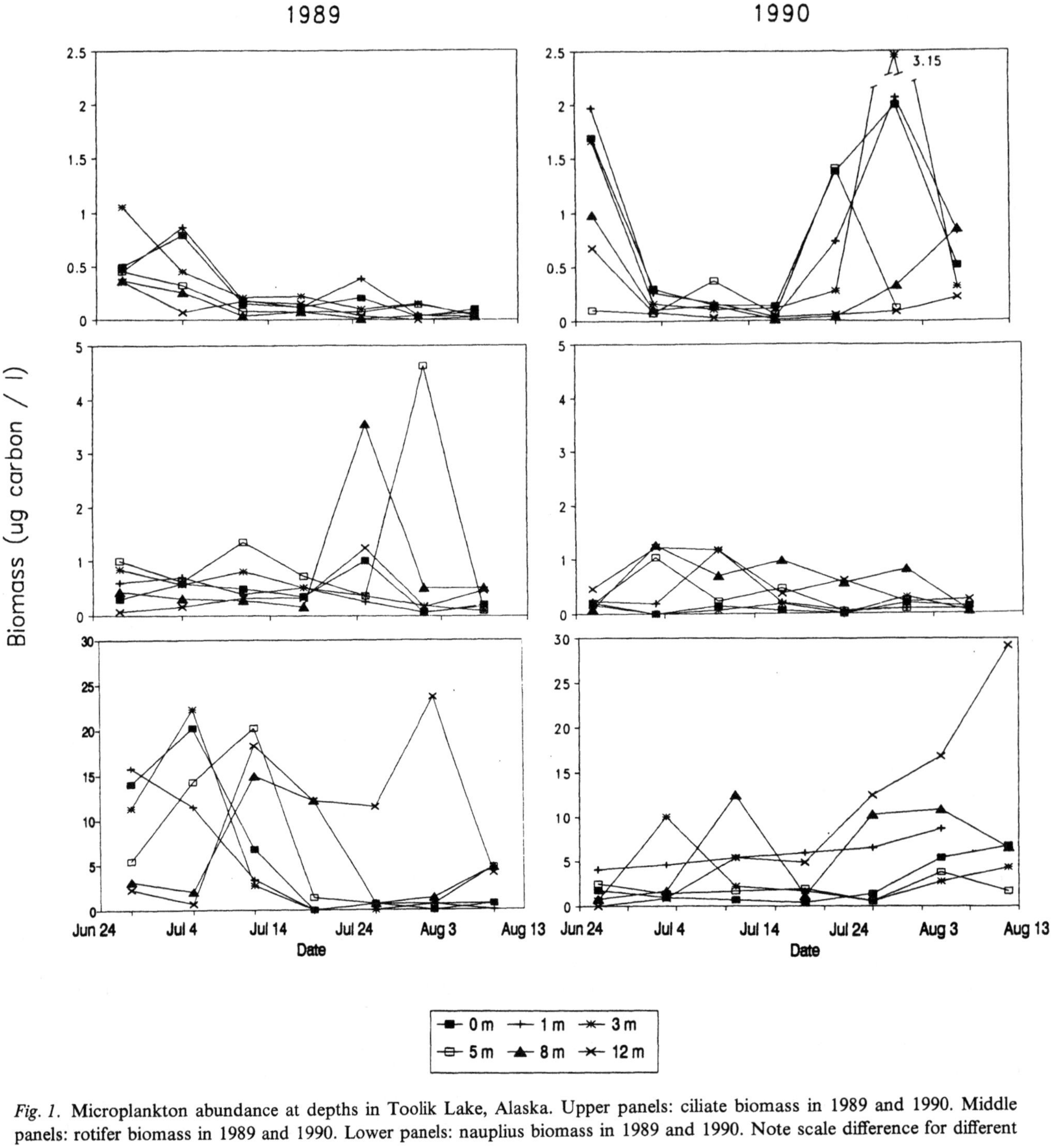

Fig. 1. Microplankton abundance at depths in Toolik Lake, Alaska. Upper panels: ciliate biomass in 1989 and 1990. Middle panels: rotifer biomass in 1989 and 1990. Lower panels: nauplius biomass in 1989 and 1990. Note scale difference for different taxa.

garis demonstrated significant negative correlations with depth ($r = -0.36$, $p < 0.01$, $n = 118$; and $r = -0.27$, $p < 0.01$, $n = 118$, respectively).

In the limnocorral experiment, total microplankton abundance appeared dependent upon trophic status. Addition of either nitrate or

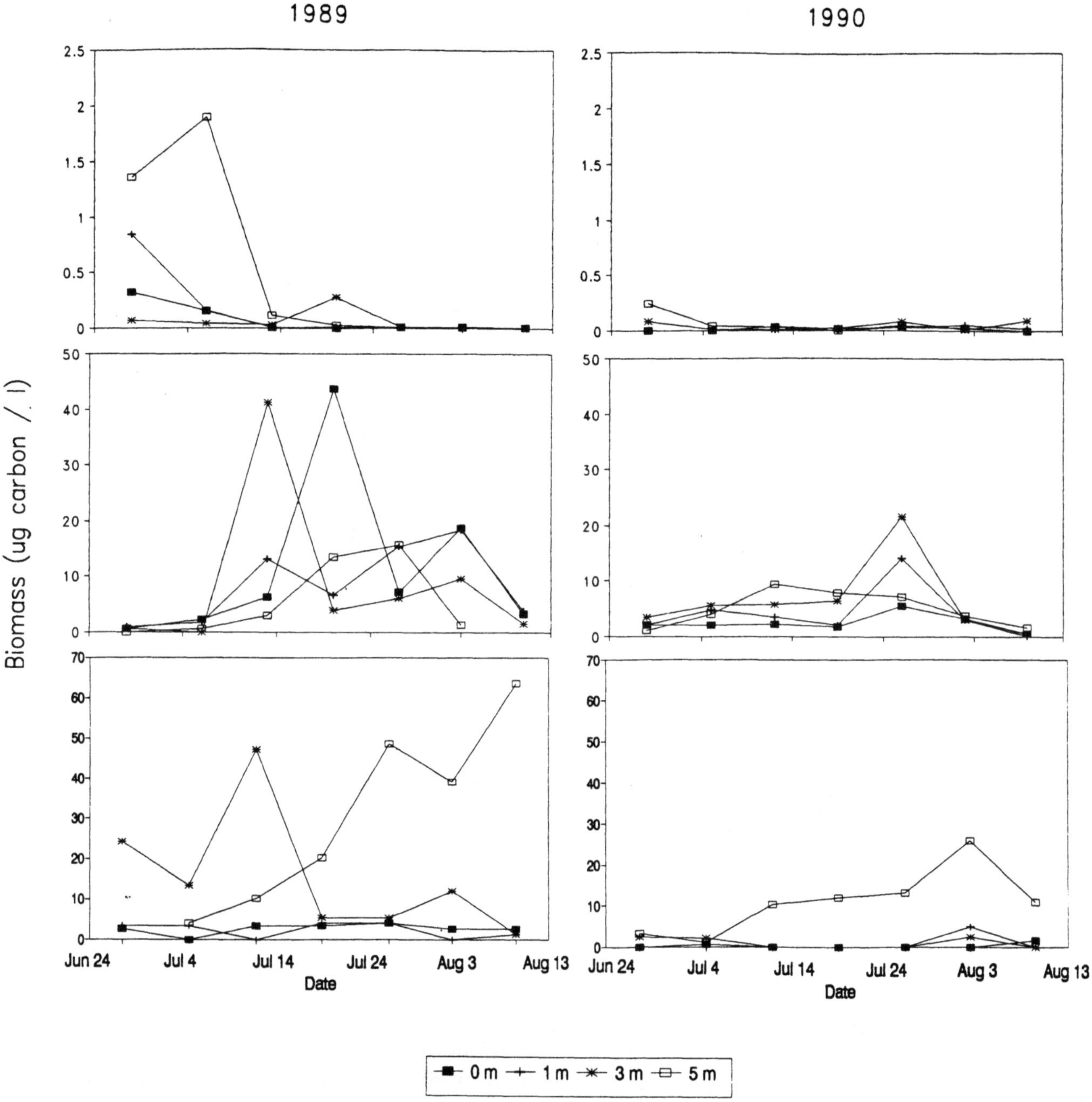

Fig. 2. Microplankton abundance at depths in Lake N-2, control side. Upper panels: ciliate biomass in 1989 and 1990. Middle panels: rotifer biomass in 1989 and 1990. Lower panels: nauplius biomass in 1989 and 1990. Note scale difference for different taxa.

ammonia resulted in dramatic increases in primary producer biomass as measured by chlorophyll *a* and was followed by increased microplankton biomass (Fig. 4). Although the relationship of microplankton biomass to chlorophyll concentration also suggested a similar dependence on primary producers, the correlation was not significant for those lakes where data on

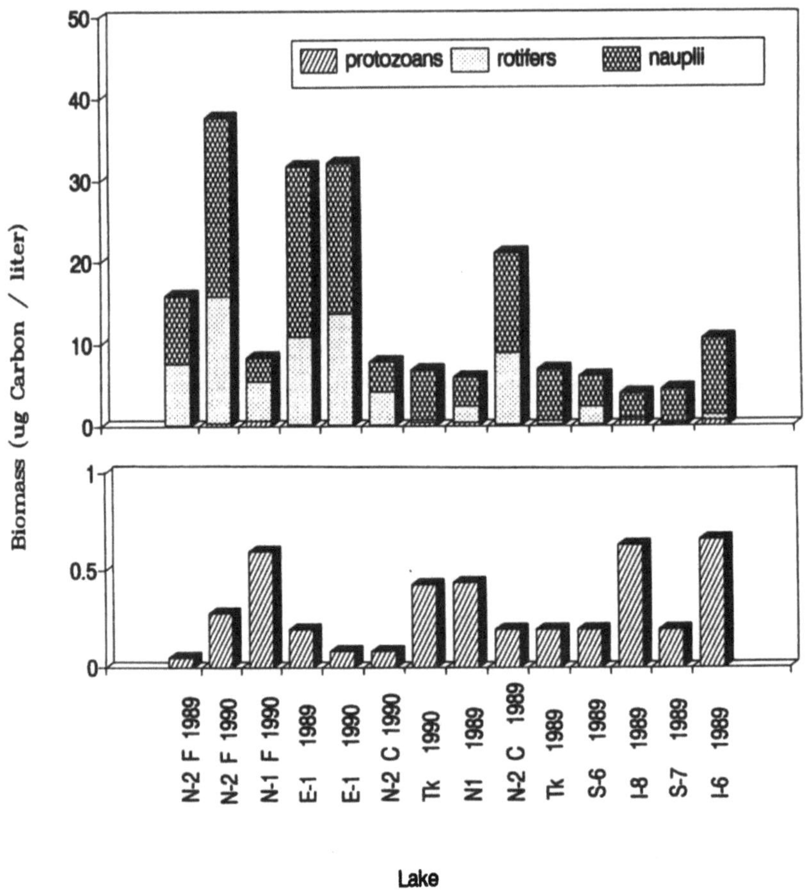

Fig. 3. Mean values for abundance of microplankton taxa in Toolik LTER lakes, 1989 and 1990. Upper, total microplankton biomass. Lower, protozoan biomass in expanded scale. Lakes are arranged in trophic spectrum from mesotrophic to oligotrophic based on chlorophyll *a* concentrations. Designation F indicates experimentally fertilized lakes and C indicates control side of N-2.

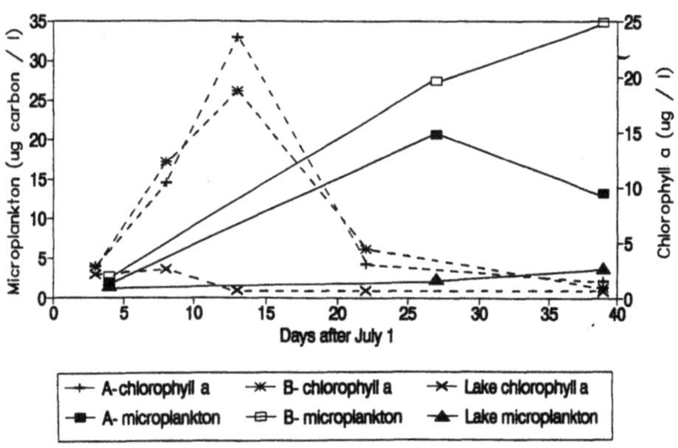

Fig. 4. Limnocorral experiment. Dashed lines represent chlorophyll *a*; Solid lines represent microplankton biomass. Limnocorral A fertilized with ammonia. Limnocorral B fertilized with nitrate. Lake samples adjacent to limnocorrals are control.

microplankton biomass and chlorophyll concentrations were available ($r = 0.337$, NS, $n = 8$).

Discussion

Previous studies provide limited information on heterotrophic microplankton in arctic freshwaters. Several studies report species lists of rotifers from arctic lakes (e.g. Moore, 1978; Chengalath & Koste, 1989), and the rotifer species reported here appear to be common to the plankton of Alaskan arctic lakes. Moore (1978) also noted the presence of ciliated protozoa, including the tintinnid *Codonella cratera*, in a series of eighteen arctic and subarctic lakes. No tintinnids were seen in lakes at the Toolik LTER site during this study.

Several other studies report biomass of microplankton in addition to identifying dominant taxa. Hobbie, Kalff, and Holmgren (unpublished data) reported six rotifer species, and biomass in the range of 0.1–0.4 mg m^{-3} for Lakes Peters and Schrader in May through July, 1968. These values represented 1% or less of the total zooplankton biomass. In Wolf Lake they found higher rotifer biomass, 1.9–5.8 mg m^{-3} over the same period, which represented up to 2% of the total zooplankton biomass. In ultra oligotrophic Char Lake, Rigler *et al.* (1974) reported annual abundance of *Keratella cochlearis*, the second most common zooplankter, to range from $0–5 \times 10^4$ individuals m^{-2} for the 27.5 m deep lake. Production of the rotifers was estimated to represent about 1% of the production of the dominant zooplankter, *Limnocalanus macrurus*. They also noted the abundance of ciliated protozoans, 0–10.3 individuals ml^{-1}, and estimated their production to exceed that of *Keratella*.

The values reported in this study for rotifer abundance and biomass are slightly higher, and in Lakes E-1 and the fertilized side of N-2, significantly higher, than most earlier studies of arctic lakes. Several factors may contribute to this. First, sampling by reverse flow filtration, the method used in this study, is less likely to damage and more likely to retain small rotifers than sampling by net tows. Second, Lakes E-1 and N-2 (fertilized) are smaller, more eutrophic systems than Char, Peters, and Schrader lakes and thus higher microplankton biomass is not unexpected.

One previous report has addressed protozoans in Toolik Lake. In that study, Hobbie & Helfrich (1988) studied heterotrophic nanoflagellates (2–20 μm in length). They found $1–2 \times 10^3$ cells ml^{-1} during July and August 1984 in Toolik Lake and control limnocorrals. They also noted that in fertilized limnocorrals, peaks of nanoflagellate abundance (up to 6.5×10^3 ml^{-1}) were found 6–7 days after peaks in bacterial numbers, suggesting a clear pattern of predator control of bacteria. These heterotrophic flagellates as well as small ciliates are taxa that were not enumerated in the current study. Preserved samples of lake water viewed in settling chambers with an inverted microscope included numerous small protozoans (e.g. 10–25 μm heliozoans) and larger myxotophic flagellates (e.g. *Dinobryon* spp.) that may be abundant but are not quantitatively retained by the reverse-flow method that was used in this study.

One goal of the LTER program is to assess the importance of bottom-up and top-down regulation in lakes (see O'Brien *et al.*, 1992). Pace (1986) presented compelling evidence for bottom-up regulation of microplankton biomass in a series of temperate Canadian lakes, and the results of this study are generally consistent with the hypothesis that biomass of a trophic level is positively correlated with bottom-up factors (food resources). For example, a comparison of the control and fertilized sides of N-2 (Fig. 3) and the limnocorral experiment (Fig. 4) indicate that bottom-up inputs lead to increased microplankton biomass, as would be predicted from temperate studies (e.g. Pace, 1986; Beaver & Crisman, 1989).

Despite general agreement of increased total microplankton biomass with increased trophic status, changes in the microplankton community structure in this study are not entirely consistent with observations of microplankton community response to changing trophic status by Pace (1986) in Quebec lakes or by Beaver & Crisman (1989) in temperate Florida lakes. Pace (1986) found little change in community structure across the trophic spectrum, and both Pace and Beaver & Crisman (1989) reported increased abundance of ciliates with increasing trophic status. In this study, the biomass of rotifers as well as the contribution of rotifers to total microplankton declined as primary producer biomass declined, while protozoans demonstrated no clear pattern (Fig. 3). It is likely, however, that chlorophyll *a*, alone, is a poor estimator of microplankton food resources in these systems. For example, the highest protozoan biomass was found in lakes N-1, Toolik, I-6, and I-8. Each of these lakes may be characterized as having small phytoplankton, probably dominated by cryptophytes, chrysophytes and centric diatoms. This compares with lower protozoan biomass in lakes N-2 and E-1, systems richer in nutrients where large phytoplankton dominate, including larger flagellates and dinoflagellates which do not represent food resource available to the small oligotrich protozoans.

The response of the microplankton community in Lake N-1 to fertilization provides additional insight into regulation of microplankton community structure. During 1989, the seasonal pattern of microplankton abundance in Lake N-1 was typical, and similar to that of Toolik Lake (Fig. 1). After whole lake fertilization of N-1 was begun in 1990, algal biomass increased and protozoan abundance showed 4-fold increase during later summer, due to both an increase in number and mean size of individual ciliates (Fig. 5). Additionally, the species composition, but not biomass, of the rotifer assemblage in Lake N-1 changed: during 1989, the rotifer assemblage was dominated by rotifers which could be termed small-particle feeders such as *Conochilus unicornis*, *Keratella* spp., and *Kellicottia longispina*, which feed mainly on small phytoplankton (Hutchinson, 1967). These species also dominated during the early summer of 1990, but by August, 'predatory rotifers', *Synchaeta* sp. and *Polyarthra vulgaris* which consume flagellates, ciliates and other rotifers (Hutchinson, 1967), comprised more than 80% of the rotifer assemblage in Lake N-1 (Fig. 6). The development of these predatory rotifer populations may be a response to the increase in protozoans which can serve as a food resource. These data also suggest interactions within the

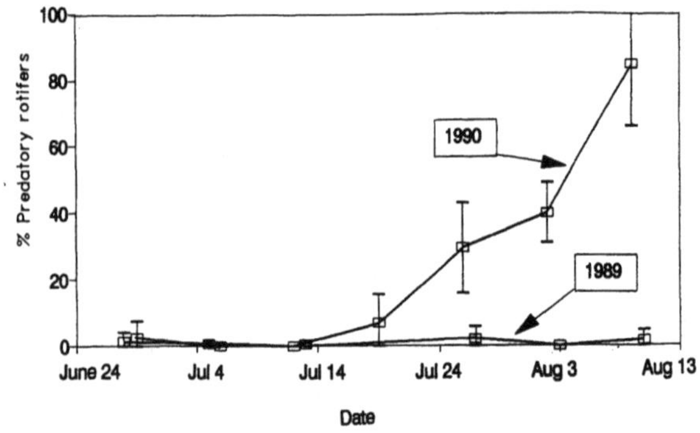

Fig. 6. Percentage of predatory rotifers (*Synchaeta* + *Polyarthra*) in Lake N-1, 1989 and 1990. Values are mean (± SD) of samples taken at 5 depths (0, 1, 3, 5, and 8 m) on each sampling date.

microbial food web which may represent a transient response as the lake system shifts toward a new equilibrium community, since predatory rotifers were not dominant in any of the other lakes studied.

Other types of regulation of microplankton also occur in these lakes. O'Brien *et al.* (1979) have noted that crustacean zooplankton community structure in arctic lakes varies considerably, and the highest abundance of rotifers was found in lakes (N-2, E-1) where the predaceous copepod *Heterocope septentrionalis* is absent. Additionally, Neill (1989) has pointed out that *Daphnia* spp. may inhibit herbivorous rotifers by interference competition, and in arctic lakes such as Lake N-1, where herbivorous zooplankton populations of *Daphnia*, *Bosmina*, and *Diaptomus* are highest, such controls may be important when food resources are limiting. Continued study of grazing interactions in control and experimentally manipulated arctic LTER lakes will aid in understanding the regulation and importance of the microbial loop in lake systems.

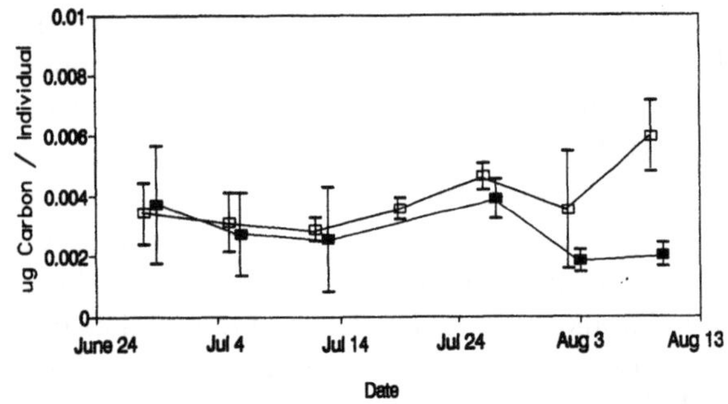

Fig. 5. Individual ciliate protozoan biomass, Lake N1, 1989 (filled square) and 1990 (open square). Values are mean (± SD) of samples taken at 5 depths (0, 1, 3, 5, and 8 m) on each sampling date.

Acknowledgements

This work was funded by National Science Foundation Grant DPP-8821679. Thanks are due to the Arctic LTER researchers for field support and to G. Kipphut, M. C. Miller, H. Kling, and W. J. O'Brien for shared data.

References

Azam, F., T. Fenchel, J. G. Field, J. S. Grey, L. A. Meyer-Reil & F. Thingstad, 1983. The ecological role of water-column microbes in the sea. Mar. Ecol. Progr. Ser. 10: 257–263.

Beaver, J. R. & T. L. Crisman, 1988. The role of ciliated protozoa in pelagic freshwater ecosystems. Microb. Ecol. 17: 111–136.

Chengalath, R. & W. Koste, 1989. Composition and distributional patterns in arctic rotifers. Hydrobiologia 186/187: 191–200.

Dodson, A. & W. Thomas, 1964. Concentrating plankton in a gentle fashion. Limnol. Oceanogr. 9: 455–459.

Hobbie, J. E., 1973. Arctic limnology: A review. pp. 127–168, In M. E. Britton (ed.) Alaskan Arctic Tundra. Arctic Inst. of North Am., Technical Paper No. 25.

Hobbie, J. E. & J. V. K. Helfrich III, 1988. The effect of grazing by microprotozoans on production of bacteria. Archiv. Hydrobiol. Beih. 31: 281–288.

Hutchinson, G. E., 1967. A treatise on limnology. Vol. II. J. Wiley & Sons, Inc. New York.

Kling, G. W., B. Fry & W. J. O'Brien, 1992. Stable isotopes and planktonic trophic structure in arctic lakes. Ecology. 72: In press.

McQueen, D. J., M. R. S. Johannes, J. R. Post, T. J. Stewart & D. R. S. Lean, 1989. Bottom-up and top-down impacts on freshwater pelagic community structure. Ecol. Monogr. 59: 289–309.

Miller, M. C., G. R. Hater, P. Spatt, P. Westlake & D. Yaekel, 1986. Primary production and its control in Toolik Lake, Alaska. Arch. Hydrobiol./Suppl. 74. 1: 97–131.

Moore, J. W., 1978. Composition and structure of zooplankton communities in eighteen arctic and subarctic lakes. Int. Revue ges. Hydrobiol. 63: 545–565.

Neill, W. E., 1988. Complex interactions in oligotrophic lake foodwebs: responses to nutrient enrichment. In Carpenter, S. R. (ed.) Complex interactions in lake communities. Springer-Verlag. New York: 31–44.

O'Brien, W. J., C. Buchanan & J. F. Haney, 1979. Arctic zooplankton community structure: Exceptions to some general rules. Arctic 32: 237–247.

O'Brien, W. J., A. E. Hershey, J. E. Hobbie, M. A. Huller, G. W. Kipphut, M. C. Miller, B. Moller & J. R. Vestal, 1992. Control mechanisms of arctic lake ecosystems: a limnocorral experiment. Hydrobiologia 240: 143–188.

Pace, M. L., 1986. An empirical analysis of zooplankton community size structure across lake trophic gradients. Limnol. Oceanogr. 31: 45–55.

Pauli, H.-R., 1989. A new method to estimate individual dry weights of rotifers. Hydrobiologia 186/187: 355–361.

Pomeroy, L. R. & W. J. Wiebe, 1988. Energetics of microbial food webs. Hydrobiologia 159: 7–18.

Porter, K. G., E. B. Sherr, B. F. Sherr, M. Pace & R. W. Sanders, 1985. Protozoa in planktonic food webs. J. Protozool. 32: 409–415.

Putt, M. & D. K. Stoecker, 1989. An experimentally determined carbon:volume ratio for marine 'oligotrichous' ciliates from estuarine and coastal waters. Limnol. Oceanogr. 34: 1097–1104.

Rigler, F. H., M. E. MacCallum & J. C. Roff, 1974. Production of zooplankton in Char Lake. J. Fish. Res. Bd Can. 31: 637–646.

Ruttner, Kolisko, A., 1974. Plankton rotifers. Biology and taxonomy. Die Binnengewässer, Suppl. Vol. XXVI, part 1.

Ruttner, Kolisko, A., 1977. Suggestions for biomass calculation of plankton rotifers. Arch. Hydrobiol. 8: 71–76.

Sanders, R. W., K. G. Porter, S. J. Bennett & A. E. DeBiase, 1989. Seasonal patterns of bacterivory by flagellates, ciliates, rotifers, and cladocerans in a freshwater plankton community. Limnol. Oceanogr. 34: 673–687.

Sherr, B. F. & E. B. Sherr, 1984. Role of heterotrophic protozoa in carbon and energy flow in aquatic ecosystems. In M. J. Klug & C. A. Reddy (eds) Current Perspectives in Microbial Ecology. Amer. Soc. Microbiol. Washington, D.C. 412–423.

Stockner, J. G. & K. G. Porter, 1988. Microbial food webs in freshwater planktonic ecosystems. In S. R. Carpenter (ed.) Complex Interactions in lake communities. Springer-Verlag, NY 69–83.

Williamson, C. E. & N. M. Butler, 1986. Predation on rotifers by the suspension-feeding calanoid copepod *Diaptomus pallidus*. Limnol. Oceanogr. 31: 393–402.

Hydrobiologia **240**: 143–188, 1992.
W.J. O'Brien (ed.), Toolik Lake – Ecology of an Aquatic Ecosystem in Arctic Alaska.

Control mechanisms of arctic lake ecosystems: a limnocorral experiment

W. J. O'Brien,[1] A. E. Hershey,[2] J. E. Hobbie,[3] M. A. Hullar,[3] G. W. Kipphut,[4] M. C. Miller,[5] B. Moller[3] & J. R. Vestal[5]
[1] *Department of Systematics & Ecology, University of Kansas, Lawrence, KS 66045, USA;* [2] *Department of Biology, University of Minnesota, Duluth, Minn., USA;* [3] *Ecosystem Center, Marine Biological Lab. Woods Hole, Mass., USA;* [4] *Departments of Biological and Geological Sciences, Murray State University, Murray, Ky., USA;* [5] *Department of Biology, University of Cincinnati, Cincinnati, Ohio, USA*

Key words: limnocorral, ecosystem control, bottom up-top down, nutrient addition, fish manipulation

Abstract

To assess the potential impact of human exploitation on arctic lakes and to determine how these ecosystems are regulated we initated a limnocorral experiment in Toolik Lake, Alaska, in the summer of 1983. The limnocorrals were 5 m in diameter and from 5-6 m in depth and were open to the sediments. In 1983 four limnocorrals were deployed in an isolated bay of Toolik Lake within a cross-classified treatment regime of high and low inorganic nitrogen and phosphorus additions and high and low free swimming fish additions. The objective of the nutrient addition was to stimulate phytoplankton growth and determine the extent to which increased plant production was passed through pelagic and benthic food chains. The objective of the fish addition was to determine the impact of fish predation on large-bodied zooplankton, especially the zooplanktivorous copepod *Heterocope*, then to study the effect of altered *Heterocope* densities on small-bodied zooplankton species population dynamics. In 1984 two more limnocorrals were deployed, one a low fish, 1 × nutrient addition treatment and the other a no fish, no nutrient treatment. The fish manipulation was changed to confining several fish in cages with the cages held in corrals for varying lengths of time.

The addition of inorganic nitrogen and phosphorus dramatically increased phytoplankton productivity. This increase in algal biomass and production greatly altered the light environment and water quality in the nutrient treated limnocorrals. The secchi disk depth in the nutrient treated limnocorrals declined each summer reaching as low as 1 m in 1985. Both oxygen content and pH increased in the nutrient treatment corrals. Corrals not receiving nutrient additions remained near lake concentrations for most water quality parameters. While phytoplankton biomass was stimulated in 1983 phytoplankton growth was not sufficient to draw down all the nitrogen and phosphorus added and these nutrients reached high levels in the last half of the summer. In 1984 phosphorus remained above 20 μg in the nutrient-treated corrals but ammonia dropped to reference levels by day 25. In 1985 both nutrient concentrations rapidly declined to reference levels.

Most pelagic components responded to the nutrient additions. Microbial production was stimulated in the nutrient treated limnocorrals and bacterial population sizes built up to nearly 8–10 times those of the reference corrals. However, microheterotrophs soon increased in abundance and apparently grazed down bacteria to reference levels. Phytoplankton population density, as estimated by chlorophyll *a* determinations, increased dramatically with nutrient addition such that each year the phytoplankton densities were higher than before. Primary productivity was also stimulated and appeared not to be light limited even when phytoplankton densities rose to high levels. In the first two years of the experiment zooplankton densities were little altered by the increased phytoplankton densities. However, by 1985 daphnid densities were quite a bit higher in the high nutrient addition limnocorrals.

The benthic community and sediment response was much less affected by nutrient addition. Overall sediment respiration increased in the nutrient treated corrals but underlying sediments seemed little affected. Decomposition of *Carex* litter was likewise little affected by nutrient addition. Benthic invertebrates were also little impacted by the nutrient addition and increased sedimentation of phytoplankton. However, the response of benthic invertebrates is difficult to assess fully in the current experiment because chironomids, a prominent component of the benthic community, failed to recruit into the limnocorrals and the corrals physically shifted during ice-out in the spring of 1984 disturbing the sediment in several corrals.

The fish additions in 1983 of free swimming grayling essentially eliminated large bodied zooplankton, especially *Heterocope septentrionalis*, from all four limnocorrals. In subsequent summers *Heterocope* were not so dramatically preyed upon but generally were found in higher densities in the low or no fish treatments. However, either when *Heterocope* were eliminated in 1983 or were in rough inverse proportion to fish density, altered *Heterocope* abundance had no obvious affect on small-bodied zooplankton abundance. The fish treatment apparently influenced the zooplankton response to high nutrient addition in 1985. In the high nutrient limnocorrals daphnid populations became very abundant, but in the high fish treatment the daphnid responding was the small-bodied *D. longiremis* while in the low fish treatment the daphnid responding was the large-bodied *D. middendorffiana*.

Thus we have considerable evidence for bottom up control of phytoplankton density and production. This increased production ultimately, but not for two years, stimulated zooplankton density increases. Increased nutrients had little effect on the benthos or sediments. Fish manipulations influenced large-bodied zooplankton but had little effect on small-bodied zooplankton. Because grayling are predominantly plankton feeders in lakes, no fish effect on benthic invertebrates was expected.

Limnocorrals thus seem good systems to study nutrient–phytoplankton interactions. They are not as suitable for benthic invertebrate studies and fish manipulations may be difficult. Most other limnocorral studies were of brief duration; however, in the present study the limnocorrals seemed to perform well over a three year period.

Introduction

Regulation of lake ecosystems

Ecologists have long been interested in the mechanisms regulating natural populations and controlling the rate of flow of energy through ecosystems. A major question has been whether control is exerted by limited amounts of nutrients available to plants (Liebig, 1874) or by mortality due to predators (Huffaker, 1958; Paine, 1966; Brooks & Dodson, 1965). Hairston *et al.* (1960) proposed that herbivores in aquatic communities are food limited whereas terrestrial herbivores are predator limited. Smith (1969) theorized that top predators must be limited by the abundance of their prey and also that top predators control prey abundance. When a population is regulated by predation, however, it cannot in turn regulate the abundance of its food. Thus, a saw-toothed pattern of abundance should occur within a food chain if trophic levels are first resource limited and then predator limited. In aquatic ecosystems it is common to refer to resource limited systems as regulated from the bottom up and to predator limited systems as controlled from the top down (Shapiro, 1990; McQueen, Post & Mills, 1986).

The idea that lake ecosystems are regulated through bottom up control began with Thienemann (1928), who noted that lakes which developed large algal blooms had unusual species of phytoplankton and were well supplied with plant nutrients, especially nitrogen and phosphorus. He coined the term eutrophic to describe such lakes. With the development of detergents and attendant polyphosphates, such eutrophication became a major water pollution problem (NAS, 1969). Controlled experiments left little doubt that nitrogen and especially phosphorus greatly stimulate phytoplankton growth in most lakes (Schindler, 1974). The extent to which this increased phytoplankton productivity is passed up the food chain has remained a significant question, especially in arctic lakes.

Some investigators (Le Brasseur & Kennedy, 1972; Hillbricht-Ilkowska & Weglenska, 1970; O'Brien & deNoyelles, 1974) have reported a direct relationship between an increase in phytoplankton biomass resulting from added nutrients and an increase in zooplankton biomass. In addition, Hall *et al.* (1970) found that benthic invertebrates as well as zooplankton increased when nutrients were added to ponds. In a series of deep large lakes, Hyatt & Stockner (1985) observed larger standing stocks of zooplankton and increased growth of salmon fry with added nutrients and Kelso (1985) found increased fish production with increased primary productivity. In contrast, in a series of tank mesocosms Drenner *et al.* (1989) found no increases in zooplankton with increased nutrients. Likewise Bare Lake in Alaska received fertilizer additions which stimulated phytoplankton but zooplankton abundance did not change (Nelson & Edmondson, 1955). Similarly, zooplankton biomass did not increase after fertilization of ELA Lake 226 (Malley *et al.*, 1977) although whitefish growth rates, condition, and fecundity increased (Mills, 1985). In fact, in ELA Lake 227 zooplankton declined after fertilization although phytoplankton primary production and biomass were greatly increased (Malley *et al.*, 1977). This decline may have been due to high pH levels caused by intense photosynthesis (O'Brien & deNoyelles, 1972).

There are a few reported cases of increased phytoplankton productivity and biomass in arctic lakes following nutrient addition. During ice melt Meretta Lake in Canada received sewage from a nearby military installation while nearby Char Lake did not. In response, Meretta Lake developed 10 times more chlorophyll than did Char Lake. Furthermore, in Meretta Lake the major crustacean zooplankter disappeared, apparently due to resting egg mortality induced by the eutrophic state of the lake (Schindler *et al.*, 1974 a.b.; Kalff & Welch, 1974; Kalff *et al.*, 1975). Several small lakes near Sagragjnac, Canada (63° N) were fertilized with nitrogen and phosphorus separately and together. Phosphorus alone did not increase phytoplankton productivity but nitrogen and phosphorus together led to sustained increases in primary productivity (Welch *et al.*, 1989).

In an experimental fertilization of a lake in northern Sweden, primary productivity and phytoplankton biomass were both stimulated by the addition of phosphorus but not by nitrogen. Initially, there was little or no effect on the zooplankton, but after two years of fertilization, *Bosmina* did begin to increase (Holmgren, 1984; Jansson, 1978).

Benthic invertebrates, especially gastropods, are often food limited; competition for food seems to control both snail growth and fecundity (Eisenberg, 1966; Brown, 1982; Cuker, 1983; Hershey, 1990). A correlation has been reported between chironomid productivity and lake primary productivity (e.g., Jonasson, 1972; Aagaard, 1978; Moore, 1978), but the mechanisms involved have not been well studied and may vary in different systems.

Thus, the evidence for bottom up control of herbivore populations in arctic as well as temperate lakes is not compelling. While added nutrients almost always stimulate increased plant production and biomass in lakes, increases in higher trophic levels are spotty. This is especially puzzling in the case of cladoceran zooplankton, which have short generation times and are commonly resource limited and thus could show dramatic and rapid increases. Such a response is not commonly observed, however.

The alternative idea, that lake ecosystems are regulated through top down control is a fairly recent concept first put forth for plankton systems by Hrbacek (1962) and Brooks & Dodson

(1965). These authors noted that the introduction of planktivorous fish into lakes dramatically altered the size and species composition of the zooplankton; most large zooplankton species were eliminated and replaced by small sizes and species. Recently it has been suggested that manipulating planktivorous fish could have a cascading effect on lower trophic levels (Carpenter *et al.*, 1985; Shapiro & Wright, 1984). Hall *et al.* (1970) showed that fish stocked in experimental ponds had dramatic impacts on both the planktonic and benthic food webs. In the wake of these findings many investigators have noted the strong correlation between the presence of planktivorous fish and small size zooplankton (Nilsson & Pejler, 1973; O'Brien, Buchanan & Haney, 1979; Post and McQueen, 1987). Dodson (1979), however, showed that numerous plankton communities did not appear to be regulated by food or by predation and thus the relationship is more complex.

Brooks & Dodson (1965) suggested that, in the absence of fish, competition was the force acting to maintain large sized species (see Hall *et al.*, 1976 for a review to that date). Dodson (1974) has further suggested that predation from invertebrate planktivores might be the more powerful force involved. Subsequent study has shown that invertebrate planktivores may indeed exclude small bodied zooplankton from lakes and ponds where such predators can become abundant (Neill & Peacock, 1980; Kerfoot, 1977; Luecke & O'Brien, 1983; O'Brien & Luecke, 1988; Lunte & Luecke, 1990). Thus, there is strong evidence that both vertebrate visual planktivores and invertebrate tactile planktivores can exert sufficient predation pressure on plankton food webs to eliminate those sizes and species most vulnerable to these predators. What is less clear, however, is the extent to which these changes in zooplankton community structure and density cascade downward and affect phytoplankton and microbial community structure and density.

In several studies (Hrbacek, 1962; Hurlbert *et al.*, 1972; Losos & Hetesa, 1973; Lynch & Shapiro, 1981) in which high densities of planktivorous fish have been added to ponds or to enclosures in ponds, zooplankton were severely reduced and phytoplankton generally stimulated (see Northcote, 1988 for a review). Hrbacek *et al.* (1978) pointed out that reservoirs with low fish stock have much lower chlorophyll concentrations than would be predicted from the phosphorus to chlorophyll correlations of Dillon & Rigler (1974) and suggested that resulting high zooplankton densities reduced chlorophyll concentrations. However, reservoirs with more normal fish stocks and zooplankton densities fit the correlation well. Similar trends have occurred in Lake Michigan as the top predator lake trout varied in density through time (Kitchell & Carpenter, 1987).

In an experiment with 3 large limnocorrals stocked with varying densities of fish, Post & McQueen (1987), and McQueen & Post (1988) found a strong negative relationship between fish biomass and zooplankton biomass. However, there was only a very weak relationship between variable zooplankton biomass and phytoplankton density. They conclude that such top down effects are unlikely to profoundly influence phytoplankton density.

In several recent studies, phytoplankton have been found to increase upon fish addition (Vanni & Findlay, 1990; Threlkeld, 1988). However, it has been questioned whether this observed increase in phytoplankton is due to a reduction in grazing pressure (Vanni & Findlay, 1990), but rather an increase in nutrient cycling by the resultant small-bodied zooplankton (Carpenter & Kitchell, 1984). Threlkeld (1987, 1988) has argued a third possibility that added fish may die and release nutrients as they decompose. Vanni & Findlay (1990) present evidence consistent with the idea that nutrient recycling from both fish and small zooplankton may stimulate phytoplankton.

Only a few studies have investigated the effects of fish removal on the entire pelagic food chain (Carpenter *et al.*, 1987). Stenson *et al.* (1978) removed fish from a small Swedish lake and found that large zooplankton soon dominated the zooplankton community, primary productivity declined by 90% and large phytoplankton increased in importance. In a more natural experiment, Benndorf *et al.* (1984) reduced planktivorous fish abundance in a small quarry pond by the addition of piscivorous fish and a definite cascading effect was seen throughout the food chain. Zooplankton biomass increased 400% with large daphnids appearing for the first time, while phytoplankton biomass in the pond remained the same although the composition of the phytoplankton community shifted predominantly to small species or ones with gelatinous coating such as *Oocystis*. A similar shift in phytoplankton community structure was also found by Porter (1976) in studying high densities of zooplankton in small enclosures in a pond and by Carpenter *et al.* (1987) in a small lake.

Experimental work on lake benthic communities has implicated predation as an important control mechanism. When fish are removed from ponds there is generally an increase in biomass of benthic insects (Ball & Hayne, 1952; Hayne & Ball, 1956; Hall *et al.*, 1970; Crowder & Cooper, 1982). The effect of fish predation is generally greatest on large invertebrates for most benthic communities (Hall *et al.*, 1970; Crowder & Cooper, 1982), but for infaunal chironomid communities, the effect is greater on smaller-bodied individuals (Werner *et al.*, 1983; Hershey, 1985a). Susceptibility to fish predation may be determined mainly by habitat structure on lake bottoms where fish are not effective predators on benthic invertebrates in habitats with dense macrophyte stands (Thorpe & Bergey, 1981; Crowder & Cooper, 1982; Gilinsky, 1984; Hershey, 1985a; Luecke, 1990).

Fish predation on snails does appear to be important in controlling size-frequency distribution and community structure (Brown & DeVries, 1985; Merrick *et al.*, 1991). However, fish predation may affect snail community structure without affecting biomass by causing shifts in the relative abundance of species. For example, fish which crush snails select prey based on shell strength (Stein *et al.*, 1975, 1984; Vermeij & Covich, 1978), thereby affecting the species composition of the community. Snail recruitment has been shown to be controlled by the feeding of mudminnows on snail eggs and juveniles (Brown & DeVries, 1985). Trout also eat snails (Macan, 1966; Smith, 1972; Johnson, 1972; Ware, 1972; Tippets & Moyle, 1978; O'Brien *et al.*, 1979a; Merrick *et al.*, 1991), especially when a pelagic forage fish is absent (Nilsson, 1972).

There is evidence that arctic plankton communities are also structured by predation. Nilsson & Pejler (1973) stocked various combinations of fish species in Swedish lakes and found that when very planktivorous fish species were stocked, small sized zooplankton communities developed. In the Alaskan subarctic, O'Brien (1975) found clearly delineated zooplankton communities; in most of the deep lakes planktivorous whitefish were abundant and only a small sized zooplankton community existed. However, in ponds where no fish occurred only the predaceous copepod *Heterocope* and other large bodied zooplankton were found. More recently, it has been well documented that *Heterocope* predation is fully intense enough to exclude small bodied species (Dodson, 1979, 1984; Luecke & O'Brien, 1983; O'Brien & Luecke, 1988).

The patterns of abundances and species distribution in Toolik Lake, Alaska, demonstrates an interesting mix of bottom-up and top-down processes. Plant nutrient concentrations are very low (Whalen & Cornwell, 1985) and phytoplankton biomass and production are also low (Miller *et al.*, 1986). Juvenile arctic grayling and small lake trout are the major vertebrate planktivores and these are thought to be relatively rare in the open waters of the lake due to the presence of large piscivorous lake trout (O'Brien *et al.*, 1979a). Probably because of the scarcity of planktivorous fish, the invertebrate predator, *Heterocope*, is also present in Toolik Lake in fair abundance; therefore the zooplankton in the lake face both large size (fish) and small size (*Heterocope*) predation with *Heterocope* predation perhaps the most intense (O'Brien *et al.*, 1979b). The smaller zooplankton in the lake do seem to be facing severe *Heterocope* predation pressure since *Bosmina longirostis* and *Daphnia longiremis* decline in abundance each year in Toolik Lake whereas in other lakes in the area, which lack *Heterocope*, populations of these two species increase throughout the summer.

The benthic invertebrate community in Toolik Lake is also structured by predation and resource limitation. The population density of chironomids is limited by sculpin predation on bare sediments but not in macrophyte beds (Hershey, 1985a). The population density of snails appears to be limited by an interaction of lake trout predation and resource competition (Merrick *et al.* 1991).

However, these controls of the aquatic populations at Toolik Lake are changing through alterations brought on by human settlement and exploitation of fish. Toolik Lake has been heavily fished during the past 12 years since the Dalton Highway was constructed and overall body size of lake trout has significantly declined. Gill net surveys in Toolik Lake in 1977 compared with surveys in 1986 carried out in the same way show a decrease in median size of Toolik lake trout (McDonald & Hershey, 1989). The medium size of Toolik trout is now below the smallest reproductive size previously observed for Toolik trout; however, condition factor has increased (McDonald & Hershey, 1989).

There is no evidence of population changes in lake trout in nearby lakes that have not been extensively fished. For example, the median size of trout in 1986 in Lake 1–8, a large lake about 1 km southeast of Toolik Lake in the same watershed, was 387 mm which is similar to Toolik trout in 1977 (McDonald & Hershey, 1989). Historical data from Alaska Fish and Game on trout from Itigaknit Lake, a large lake 3 km northwest of Toolik Lake show even larger trout of 415 mm fork length in 1975 and 454 mm fork length in 1979. In 1987 median fork length of trout from Itigaknit was 468 mm (McDonald & Hershey, 1989). Thus the median size of trout in Toolik Lake seems to have declined but in more remote lakes no such change has occurred. Such a reduction in overall body size and possibly population size could have considerable implications for food web dynamics in Toolik Lake.

To study the control mechanisms operating in Toolik Lake and the ramifications of altering these controls, a limnocorral manipulation experiment was performed beginning in 1983. In 1983 four large limnocorrals were constructed of reinforced polyethylene and deployed in a small bay in Toolik Lake and two more corrals were added in 1984. Two general treatments were applied:

Table 1. Treatments applied to specific limnocorrals.
New water enclosed

		No fish	Low fish 7 fish days per week	High fish 42 fish days per week
1983	No nutrients (0)		A	C
	Low nutrients			
	High nutrients (+10)		B	D

New water enclosed

		No fish	Low fish 4.7 fish days per week	High fish 21 fish days per week
1984	No nutrients (0)	F	C	A
	Low nutrients (+1)		E	
	High nutrients (+10)		B	D
1985	No nutrients (0)	F	A*	C
	Low nutrients (+1)		E	
	High nutrients (+5)		B	D

* New water enclosed

added nutrients and added planktivorous fish predators. The treatment design is summarized in Table 1. The objective of the nutrient additions was to stimulate phytoplankton growth and determine the extent to which any increased phytoplankton production was passed up through the planktonic and benthic food chains. The objective of the fish additions was to alter the density of *Heterocope*, a planktivorous copepod thought to be a major predator in Toolik Lake (O'Brien, *et al.*, 1979b), and determine the extent to which release from this predator cascades down through the planktonic food chain. These treatments were maintained for 3 years. In 1986 the fish treatment was discontinued and the nutrient treatments were changed considerably so that the results from the 1986 experiment are not included here.

Site description of Toolik Lake

Toolik Lake (68°38′00″N, 149°36′15″W) is a complex kettle lake (Fig. 1) and the largest of a number of lakes set in a terminal moraine of the Itkillik II glaciation, a late Wisconsin local glaciation. It is located in the foothills of the Brooks Mountains at 720 m elevation, some 20 km north of the mountains and just off the Dalton Highway. The lake is about 12400 years old (Colinvaux, personal communication). The morainal material around the lake is composed mostly of Devonian conglomerates of the Chandler formation and Cretaceous sandstones (Keller *et al.*, 1961). These moraines extend into the lake itself where they form shoals which divide the lake into several rather discrete basins (Miller *et al.*, 1986) (Fig. 1). Tussock and upland heath tundra dominate the vegetation in the watershed and ground cover is nearly 100% (Miller *et al.*, 1986).

The lake has a surface area of 149 ha and a volume of 10×10^6 m^3. The maximum depth is 25 m, the mean depth 7 m and only 10% of the lake is deeper than 10 m. The Toolik watershed covers 65 km^2, a ratio of catchment to surface area of 43:1. The major inlet drains 75% of the watershed including a series of 12 lakes upstream of Toolik Lake. A secondary inlet which drains another 7% of the watershed and several ephemeral streams account for the remainder of the inflow (Whalen & Cornwell, 1985). The major inflow comes in the spring from snowmelt from mid-May to early July (Hobbie *et al.*, 1983). Ice-out in the lake has varied from 23 to 30 June over the last 10 years. The lake level, which fluctuates by as much as 59 cm, is generally highest in the late spring at the time of major inflow, and declines throughout the summer.

The average monthly summer air temperature varies from 5 to 14 °C while the average annual temperature is -10 °C. The summer rainfall varies from 30 to 40 mm with the annual precipitation varies from 200–250 mm. Most of the precipitation occurs as snowfall. Summer solar radiation ranges from 10 to 45 cal cm^{-2} day^{-1}.

The lake is dimictic despite its northern location and the epilimnetic temperature varies from 12°–18 °C during early July. The lake first stratifies in July when the thermocline is located at 5 m. By August, it is at 7 m and the lake mixes in late August. The ice cover forms in early October and may be up to 1.5 m thick and lasts for 8.5 months. The lake is fairly clear with summer secchi disk readings around 6–7 m.

The chemistry of Toolik Lake is typical of oligotrophic lakes. Oxygen is generally near saturation even in the hypoliminion during summer stratification and decreases by less than 50% during the long arctic winter. The alkalinity is 0.4–0.5 meq^{-1} and summer pH is around 7.6 but water pH may be slightly lower in the winter (Whalen & Cornwell, 1985). Plant macro-nutrients are extremely dilute because of the low annual loadings of total dissolved nitrogen and phosphorus of 290 mmols N m^{-2} and 4.64 mmols P m^{-2}, respectively. Average lake concentrations of nutrients are 6.4 $\mu g\,l^{-1}$ NO_3-N and 2.4 $\mu g\,l^{-1}$ NH_4-N and 7.1 $\mu g\,l^{-1}$ total phosphorus (Whalen & Cornwell, 1985).

The low levels of plant nutrients lead to low levels of phytoplankton biomass and productivity. Photic zone chlorophyll averages around 1.3 $\mu g\,l^{-1}$ (Whalen & Cornwell, 1985) and phytoplankton primary productivity averages less than 14 g C m^{-2} yr^{-1}, ranging from 16 to 228 mg

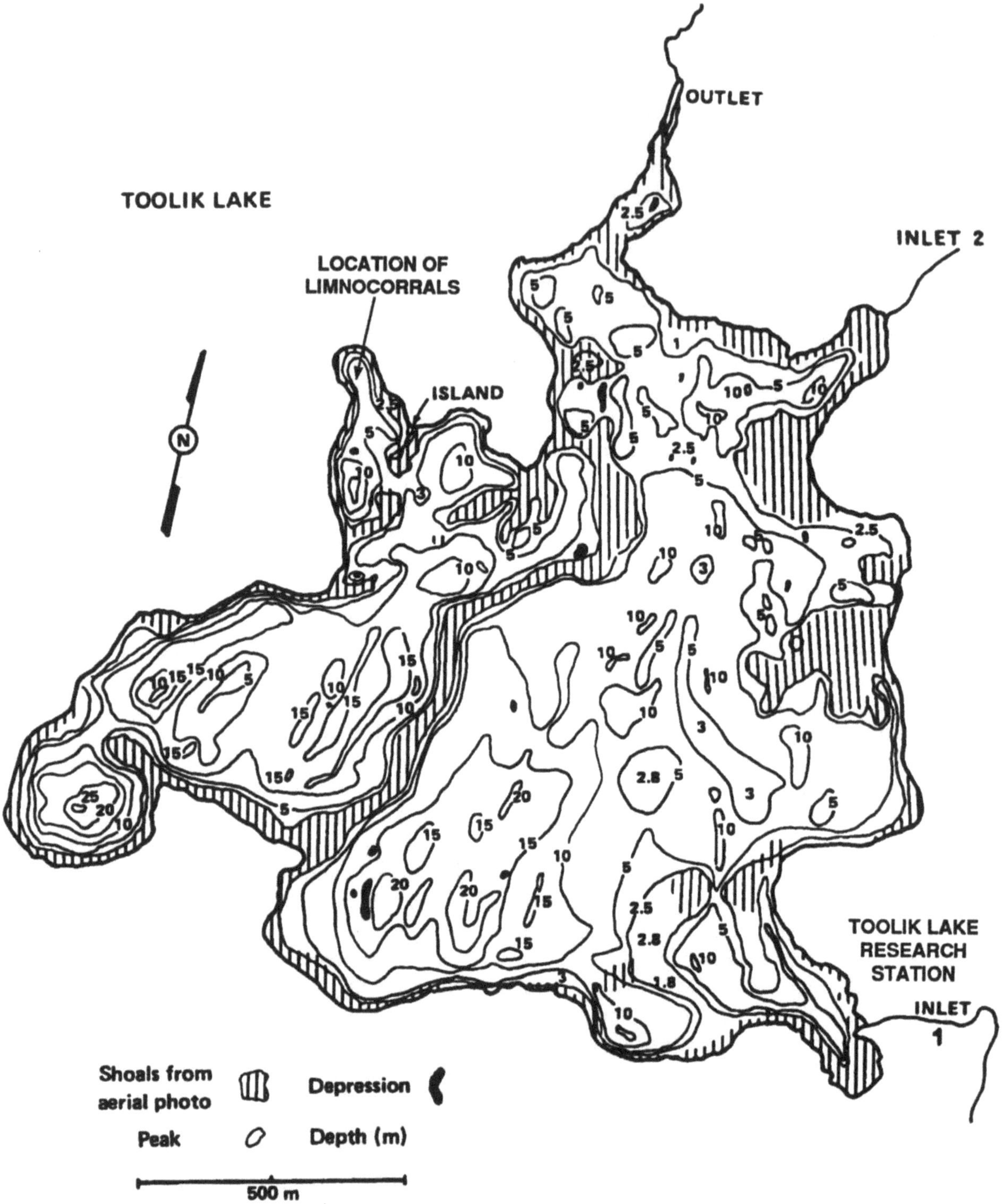

Fig. 1. Morphometry of Toolik Lake and location in which the limnocorrals were emplaced.

C m^{-2} day^{-1} during summer months (Miller *et al.*, 1986). Productivity tends to decline through the summer with only one summer out of the seven sampled showing an increase as the thermocline breaks down. Unlike many northern oligotrophic lakes, Toolik Lake does not develop a hypolimnetic phytoplankton maximum and primary productivity is always greatest near the surface.

The phytoplankton in Toolik Lake are sparse, around 2×10^6 cells l^{-1}, and small, generally less than 6 μm in diameter. Most of the algae are flagellates belonging to the Chrysophyceae and Cryptophyceae. There are some planktonic centric diatoms (*Cyclotella compta* and *C. bodanica*) which are also quite small (Miller *et al.*, 1986). Bacteria numbers are also low, commonly 0.5×10^6 ml^{-1} (Hobbie *et al.*, 1983).

The common species of zooplankton are *Bosmina longirostris*, *Daphnia longiremis*, *Diaptomus pribilofensis* and *Cyclops scutifer*. The densities of these species in Toolik Lake are quite low. *Bosmina* populations occasionally reach 10 l^{-1} but are commonly 1–5 l^{-1} with the other three species even less abundant. An omnivorous copepod, *Heterocope septentrionalis*, can be a very important predator on these herbivorous species but is rarely more abundant than 0.3 l^{-1}. Typically, *B. longirostris* & especially *D. longiremis* decline in population density by nearly a factor of 10 during the short arctic summer. Two other species, *Daphnia middendorffiana* and *Holopedium gibberum* are large bodied but uncommon, often less than 0.1 l^{-1} (O'Brien *et al.*, 1979a). Rotifers are rare and are of no quantitative significance in Toolik Lake (Rublee, 1992).

The benthic environment of Toolik Lake may be divided into rocky littoral habitats and soft sediment habitats. The rocky habitat occurs around the shore and between the basins of the lakes, comprising approximately 25% of the bottom of the lake. Areas of soft sediment make up the rest of the lake bottom. The shallower, soft sediments are patchily covered with beds of diminutive macrophytes (*Nitella flexilis*) and mosses.

The benthic invertebrate fauna of Toolik Lake is dominated by chironomids and molluscs. The chironomid community is comprised of at least 27 species, spread among 21 genera (Hershey, 1985b). Densities of larval chironomids vary from 8300–20 000 individuals m^{-2} in the macrophytes and from 3800–10 700 m^{-2} in the bare soft sediments (Hershey, 1985a, 1985b). The molluscs include two genera of clams (*Pisidium* and *Sphaerium*) and four species of snails (*Lymnaea elodes*, *Valvata lewisi*, *Gyraulus* sp. and *Physa* sp.) Larval caddis flies also occur but are not abundant.

Five species of fish are present in Toolik Lake. Lake trout, *Salvelinus namaycush*, feed on zooplankton when small but feed on snails, other benthic invertebrates and other fish as adults. Round whitefish, *Prosopium cylindraceum*, feed on zooplankton when they are juveniles but feed on benthos when large (O'Brien *et al.*, 1979a). Arctic grayling (*Thymalus arcticus*) feed on zooplankton when small but become omnivorous when longer than 15 cm (O'Brien *et al.*, 1979a). The slimy sculpin (*Cottus cognatus*) is quite common in the rocky littoral and soft-sediment zones and feeds entirely on benthic invertebrates (McDonald *et al.*, 1982; Cuker, 1981; Hershey, 1985a; Hershey & McDonald, 1985). The burbot (*Lota lota*) is also present and has been observed to feed on sculpin (Hershey, pers. obs.) and lake trout (McDonald, pers. comm.).

Methods

Limnocorral construction and treatments

Limnocorral construction

The limnocorrals have three major structural components: the reinforced polyethylene plastic tubes, the flotation collars, and an attachment beam. The tubes were made of two layers of 6 mil translucent polyethylene reinforced with fiberglass mesh. This material was formed into a cylinder 15.2 m in circumference and 6 m deep. Two pockets, extending around the full circumference of the corrals, were sewn at one end, one at the very bottom and another 1 m up from the bottom.

Each of these pockets was threaded with 15.2 m of chain to weight the bottom of the tubes.

The flotation collars were constructed from fabricated beams. Each of these beams was constructed by sandwiching a layer of 5 cm thick, dense Styrofoam between two 5 × 20 cm boards 4.8 m long with the layers glued and bolted together. The beams were formed into a square 4.8 m on a side and then into an octagon by bridging the corners with smaller beams (Fig. 2).

The attachment beam was constructed from three thicknesses of 5 × 20 cm boards bolted together to form a 13.4 m long beam kept afloat with six 210 liter oil drums, two located at each end and two in the middle. The beam was held in position in 5–6 m deep water with eight anchors. To hold the limnocorrals in place at 'ice out', each corner of the limnocorral array was anchored to shore by 1 cm steel cable attached to reinforcing rod driven into the ground at the water line. To further reduce movement at 'ice out' and prevent ice flows from intruding from the main basin of the lake two ice booms were constructed from 10 and 12 wooden beams, similar in shape and size to railroad ties, attached to a 1 cm steel cable and anchored to the shore as described for the limnocorral cable.

The limnocorral array was situated in a small bay in Toolik Lake (Fig. 1). The bay was chosen because it is protected from wind on three sides by the steep rising shoreline and from major wave action by a small island and shoal which cuts most of the way across the mouth of the bay.

In 1983, four flotation collars were constructed and fastened to the attachment beam, two along each side with a 3.6 m space between them (Fig. 2). This arrangement allowed some movement of the flotation collars while the attachment beam was held rigidly in place. To put the corrals in the lake, each tube was gathered up into a large circle of material and the top clamped to the eight sides of the flotation collar with three 'C' clamps per side. Ropes were run underneath the flotation collar and the gathered tube and as the rope was slowly played out from each corner of the flotation collar the tube was slowly lowered. A diver with SCUBA gear observed the lowering and

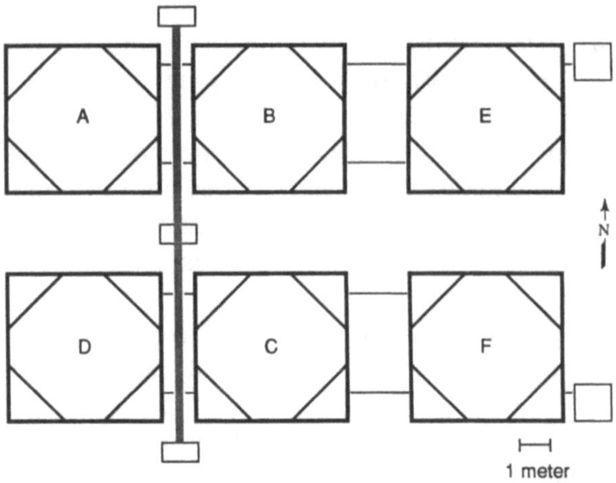

Fig. 2. Diagram of the limnocorrals and attachment beam.

pushed the bottom of the tube into the sediment once the limnocorral was in place. Within a few days of deploying the limnocorrals all or most of the contained zooplankton had died. The corrals were raised, left raised for 2 days and then redeployed after which there was no dramatic zooplankton mortality. Lund & Reynolds (1982) discuss at some length the choice of corral material to minimize such effects. They point out that plastics may have various potentially toxic additives which might not even be known to the suppliers and hence they used butyl rubber (Esso butylite).

The limnocorral array shifted during ice out in the spring of 1984. This movement disturbed the sediment and collapsed one limnocorral. Therefore in late June of 1984, all four corrals were raised, scrubbed and lowered again. Two more limnocorrals and flotation collars were constructed in 1984. The only significant difference in construction was that the flotation collars were constructed so that each side of the octagon was of equal length so as to maximize the volume of water enclosed. To further increase the volume of water enclosed, the new tubes, as well as the original four tubes, were lowered over a rigid circle of plastic pipe 15.2 m in circumference.

In the early spring of 1985 a sampling hole was drilled in the ice cover of corral A. Surprisingly, this allowed water to flow out of the hole and corral A collapsed. In June of 1985 corral A was raised and lowered again as described for the 1984 deployment. However, each spring after the ice has melted there has been a lens of dilute water, approximately 1 m thick, on the top of each corral. This water was mixed into the whole corral with an air-lift pump to simulate lake turnover. The limnocorrals have needed little maintenance during the three years of use.

Limnocorral treatments

Nutrients. The rates of nutrient addition to the limnocorrals approximated whole multiples of the natural loading rates of nitrogen and phosphorus to Toolik Lake (Whalen & Cornwell, 1985). The 10 × nutrient addition levels for 1983 and 1984 were 2.77 g N corral^{-1} day^{-1} and 0.465 g P corral^{-1} day^{-1} (Table 1). This is equivalent to 70 μg $N\text{-}NH_3$ l^{-1} day^{-1} and 4.9 μgs $P\text{–}P0_4$ l^{-1} day-1 for the estimated corral volume of 92 m^3. Solutions of ammonium chloride and potassium dihydrogen phosphate were used as sources for N and P. Nutrient solutions were added to the surface water of each treated corral every day of treatment with no attempt to mix the solutions evenly throughout the corrals. In 1983, there were two high nutrient treatments (10 ×), corrals B and D and two control nutrients (no nutrients added), corrals A and C. In 1984, there were two high nutrient (10 ×) corrals B and D, one low nutrient (1 ×) corral E and three control nutrient (no nutrient added) corrals A, C, and F (Table 1). These treatments were continued in 1985 except that the high treatments were reduced from 10 × to 5 × . Nutrient additions were made daily beginning on 11 July in 1983, 5 July in 1984, 2 July 1985 and ended in mid-August each year for a 36-day, 40-day and 44-day period in 1983, 1984 and 1985, respectively (Table 2). These beginning dates are subsequently referred to as day 1 for each year and for year to year consistency time is marked from these dates.

Fish. The inclusion of fish in this experiment was intended to regulate *Heterocope* densities and represent a treatment important in itself. However, there was no intention to study fish growth. A realistic study of fish can be difficult in limnocorral experiments since even in more productive lakes fish may represent an important source of nutrients (Rudd *et al.*, 1980, Andersson *et al.*, 1978).

In the summer of 1983, arctic grayling were directly added to each limnocorral. The fish additions began on 14 July with one fish added to the two low fish treatment limnocorrals and two fish added to each of the two high fish treatments (Table 1). Over the next six days four more fish were added to each of the high treatment corrals to bring the final fish treatment levels to one fish in each low treatment and six fish in each high treatment. This resulted in a low fish treatment of 7 'fish days' per week and 42 'fish days' per week in the high fish treatment (Table 1). A fish day is one fish in a corral for a 24-hour period. This was roughly 5 g m^3 in the high fish treatment and this was halved in subsequent years.

To better monitor fish survival and reduce predation intensity, the fish were confined to cages in the summers of 1984 and 1985. Three 1 m^3 cages were constructed with a frame made of 2.5 cm PVC plastic pipe and enclosed with 0.5 cm mesh net. The treatments consisted of three fish per

Table 2. Nutrient addition concordance and calendar date for treatment day from 1983 through 1985.

Date	Year (days elapsed)		
	1983	1984	1985
July 1			
July 2			1
July 5		1	4
July 10		6	9
July 11	1	7	10
July 15	5	11	14
July 20	10	16	19
July 25	15	21	24
July 30	20	26	29
August 1	22	28	31
August 5	26	32	35
August 10	31	37	40
August 14	35	41	44
August 15	36		45
August 16	37		

cage with the cage left in the two high fish treatments. Cages containing 2 fish were moved among the three low fish treatments on a schedule of two days in a limnocorral out of every six days. This resulted in a treatment level of 21 'fish days' per week in the high treatment and 4.7 'fish days' per week in the low treatment level (Table 1). In 1985 there was some fish morbidity and mortality in limnocorral B so that several fish had to be replaced.

Limnocorral volume and leakage estimates
To assess whether the corrals successfully isolated water from the rest of the lake and to accurately estimate the volume of the corrals, a known amount of sodium chloride (2 kg) was added to each corral. This was done in August of 1983 and in July of 1984 and 1985. Sodium was chosen as the tracer because of its low concentration in Toolik Lake (less than 1 mg l^{-1}1) and because of its negligible chemical and biological activity. After the salt had mixed completely, typically after 3-5 days, sodium samples were collected several times per week. Sodium concentrations were determined by atomic absorption.

Results

Limnocorral volume and leakage estimates

Measurement of the dilution of added sodium in each corral provided an estimate of realized corral volumes (Table 3). In most cases, the actual volume of water enclosed by the corrals was less than the potential 92 m^3 which would be enclosed with a cylinder 5 m in diameter and 5 m depth. The discrepancies are partly due to the octagonal shape of the collars; a calculation based on this shape gives an estimated volume of 86.5 m^3 for a 5 m depth. The remaining discrepancies were due to inward bowing of the tubes or corral depths somewhat greater than 5 m, especially in the case of corrals E and F.

Exchange of water between the corrals and the lake can also be estimated by any change through time in the dilution of sodium in the corrals. During August 1983, there was no dilution of the sodium in any of the corrals. In 1984 there was some dilution in corral A near the end of July caused by detachment of a section of the corral top from the supporting collar. The decline in sodium concentration indicates that perhaps one third of its volume was exchanged with the lake. There may also have been minor leakage problems with corrals C and F. None of these corrals had added nutrients. In 1985 corral A obviously was leaking such that the volume estimate is likely in error as well (Table 3).

Table 3. Corral volumes, in m^3, estimated from sodium data.

	Corral					
	A	B	C	D	E	F
New water enclosed 1983	57	60	63	55	–	–
New water enclosed						
1984	67	72	62	62	72	72
1985	44*	59	63	76	92**	106**

New water enclosed.
**Possible 15% exchange of water during the winter.

Volumes calculated from sodium measurements during 1985 have a greater uncertainty than volumes calculated for 1983 and 1984. This arises because the corrals were not raised and lowered in 1985, as occurred at the start of the 1983 and 1984 seasons. Volumes calculated for 1985 assumed that all sodium had been retained by the corrals over the previous winter. Loss of sodium during the winter would result in anomalously high calculated volumes for the following year. This is the probable explanation for the apparent increase in volume for corrals E and F from 1984 to 1985. The 1984 volume estimates for corral E and F are probably more accurate. Despite uncertainties in some corral volumes, the sodium was very useful in monitoring leakage of the corrals during the summer season. We are sure that the two high nutrient corrals, B and D, did not exchange significant amounts of water with the lake during 1983 through 1985.

Limnocorral measurements

Physical factors

The temperature of the limnocorrals and bay was measured weekly at meter intervals with a YSI Model 52 oxygen/temperature probe calibrated at the beginning of the field season against a NBS mercury thermometer. Light extinction in the limnocorrals was measured with a secchi disk and a LiCor LI-188B digital light meter. The light meter measurements were made at the surface of the lake, and at depths of 0.1, 0.2, 0.5, 1, 2, 3 and 4 meters.

Chemical factors

Water quality. The oxygen concentration of the limnocorrals and bay was measured weekly at meter intervals using a YSI Model 52 oxygen/temperature meter. The meter was air calibrated. In 1983, water samples were collected for chemical measurements at depths of 1 m and 4 m with a submersible pump. In 1984 and 1985, a depth-integrated, composite water sample was taken from each corral by raising a submersible pump from near-bottom to the surface at a constant rate.

The pH of the water samples was measured with a Corning Model 610 meter. The alkalinity of the water samples was determined by titrating the water with a strong acid to a pH of 4.6 as measured with the pH meter.

Nutrients. For the soluble phosphorus and ammonia measurements, freshly collected water was filtered through a glass fiber filter and 50 ml aliquot poured into an acid-rinsed bottle. The specific methods are colorometric techniques modified from Solarzano (1969) for ammonia and from Strickland & Parsons (1972) for phosphorus. In 1983 the samples were frozen and later analyzed with an Autoanalyzer II in the early fall. In 1984 and 1985 the samples were analyzed in the field using a Hitachi 100–60 semi-automated spectrophotometer.

Particulate C, N, and P were determined from water filtered through Gelman A-E glass fiber filters. Particulate C and N were determined with a Perkin-Elmer Carbon Nitrogen analyzer, and particulate P was determined by extraction of soluble PO_4 from ashed filters (Stainton *et al.*, 1977). Sedimentation of C, N, and P was determined with 46 cm^2 sediment traps suspended near the bottom of each corral. Carbon, N and P were determined on subsamples of sediment trap material by the same methods used for particulate C, N,and P. Natural, long-term rates of sedimentation in the bay were determined by the Pb-210 method (Chanton *et al.*, 1983).

Planktonic components

Microbes. Samples for estimates of planktonic bacteria were taken with a submersible pump or grab sample at a depth of 1 m and placed in clean polyethylene bottles. Samples were either immediately processed or preserved with 5% formaldehyde. Samples were taken weekly in 1983 through 1985 but daily from some corrals in 1984.

In 1983 through 1985, bacteria were stained with fluorochrome 2,6-diamidino 4-phenylindole (DAPI, Sigma) (Porter & Feig, 1980) and counted using epifluorescence microscopy (Hobbie *et al.*, 1977). In 1984 and 1985, some samples were counted both in the field and in the lab during the subsequent autumn to assure adequacy of preservation.

In 1984 and 1985, heterotrophic microflagellates were counted using the method of Caron (1983) as modified by Hobbie & Cole (1984). Samples were double stained with DAPI and primulan and counted within two days of preservation so that autofluorescence of chlorophyll could be used to distinguish between autotrophs and heterotrophs.

Incorporation of tritiated thymidine was used to measure bacterial DNA production and thus to infer bacterial growth activity (Fuhrman & Azam, 1982). Subsamples (15 ml) from each limnocorral were incubated in triplicate with 5 nM of 20 Ci/n mol^{-1} ^3H-thymidine (New England Nuclear) for 0.5 to 1.5 hrs depending on the temperature. One subsample from each site was killed with formaldehyde to control for abiotic adsorption of label. After incubation, the sample was filtered and the acid soluble materials extracted using 5% cold TCA. The level of radioactivity on the filter (mostly DNA) was determined with a scintillation counter. Planktonic bacterial production was also estimated using the rate of incorporation of ^{14}C-labelled acetate into microbial lipids (White *et al.*, 1977). Corral water was taken from a depth of 1 m. Subsamples, 1.9 ml, were incubated in 20 m glass scintillation vials in the dark with 0.1 ml of ^{14}C-UL-Acetate (2.5 μ Ci-ml^{-1}). After 2 hr at 10 °C, incorporation was halted by adding 2.5 ml chloroform and 5.0 ml methanol to extract the lipids. The amount of radioactivity in the lipid fraction was determined with liquid scintillation.

During 1985, differential filtration was used to separate planktonic production by the large particulate fraction (algal) from the small particulate fraction (bacterial) in the water. Water was collected from 1 m depth from each corral and 900 ml was filtered through a Whatman glass fiber filter (GFB, 1.2 μm nominal pore size). Some of the filtrate was then filtered through a 0.22 μm membrane filter to retain the free living bacteria. These filters, the GFB filtrate and unfiltered corral water were assayed for bacterial production using acetate incorporation, as above. In addition, the photosynthetic incorporation of $H^{14}CO_3$ into microbial community lipids was measured by incubation of an unfiltered water sample (1.9 ml) with ^{14}C-bicarbonate (0.1 ml; 10 μ Ci ml^{-1}; 180 mCi $mmole^{-1}$). The samples were incubated under 85 μ E m^{-2} sec^{-1} PAR illumination for 2 hours, and then the lipids were extracted as described above.

Total viable microbial community biomass was estimated in 1985 by filtering 900 μl of water from each corral through a glass fiber filter (GF/B) and the filtrate through a 0.2 μm phosphocellulose filter. The filters were extracted for lipids (Bligh & Dyer, 1959) and the quantity of lipid phosphate determined colorimetrically after degestion with perchloric acid (White *et al.*, 1979).

Phytoplankton. Phytoplankton abundance was estimated weekly by chlorophyll determinations. Corral and bay water used for these determinations was collected in the manner described for chemical measurements, and filtered through a glass fiber filter. The filters were then extracted in 90% acetone for 24 hours in the dark. The chlorophyll was determined using a Turner 111 Fluorometer. The fluorometer was calibrated against spinach chlorophyll in which the amount of chlorophyll was calculated by its absorbance in 90% acetone at 750, 650, 645 and 665 nm wavelengths using a Hitachi double beam spectrophotometer.

Direct counts of phytoplankton were made with algal cells preserved in Rodhe's Lugols iodine and settled in Utermohl Cells (25–30 ml) for 24 hours. Algal cells were examined using a phase contrast inverted microscope and species were identified using keys by Skuja (1958), Huber-Pestalozzi (1968), Prescott (1962), and Fogged (1981).

Phytoplankton primary productivity was measured weekly using the carbon-14 method modified from Wetzel & Likens (1979). In 1983 and 1984 water was pumped from depths of 0, 1, 2, 4 m directly into incubation bottles and 0.2 ml of ^{14}C-HCO_3 (16.2 or 20.0 μCi ml^{-1}) was added to each bottle. The bottles were then suspended at the depth the water was collected and incubated for 24 hours. In 1985 an integrated composite water sample was collected from depths 0 through 4 m and incubated at 1 m in the respective corral and at constant light intensity in a lakeside incubator. Subsamples were taken from each bottle and filtered onto a 0.45 μm membrane filter. Filters were dried and counted with liquid scintillation in a toluene-POPOP-PPO cocktail using the channel ratio method to determine quench. Total primary productivity was calculated using the formulae in Wetzel & Likens (1979). Total inorganic carbon was calculated from pH, temperature, and alkalinity measurements.

Zooplankton. Zooplankton were sampled by vertical plankton tow from the limnocorrals and from the bay. A 30 cm diameter plankton net of 153 μm mesh netting was towed from 4.5 m in 1983 and 1984 and 4.0 m in 1985 to the surface. In the limnocorrals, tows were taken using a small floating frame that could be variously positioned in the limnocorral. A comparable tow was taken in the bay from a boat. All plankton tows were made in duplicate. All zooplankton samples were preserved with a mixture of 100% ethyl alcohol and 4% formalin. Samples were taken every week in 1983 and twice a week in 1984 and 1985. Zooplankton containers were subsampled (1 to 3 ml from an approximately 50 ml sample) and animals identified as to species and counted.

Benthic components

Sediment chemistry. Chemical fluxes between the sediment and overlying water were determined using plexiglas benthic chambers in the corrals and Toolik Lake. The chambers were constructed of clear or opaque plexiglas open on the bottom and cover 0.2 m^2 of sediment. They enclosed 20-60 l of water which was mixed by a 1 rpm battery driven stirrer. The chambers were deployed on the sediment by a SCUBA diver in the lake but were simply lowered to the sediment in the limnocorrals. Samples were withdrawn through tubes extending to the surface of the water from sampling ports on the chamber top. Oxygen and inorganic nitrogen concentrations were measured for various periods of time using a micro Winkler technique for oxygen and modified techniques of Stainton *et al.* (1977) for nitrogen measurements.

Benthic microbes and decomposition. Sediment bacterial activity was measured as the ^{14}C-acetate incorporation into microbial lipids as described by McKinley *et al.* (1982). In 1983 sediment cores were obtained in 5 cm (OD) core tubes by SCUBA divers. The top 4 cm of the sediment was removed, mixed with filtered lake water (0.22 μm), and assayed for activity. Portions of this mixture were also dried at 60 °C overnight to a constant weight. Because of the movement of the limnocorrals during ice out in 1984 bay sediment in 10 cm diameter core tubes placed in plastic trays were put in each limnocorral. The sediments were subsequently sampled by a SCUBA diver using a large bore syringe and sediment microbial activity was assayed as above.

Carex decomposition was studied using fresh green *Carex* leaves collected from the margins of Toolik Lake. The leaves were cut into 5 cm segments, and oven dried at 65 °C. Approximately 300 mg of this dried litter was sewn into 5 x 10 cm fiberglass bags (1.0 mm mesh). In 1983 and 1984, the bags were placed on the sediment surface but in 1985 the bags were suspended at 1 m. After 31 days the bags were retrieved and gently washed in lake water, brushed to remove any surface materials, and the contents oven dried and weighed (Federle & Vestal, 1980). Heterotrophic activity of the decomposer community on the *Carex* leaves was measured using incorporation of ^{14}C-labelled acetate (McKinley *et al.*, 1982).

Benthic algae. Epibenthic algal biomass was esimated from chlorophyll extracts of the top layer of corral sediment. Sediments were sampled using a K-B corer and a small subsample taken of the top 1 cm of sediment. This subsample was mixed in a blender and filtered onto a glass fiber filter. The chlorophyll on the filter was extracted in a manner similar to the procedure used to determine phytoplankton chlorophyll concentration.

Epibenthic algal primary production was measured at the beginning and near the end of the summer experimental period in 1983. The top 1 cm of the core sediment was removed and placed in a flask and brought up to 100 ml with filtered lake water. One milliliter of this solution was added to 100 ml of filtered lake water, 0.2 ml of carbon-14 added, and incubated at 0.25 meter.

Benthic invertebrates. To study benthic invertebrates in the limnocorrals, three initial sediment samples were taken from each limnocorral and the bay station on 7 July 1983 and this was repeated a month later. The samples were collected by a SCUBA diver using a hand-held plastic core tube 10 cm in diameter. Samples were passed through a 100 μm net, which retains larval chironomids in instars II-IV,and instar I for many species (Hershey, 1985b). Animals were preserved in 95% ethanol. Animals were sorted with sugar flotation (Anderson, 1959) and identified using a phase contrast microscope (Hershey, 1985a).

In 1984, sediment core tubes were collected by a diver from a nearby undisturbed area, and nine cores placed in plastic trays in each limnocorral and at the bay station. This procedure was necessary because of disturbances of the sediments caused by a shift of the corrals during ice out of 1984. Three of these core tubes were removed from each limnocorral and the bay station after 1 year (summer 1985) and processed as described above. Three samples not enclosed in core tubes were also collected from the bay station in order to determine the effect of the core-tube enclosure.

Limnocorral response to nutrient addition

Physical factors

The water temperature in the limnocorrals averaged between 11 and 13 °C during most of the summer months and there was no thermal stratification in any of the limnocorrals. Generally, the temperature of the limnocorrals differed by less than 1 °C from each other or from the bay. The Secchi disk depth was at or near the bottom (5 m) for the limnocorrals prior to treatment and around 6 m in the bay. This transparency level continued in 1983 except in the two nutrient treated corrals where the Secchi disk depth began to decrease by day 5 and declined to 2 m by day 15. There was a similar pattern in 1984 except that Secchi disk depths in the two high nutrient addition corrals decreased to almost 1 m by late July. Secchi disk depths declined in all other corrals as well. In 1985 non-treatment corral Secchi depths were more similar to the bay than in 1984 and treatment corral Secchi depths were the shallowest yet observed. These general trends are shown by the average Secchi depth in the last 2 weeks of each year (Fig. 3). The difference between bay and control corral Secchi depths is due to the shallower depths of these corrals and not the water transparency.

The light extinction patterns as measured with a light meter generally matched the changes in Secchi disk depth, increasing as Secchi disk depth decreased. Prior to treatment in 1983, the extinction coefficients were similar in corrals and the

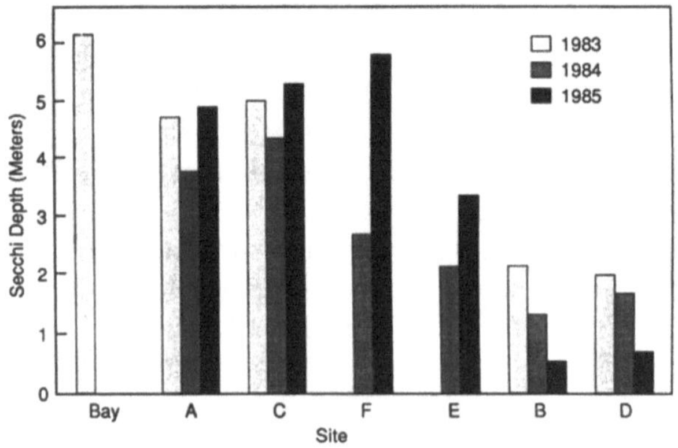

Fig. 3. Average Secchi disk depth of the last 2 measurements of each summer in the limnocorrals for all three summers.

bay, averaging about 0.6 m^{-1}. However, light extinction seems to be somewhat less sensitive to changes in biologically induced turbidity than does Secchi disk depth. In 1983, the Secchi disk depth began decreasing after the fifth day of nutrient addition but the light extinction coefficient did not begin to increase in these same corrals until day 10 or 12. A similar pattern was observed in 1984 and 1985.

Chemical factors

Water quality. The oxygen concentration in the limnocorrals and bay was quite similar through the early part of the summer of 1983, a little above 100% saturation. By day 8, however, the oxygen in the high fertilized corrals began to increase, and finally reached 140% of saturation while the nutrient control corrals and the bay declined to 95% of saturation. This general pattern recurred in 1984; the high nutrient treated corrals diverged soon after treatment and reached 150% saturation by early August. Corral E showed a slight increase to 110% by late July while the nutrient control corrals declined to 90% saturation. This pattern was again repeated in 1985. The oxygen data clearly reflects the levels of primary productivity in the fertilized and reference corrals.

In 1983 the pH of the nutrient control limnocorrals and bay remained very similar for most of July, ranging between 7.2 and 7.6. However, by day 12 in the nutrient addition corrals, the pH

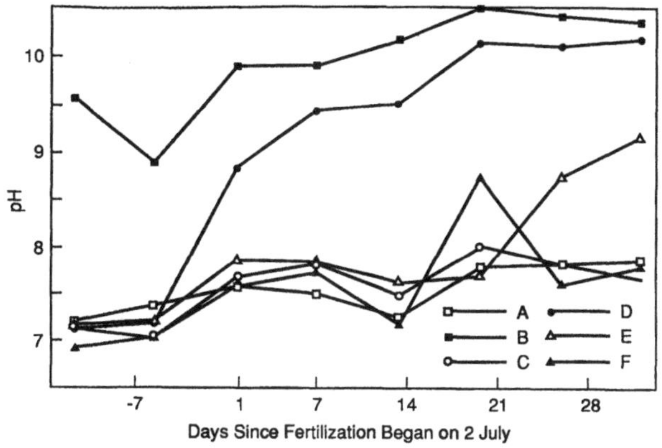

Fig. 4. The pH values in the limnocorrals throughout 1985.

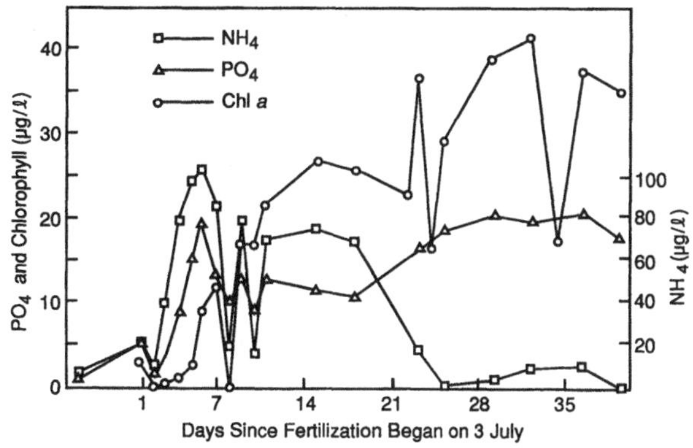

Fig. 6. Soluble ammonia, soluble reactive phosphorus and chlorophyll concentrations in corral B in 1984.

had risen dramatically to a high of 9.6. This same pattern occurred in 1984 when the pH in both B and D rose above pH 10 by early August. The pH in corral E did not go above that in the nutrient control corrals. In 1985, the high pH's in the nutrient treated corrals were even more pronounced than in past years (Fig. 4). All the corrals began the summer with a pH around 7 to 7.2 with the notable exception of corral B which had a pH of 9.6. The pH rose throughout the summer in corral D to a high around 10 while corral B reached a pH of 10.5 in late July. Unlike 1984, the pH in corral E did diverge from the nutrient control corrals by late July reaching a pH near 9.0. Like the oxygen concentrations, the pH levels are also clearly related to the different levels of primary productivity in the fertilized and reference corrals.

Nutrients. Just as in Toolik Lake (Whalen & Cornwell, 1985) bay and corral ammonia, nitrate and nitrite levels were nearly below detection. Once nutrients were added to the corrals in 1983, the ammonia level rose dramatically; by day 4 the ammonia levels in both corrals had risen to over 380 $\mu g\ l^{-1}$ and remained around this level throughout the rest of the summer. The reference corrals and bay, however, never had more than 20 $\mu g\ l^{-1}$ ammonia.

In 1984, the levels of ammonia in the nutrient treated corrals behaved very differently (Fig. 5). As in 1983, levels of ammonia and total dissolved inorganic nitrogen remained very low in the reference corrals, but ammonia levels rose dramatically in the treated corrals. However, at day 10

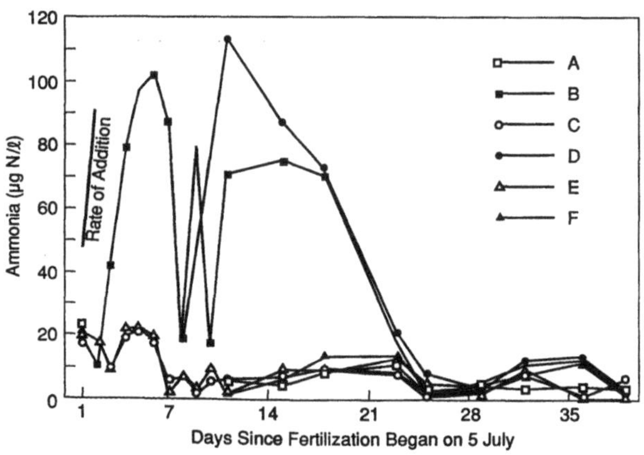

Fig. 5. Soluble ammonia in the limnocorrals in 1984.

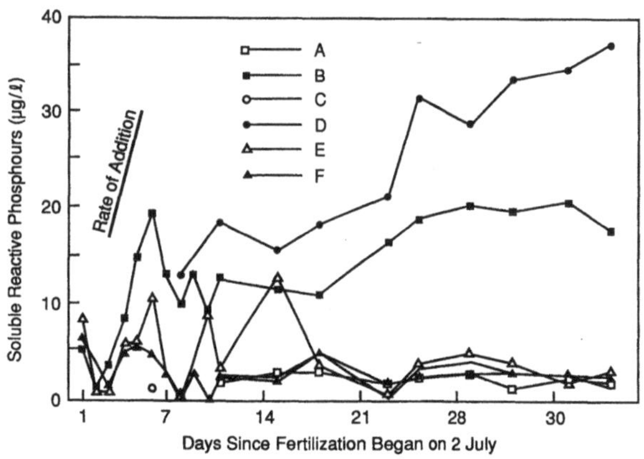

Fig. 7. Soluble reactive phosphorus in the limnocorrals in 1984.

the ammonia levels in B and D began to decline and by day 25 the ammonia levels in these corrals were similar to nutrient control levels. The decline of ammonia in corral B coincided with an increase in phytoplankton growth (Fig. 6). Ammonia levels in corral E never showed any increase over reference levels.

In 1985, the ammonia pattern was much like that of 1984. The treated corrals developed initial high values of ammonia which quickly declined to control levels. This decline was associated with a dramatic increase in a small coccoid green alga. In 1985, however, ammonia levels increased in late July in corrals D and E but not in B. Nitrate and nitrite were generally low in all corrals although there was 118 $\mu g\ l^{-1}$ of nitrate and nitrite in corral D early in the summer before nutrient additions.

Soluble reactive phosphorus (SRP) levels are quite low in Toolik Lake (Whalen & Cornwell, 1985) and the nutrient control corrals reflect this.

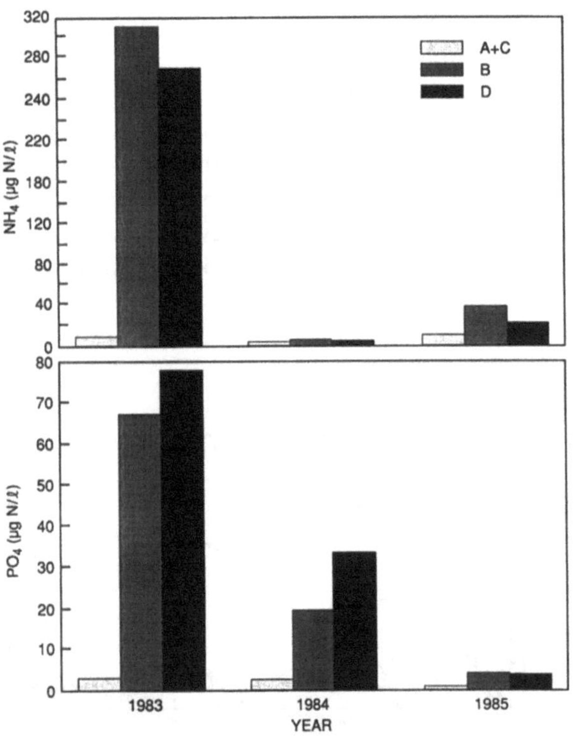

Fig. 8. Average ammonia and phosphorus concentration in the limnocorrals for all three summers. The values represent the average of the last half of each summer.

However, in 1983, phosphorus levels increased in corrals B and D almost at the rate of addition from day 0 to day 12 when they leveled off around 90 $\mu g\ l^{-1}$. In 1984 the phosphorus concentrations also climbed in the treated corrals, but at a much slower rate (Fig. 7). In corral B the phosphorus concentration increased initially, then leveled off and then increased again (Fig. 6). By the end of the summer, SRP had risen to 35 $\mu g\ l^{-1}$ in D and 20 $\mu g\ l^{-1}$ in B. The initiation of this secondary accumulation of phosphorus in the high nutrient corrals in 1984 coincided with the beginning of the severe decline in ammonia levels (Fig. 6). In 1985 the soluble phosphorus concentration in the

Table 4. Particulate carbon and nitrogen concentrations for the limnocorrals, limno bays, and central Toolik Lake, for 1983.

Corral/Depth (m)	Particulate C (nMoles/L)	Particulate N (mMoles/L)	C/N
9 August			
A/1	16.4	0.9	18.5
A/4	19.6	1.6	12.5
B/1	121.6	17.9	6.8
B/4	70.3	10.6	6.6
C/1	16.9	1.3	13.2
C/4	18.1	2.0	9.0
D/1	–	–	–
D/4	113.4	16.4	6.9
Bay/1	19.0	1.4	13.1
Bay/4	18.6	0.9	20.9
Center/1	17.2	1.4	11.9
Center/4	21.6	1.3	16.3
16 August			
A/1	19.6	1.3	15.1
A/4	30.5	2/9	10.9
B/1	99.2	12.9	7.7
B/4	50.3	6.5	7.7
C/1	20.8	1.3	16.0
C/4	23.0	1.6	14.4
D/1	121.3	18.7	6.5
D/4	131.3	14.8	8.9
Bay/1	19.6	1.3	15.1
Bay/4	19.8	1.4	14.1

Table 5. Particulate carbon, nitrogen, and phosphorus data for limnocorrals in 1984. Concentrations were determined from integrated water samples which represent average concentrations within an entire corral.

Corral	Particulate C (mMoles/L)	Particulate N (mMoles/L)	Particulate P (mMoles/L)	C/N	N/P
A 4 Aug	17.1	1.7	0.19	10.2	8.6
A 11 Aug	21.0	2.1		10.2	
B 4 Aug	150.5	22.0	0.40	6.8	54.0
B 11 Aug	176.9	30.4		5.8	
C 4 Aug	17.2	2.1	0.19	8.3	10.7
C 11 Aug	25.9	3.0		8.5	
D 4 Aug	136.2	–	1.7	–	17.9
D 11 Aug	151.8	25.0		6.1	
E 4 Aug	31.1	4.0	1.06	7.8	3.7
E 11 Aug	43.7	6.7		6.5	
F 4 Aug	16.9	2.1	0.63	8.0	3.3
F 11 Aug	22.2	3.4		6.6	
Bay 4 Aug	14.3	1.4	0.28	10.4	5.0
Bay 11 Aug	15.5	1.6		9.6	

corrals was much more variable than in past years. However, the concentration of SRP was also almost 25 times higher in the high treated corrals in 1983 than in 1985.

In 1983, both SRP and ammonia were very high. In 1984, SRP was lower than in 1983, but ammonia concentrations were similar to the control corrals. In 1985, both SRP and ammonia were very low. This dramatic change in the behavior of SRP and ammonia concentrations in the high nutrient corrals over the three years of the experiment is shown in Fig. 8. This figure depicts the average SRP and ammonia concentrations for the last half of each summer when these nutrients may have approached equilibrium.

Concentration of particulate C, N, and P were much higher in the nutrient addition than in the control corrals during 1983 (Table 4). The concentrations of particulate carbon and nutrients in 1984 correlated well with the amounts of nutrient added; corrals B and D were highest, E was next, and A, C, and F were lowest and similar to the concentrations in the bay (Table 5). Much of the ammonia nitrogen that disappeared from the dissolved phase in 1984 appears as particulate nitrogen suspended in the water column (Table 5). For example, the quantity of particulate N in corral B on 11 August is 420 $\mu g\ N\ l^{-1}$ (Table 5) while the maximum N as ammonia was only 70 $\mu g\ N\ l^{-1}$ on July 20 (Fig. 5). In 1985 particulate C, N, and P again reflected the nutrient treatment levels. However, particulate carbon densities were higher at all treatments as well as in the bay (Tables 6 and 7).

The sedimentation rates of particulate C, N and P in 1984 did not follow the nutrient addition scheme closely, but, again, corrals B and D were highest and A and C lowest (Table 8). Both E and F were in the middle of the range. This was not expected for F which did not receive nutrient addition. The high sedimentation rate for the bay was also unexpected and may reflect the oversampling that occurs when sediment traps are used in turbulent environments. Due to the position of corral F, being furthest out in the bay, it too may have been exposed to conditions causing over sampling.

These nutrient data can be used to calculate a nutrient budget for corrals B and D for 1984

Table 6. Particulate carbon, nitrogen, and phosphorus concentrations for the limnocorrals and limnobay, 1985. Samples represent integrated depths, as in Table 5.

Corral	Particulate C (mMoles/L)	Particulate N (mMoles/L)	Particulate P (mMoles/L)	C/N	N/P
A 9 July			0.09		
A 16 July	–	–		–	
A 23 July	21.1	2.6	0.19	8.2	13.3
A 30 July	39.5	3.8		10.4	
A 5 Aug		2.9	0.08		35.9
A 12 Aug	33.7	3.7	0.08	9.2	46.2
B 9 July			0.76		
B 16 July	233.4	19.1		12.2	
B 23 July	125.2	14.0	0.17	8.9	84.3
B 30 July					
B 5 Aug	145.2	12.7	1.70	11.4	7.4
B 12 Aug	245.1	24.3	1.63	10.1	14.9
C 9 July			0.11		
C 16 July	32.0	2.4		13.3	
C 23 July	12.5	2.1	0.15 ± 0.4	6.0	14.1
C 30 July	31.2	3.7		8.4	
C 5 Aug	32.5	2.4	0.06	13.7	38.4
C 12 Aug	28.6	2.9	0.13	9.8	22.5
D 9 July			0.18		
D 16 July	119.6	13.2		9.1	
D 23 July	128.4	23.6	1.52	5.5	15.5
D 30 July					
D 5 Aug	157.9	16.8	1.83	9.4	9.1
D 12 Aug	284.4	42.2	2.25	6.7	18.8
E 9 July			0.11		
E 16 July	35.9	3.8		9.4	
E 23 July	36.1	5.4	0.26	6.6	20.8
E 30 July	46.6	5.1		9.1	
E 5 Aug	35.5	5.3	0.19	6.7	28.1
E 12 Aug	74.9	10.2	0.26	7.3	38.9
F 9 July					
F 16 July	25.2	1.8		14.0	
F 23 July	26.8	4.7	0.23 ± 0.13	5.7	20.6
F 30 July	25.6	3.0		8.6	
F 5 Aug	23.4	2.1	0.06	10.9	37.5
F 12 Aug			0.09		
Bay 9 July			0.08		
Bay 16 July	27.3	2.6		10.5	
Bay 23 July			0.15		
Bay 30 July	33.7	3.7		9.2	
Bay 5 Aug	28.4	2.4	0.07	11.9	36.1
Bay 12 Aug	46.5	3.3	0.12 ± 0.04	14.0	27.8

Table 7. Summary of particulate C, N, and P throughout the study.

		Particulate C mmoles/P	Particulate N mmoles/P	Particulate P mmoles/P	N/P
Bay	83	19.2	1.25		
	84	14.9	1.5	0.28	5.4
	85	34.0	3.0	0.11	27.3
A	83	21.5	1.65		
	84	19.1	1.9	0.19	10.0
	85	31.4	3.3	0.11	30.0
C	83	19.7	4.0		
	84	21.6	2.6	0.19	13.7
	85	27.4	2.7	0.11	24.4
F	84	19.5	2.8	0.63	4.4
	85	25.3	2.9	0.13	22.3
E	84	37.4	5.4	1.1	4.9
	85	45.8	6.0	0.2	30.0
B	83	85.2	12.0		
	84	164.0	26.2	0.4	65.5
	85	187.0	17.5	1.07	16.4
D	83	122.0	16.6		
	84	144.0	25.0	1.7	14.7
	85	172.0	24.0	1.4	17.1

Table 8. Sedimentation of C, N, and P in limnocorals 13 July–10 August, 1984.

Corral	Particulate C	Particulate N millimoles m^{-2} day^{-1}	Particulate P
A	3.6	0.4	0.07
B	16.6	2.4	0.17
C	4.2	0.4	0.08
D	16.6	2.5	0.10
E	8.2	0.9	0.07
F	8.3	0.8	0.03
Bay	16.3	1.9	0.11

Note: The long-term accumulation rates of C and N in limnocorral bay have been determined to be 583 millimoles $m^{-2} yr^{-1}$ and 43 millimoles $m^{-2} yr^{-1}$.

(Table 9). As shown we can account for 53–60% of the added nitrogen but only 30–50% of the added phosphorus. Most of the nutrient remained in suspension in particulate form.

Table 9. Nutrient budgets (moles) for corrals B and D calculated as the sum of rates since 5 July and as the concentration on 14 August, 1984.

	Corral B		Corral D	
	N	P	N	P
Attached algae	0.8	0.04	0.9	0.06
Sedimentation	0.5	0.04	0.6	0.03
Suspended particulates	2.2	0.03	1.6	0.11
Dissolved	0.1	0.02	0.1	0.02
Total	3.6	0.13	3.2	0.22
	(60%)	(30%)	(53%)	(50%)

The percentages in ( ) show how the totals compare to the amount of nutrient that was added as fertilizer.

Planktonic components

Microbes. There was little fluctuation of bacteria numbers in the control corrals throughout the summer of 1983. Planktonic bacteria numbers increased from control levels of $1\text{-}2 \times 10^6\ ml^{-1}$ to over $8 \times 10^6\ ml^{-1}$ in the nutrient treated corrals by day 14 (Fig. 9). However, two weeks later, bacteria numbers had declined to reference levels and remained near reference values the remainder of the summer. Bacterial production also showed a marked response to nutrient addition, rising some 2–3 fold in these same corrals within the first week of nutrient addition and generally remaining higher than bacterial production in the nutrient control corrals. Interestingly, this peak in production occurred as bacteria numbers were declining in corral B.

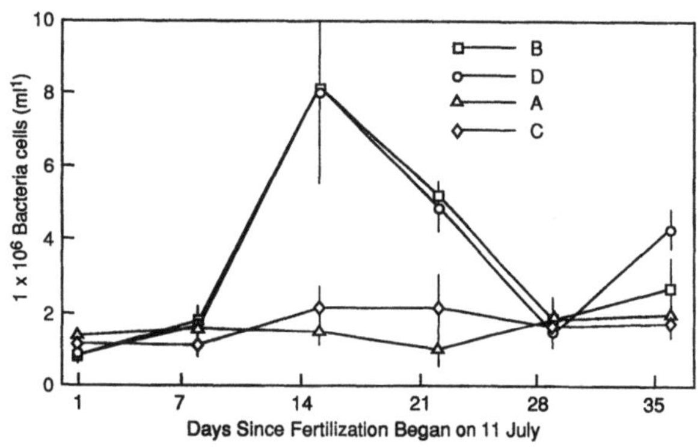

Fig. 9. Bacterial abundance in the limnocorrals in 1983.

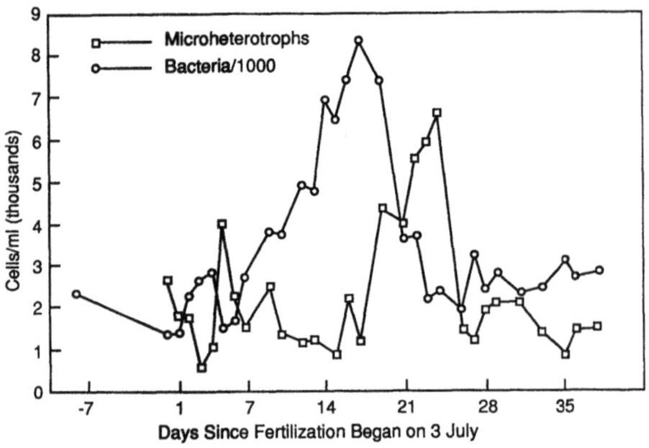

Fig. 10. Bacterial and microheterotroph abundance in corral B in 1984.

In 1984, bacteria population dynamics and production had a pattern quite similar to that observed in 1983 but the density peak and decline was delayed somewhat in corral B, beginning around day 20 (Fig. 11). In 1984, microheterotroph (protozoan) densities were measured for the first time and the abundance of these organisms increased in corral B, but only after bacterial densities had risen (Fig. 10). Once the bacteria densities declined, the microheterotrophs did as well. Interestingly, the peak of bacterial production occurred at the time of the rapid decrease of bacterial numbers (Fig. 10). Thus, microheterotroph grazing may well have caused the decline of bacteria numbers in 1983, since bacterial production also lagged behind bacteria numbers in 1983 as in 1984.

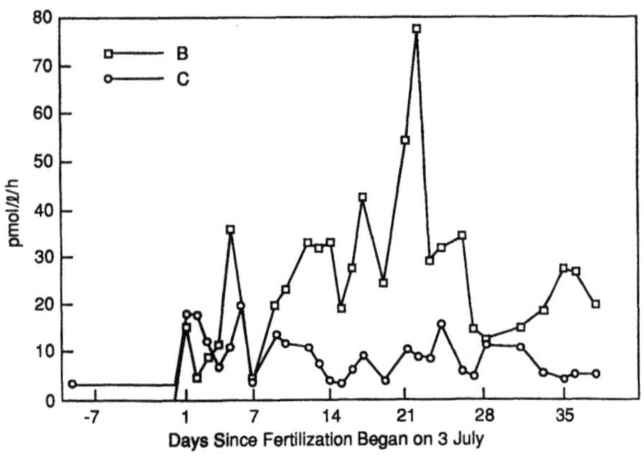

Fig. 11. Bacterial production in corral B and C in 1984.

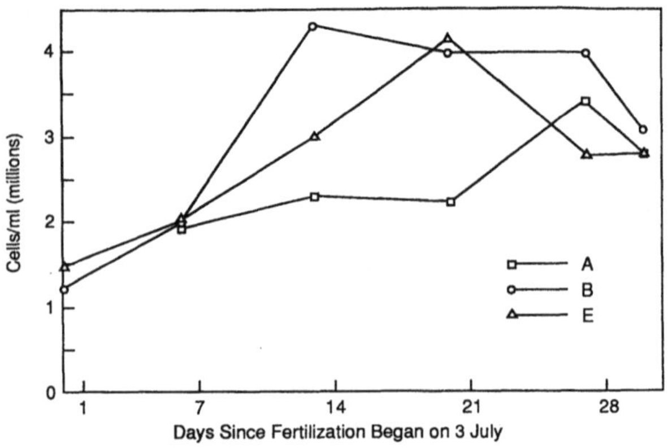

Fig. 12. Bacterial abundance in corrals A, B and E in 1985.

In 1985, bacteria numbers again showed comparable responses to nutrient addition. In 1985, the nutrient addition was halved to a 5 × addition and peak bacteria numbers in corral B on day 13, 4×10^6 ml^{-1}, was about half the earlier peaks in 1983 and 1984 (Fig. 12) and bacteria numbers also increased in corral E. There was no dramatic decline in bacteria numbers as occurred in previous years, but microheterotroph densities were just as high or higher than in 1984. Bacteria production in corrals B and D also increased to levels close to those observed in the past.

In 1984, the microbial incorporation of acetate into lipids, another measure of heterotrophic activity, also showed a response to added nutrients. By day 14, acetate incorporation in corrals B and D was twice that of the reference corrals with corral E intermediate (Fig. 13). By day 35, acetate incorporation in corrals B and D was almost 10 times that in the reference corrals. A comparison of Figs 11 and 13 shows that both bacterial production and acetate incorporation show the same general patterns. Both measures also show increase as phytoplankton production increased during the summer.

In 1985, differential filtration was used to separate the photosynthetic and organic incorporation activity of the large size particulate components (mainly algal) from that of the small size particulate components (mainly bacterial). Total photosynthetic activity was very high in the nutrient enriched corrals. However, incorporation of bicarbonate by the small size fraction was low

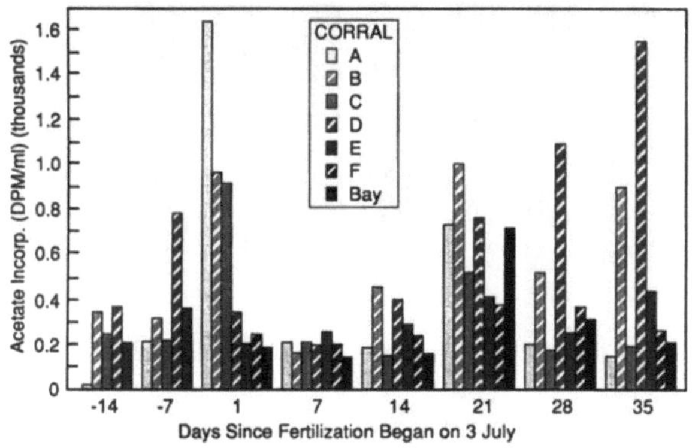

Fig. 13. Microbial acetate incorporation in limnocorrals throughout 1984.

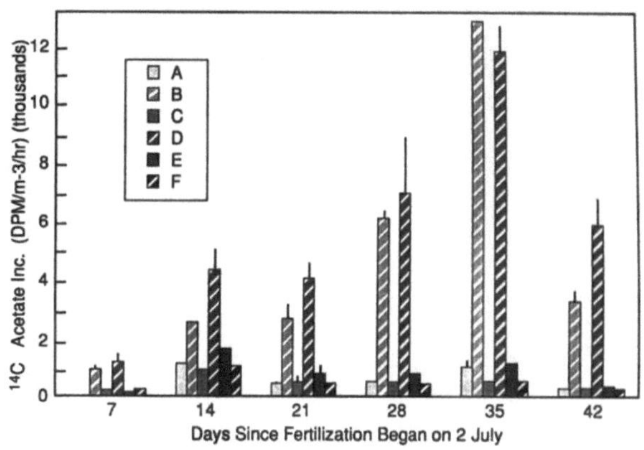

Fig. 15. Microbial acetate incorporation in limnocorrals throughout 1985.

(Fig. 14) and there was no difference in uptake between the fertilized and unfertilized corrals. Total acetate incorporation mirrored the total algal production in corrals B and D (Fig. 15). However, when the heterotrophic uptake due to the large particulate component was removed, heterotrophic activity due to bacteria was shown to be low in corrals B and D throughout the summer and there was no pattern of the effects of nutrient stimulation in corrals B and D similar to the unfiltered samples.

If one considers the percentage of attached microbes in the fertilized corrals, especially corral D, versus the reference corrals, especially corral F and Bay station, it is apparent that most microbial activity in the fertilized corrals is attached to large particles. The microbes were attached to either living or dead particulate matter, most likely algae (Fig. 16).

The total biomass of the plankton community (both bacteria and phytoplankton) was estimated by measuring the quantity of lipid phosphates. As can be seen from Fig. 17, there was a dramatic increase in biomass in the highly fertilized corrals (corral D) compared to low fertilized corrals (corral E) or a reference corral (corral F). This shows that the algae were alive and growing until late in the summer. Bacterial production determined by this method (the 0.2–1.0 μm fraction) was below detection indicating that free living bacteria did

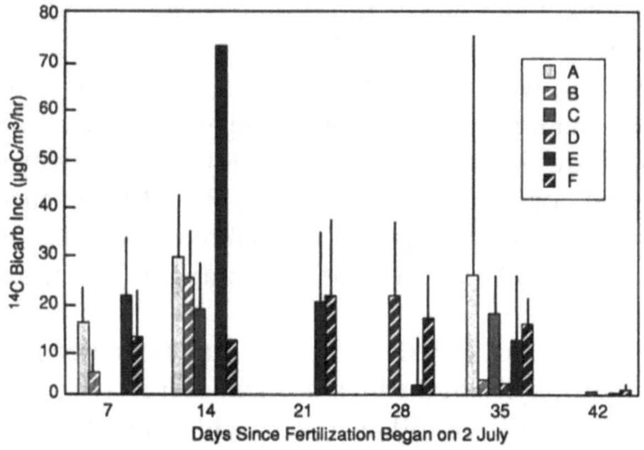

Fig. 14. Bicarbonate uptake by small particulate fraction (mainly bacteria) in limnocorrals in 1985.

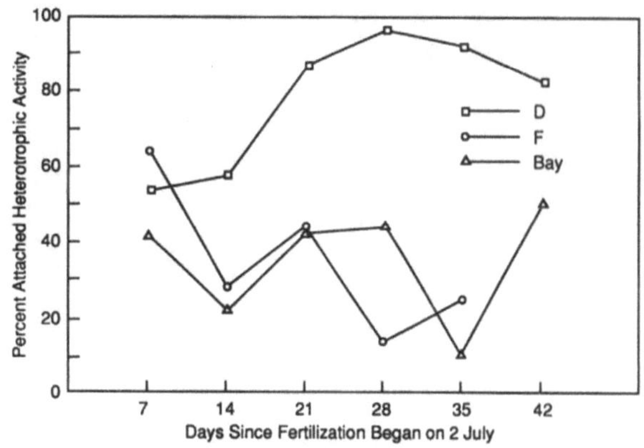

Fig. 16. Percent attached heterotrophic activity in limnocorrals D and F and the Bay station in 1985.

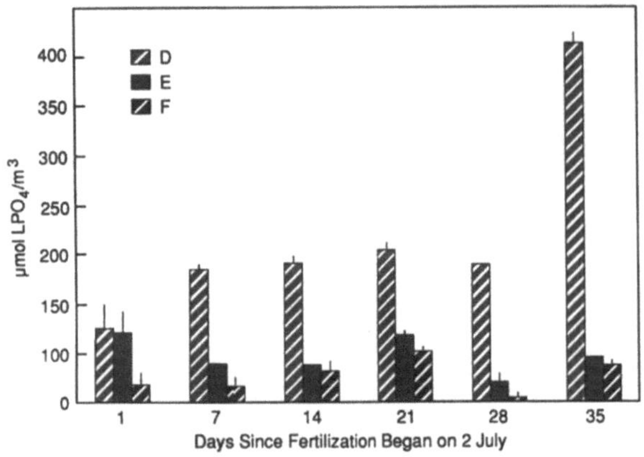

Fig. 17. GFN filter lipid phosphates in limnocorrals D, E, and F in 1985.

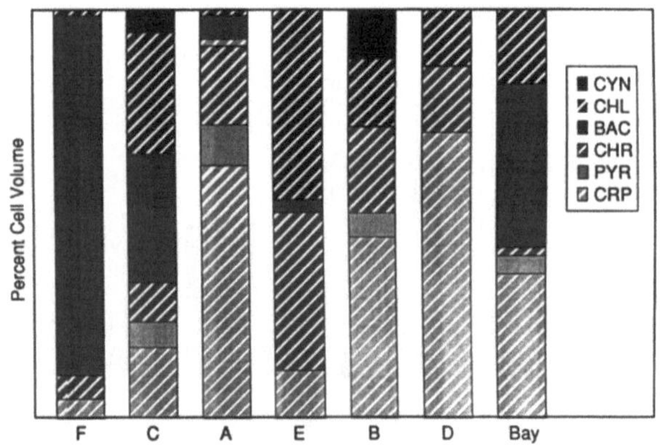

Fig. 19. Phytoplankton composition in the corrals and Bay station on 8 August in 1984.

not comprise a major portion of the planktonic microbial community.

Phytoplankton. The phytoplankton community in Toolik Lake is distinctive and is dominated by small flagellates (Cryptophytes, Chrysophytes) and centric diatoms. In general this community was retained in the reference corrals; however, the phytoplankton community in the fertilized corrals changed by August 1983 (Fig. 18). In 1984, the Toolik Lake phytoplankton community remained as before but the reference corrals differed more than in 1983 (Fig. 19). Corral C had a phytoplankton composition much like the lake but corral A contained large Cryptophytes and Chlorophytes more like the nutrient addition corrals. The high nutrient corrals were initially dominated by large Cryptophytes (*Crytomonas erosa* especially) but later shifted to Chrysophytes (*Malomonas mirailis*) and a small coccoid green alga (*Pteromonas anqulosa*). This latter species seemed to be responsible for the difficulty in obtaining complete chlorophyll extractions in corrals where it was abundant. The phytoplankton community in corral F was unusual in that it was 80% diatoms. By 1985, however, the phytoplankton community had greatly changed in all limnocorrals.

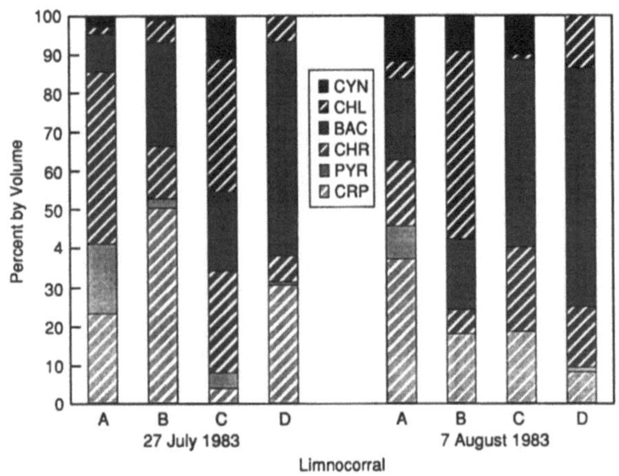

Fig. 18. Phytoplankton composition in the four corrals on 27 July and 7 August 1983.

The best estimates of total phytoplankton biomass in the limnocorrals were made by measuring chlorophyll *a* concentrations. Phytoplankton chlorophyll *a* concentrations are quite low in Toolik Lake, never more than 5 $\mu g\ l^{-1}$ and averaged 1.3 $\mu g\ l^{-1}$ over recent summers (Whalen & Cornwell, 1985; Miller *et al.*, 1986). In 1983 this changed dramatically in the limnocorrals with the addition of nutrients. By day 17, the chlorophyll concentrations in corrals B and D reached over 30 $\mu g\ l^{-1}$ (Fig. 20) The reference corrals had chlorophyll *a* concentrations comparable to those in the lake.

Again in 1984, there was a dramatic increase in chlorophyll *a* in the high nutrient corrals (Fig. 21). By day 10, 28–30 $\mu g\ l^{-1}$ of chlorophyll *a* had developed in corrals B and D. But as in 1983, there was little more chlorophyll *a* accumulation

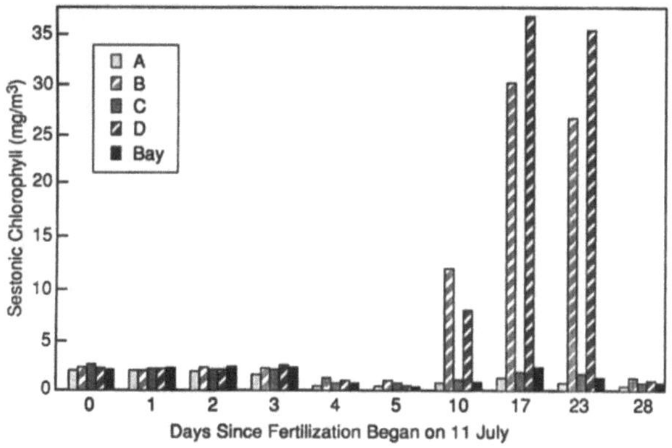

Fig. 20. Chlorophyll concentrations in the limnocorrals and Bay in 1983.

during the remainder of the summer with the exception of over 60 $\mu g\ l^{-1}$ chlorophyll which developed in corral D by day 28. These chlorophyll concentrations were 10 to 15 times greater than the reference corrals. Chlorophyll concentrations in corral E were only slightly, but consistently, greater than those of the controls.

The pattern of chlorophyll *a* concentration in the corrals in 1985 was much the same as in 1983 and 1984, although large concentrations of chlorophyll *a* developed more rapidly in corrals B and D than in past years (Fig. 22). Chlorophyll analysis was somewhat hindered with the advent of a phytoplankton species in corral B which was quite resistent to chlorophyll *a* extraction with acetone. The values given in Fig. 22 for corral B are esti-

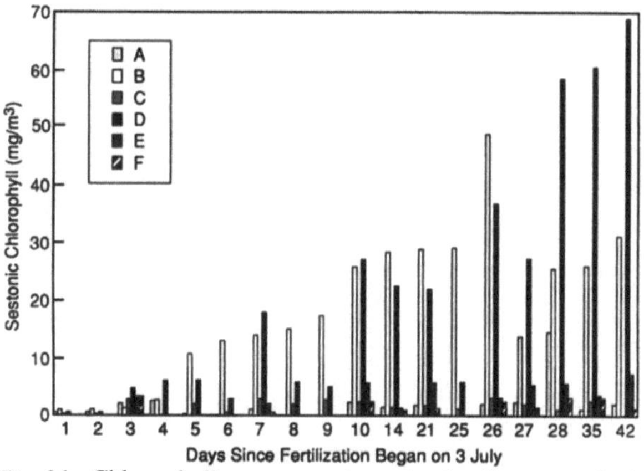

Fig. 21. Chlorophyll concentrations in the limnocorrals in 1984.

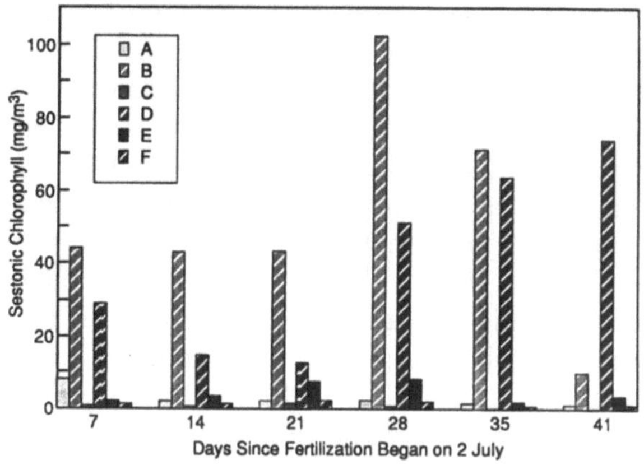

Fig. 22. Chlorophyll concentrations in the limnocorrals in 1985.

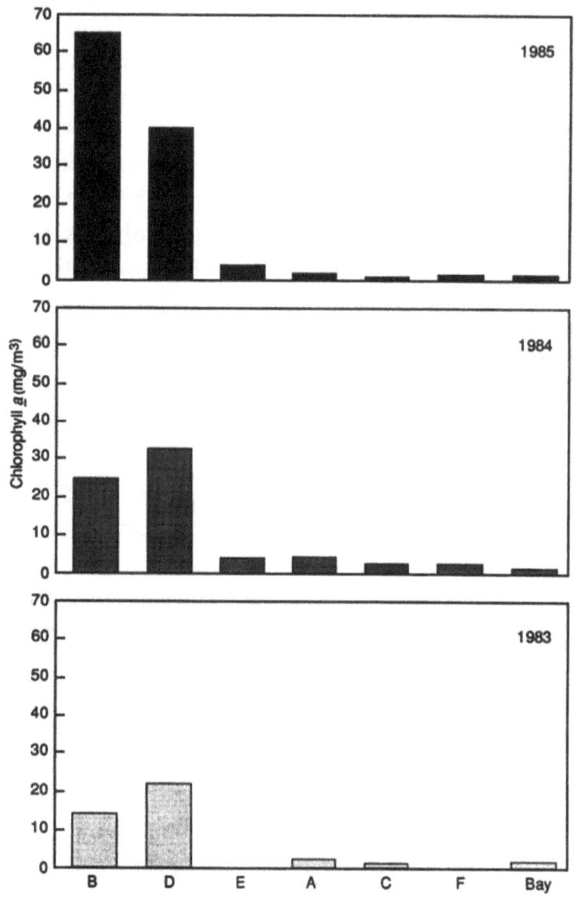

Fig. 23. Average chlorophyll concentrations for all three years of the experiment.

mates based on regressions of in vivo fluorescence and chlorophyll *a*.

Summer average chlorophyll *a* concentrations show a very consistent relationship to treatment (Fig. 23). In 1983, planktonic chlorophyll *a* concentrations in limnocorrals B and D averaged about 10–12 times that of the reference corrals or bay. In 1984, B and D have average chlorophyll *a* concentrations nearly 15 times that of the controls with E having just about twice the average amount of the controls. In 1985, B and D averaged 21 times higher than the controls. Thus, there was a successive increase in chlorophyll *a* concentrations of the fertilized corrals.

These average chlorophyll *a* concentrations may be compared to the response Ahl (1975) found for phosphorus loading versus average chlorophyll *a* concentration in northern temperate lakes (Fig. 24). The average chlorophyll *a* of Toolik Lake and the reference corrals generally fall below that predicted while the average chlorophyll *a* of the fertilized corrals all lie above that predicted for north temperate lakes.

The pattern of planktonic primary productivity in the limnocorrals is complex. In 1983, the corrals to which nutrients were added always had higher primary productivity than the non-nutrient corrals (Fig. 25) but productivity did not increase steadily as nutrients were added. On day 10, productivity was fully 12 times greater in nutrient-treated corrals than reference corrals.

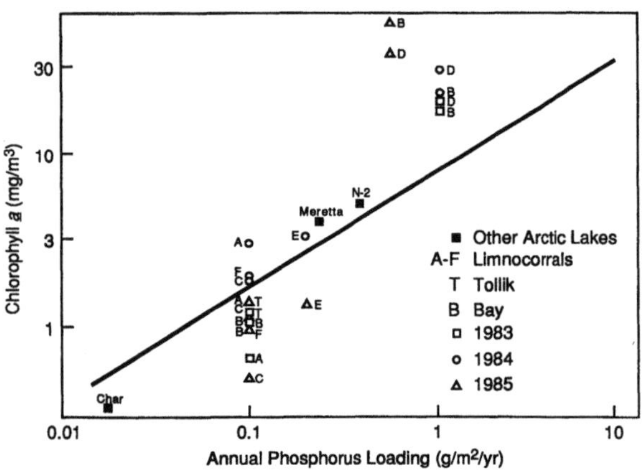

Fig. 24. Chlorophyll concentration versus phosphorus loading for Toolik Lake, lake N-2, and all three years of the limnocorrals. The line is from Ahl (1975) and several other arctic and subarctic lakes are also plotted.

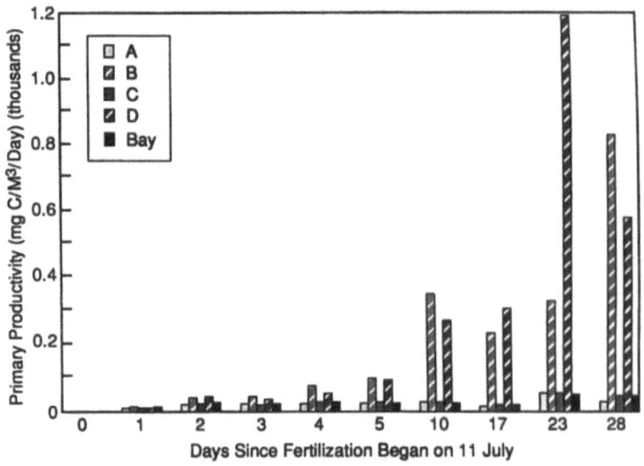

Fig. 25. Planktonic primary productivity in the corrals in 1983.

The productivity pattern in 1984 was more straightforward; where it increased almost linearly in corrals B and D throughout the summer (Fig. 26). Furthermore, corral E which received 1+ nutrient addition, showed a small but consistently higher level of productivity than reference corrals. The same general trend occurred in 1985 (Fig. 27); production increased early in the summer and peaked later. However, this was the first year any carry over effect was obvious and there was a dramatic drop in productivity on day 13 in the nutrient-treated corrals.

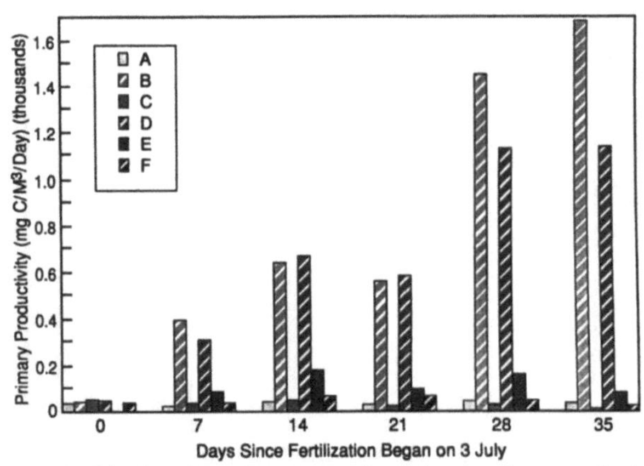

Fig. 26. Planktonic primary productivity in the corrals in 1984.

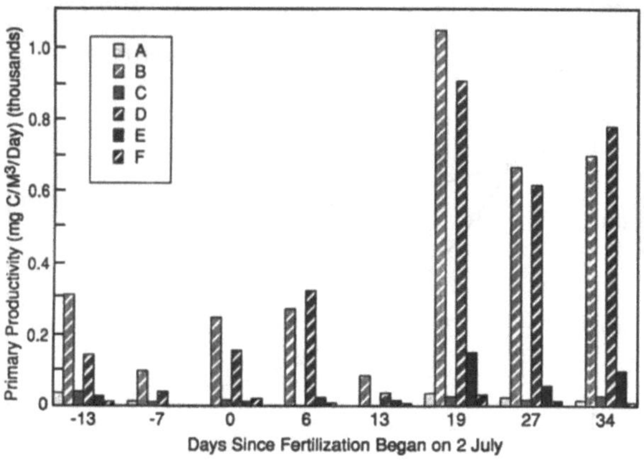

Fig. 27. Planktonic primary productivity in the corrals in 1985.

Zooplankton. The response of herbivorous populations to the limnocorral treatments was complex. In 1983, *Bosmina* densities in the reference corrals increased but there was no increase in the high nutrient corrals (Fig. 29). There was, however, a brief increase of *Bosmina* in corral B in 1984, so high nutrients did not have a negative effect. In 1985, *Bosmina* densities were much reduced in all the limnocorrals.

The numbers of *Daphnia longiremis* in 1983 showed a decline in all the limnocorrals and bay with a late summer rebound in both the high nutrient addition corrals (Fig. 28). In 1984 there was a very rapid decline in *D. longiremis* numbers in the limnocorrals; however, once again the populations in the high and medium nutrient treatment corrals rebounded to around $0.1\ l^{-1}$ by the middle of the summer and remained at these densities. Thus, there was a modest response of *D. longiremis* to nutrient addition. In 1985 *D. longiremis* greatly increased in corral D a high-fish treatment in 1985. However, at the end of the summer *D. middendorffiana* showed a tremendous increase to over $9\ l^{-1}$ in corral B, a low-fish treatment.

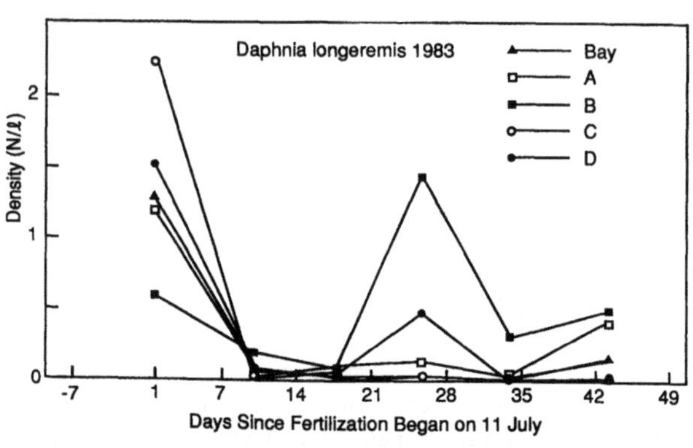

Fig. 28. Densities of *Daphnia longiremis* in the corrals in 1983.

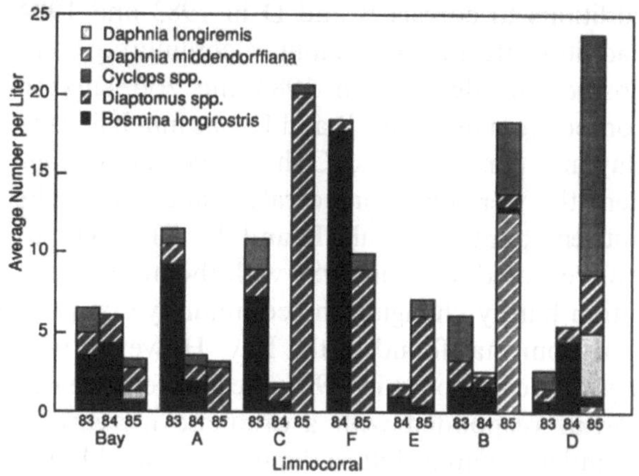

Fig. 29. Zooplankton community composition for all limnocorrals and bay throughout the study. The numbers represent the average of densities from early, mid and late summer.

The response of the zooplankton community to the limnocorrals and the various treatments showed some quite consistent patterns (Fig. 29). The two nutrient reference corrals in 1983 were very similar to each other and fairly similar to the Bay zooplankton community. *Bosmina* was the dominant zooplankton species in these corrals in 1983 as it was in the Bay. However *Bosmina* densities greatly declined in these corrals in 1984 and *Bosmina* was almost completely absent in 1985. This is a consistent pattern that was also found in corrals E and F. Even though corral F had tremendous densities of *Bosmina* in 1984, the first year corral F was in place, this species was essentially gone from that corral in 1985. While *Bosmina* was much less abundant in corral E in1984 it was lower still by 1985. In all these cases it is *Diaptomus pribilofensis* which takes over dominance as *Bosmina* declines. There is a very consistent shift in the reference corrals and in E and F to copepods, especially diaptomids, but *Cyclops* copepods are the next most abundant species.

It is very interesting that the summer nutrient

additions to corrals B and D in 1983 and 1984 had little effect on zooplankton community composition or density. In 1983 the zooplankton composition in corrals B and D was much like the Bay and corrals A and C; however, the average densities were higher in corrals A and C than the nutrient treated corrals B and D. This pattern was repeated again in 1984 with the nutrient addition hardly changing the community composition from that found in the Bay. However, with a year's carry-over in 1985 the nutrient treated corrals responded with a dramatic increase in abundance and a shift to daphnids. Corral E was little affected and did not show this shift to daphnids.

Interestingly if we use these data to get some indication of species diversity it is the nutrient treated corrals which maintain the greater diversity. By 1985 the nutrient control corrals A and F had but 2 species, both copepods, with corral C having just a trace of *Bosmina* so just barely 3 species. However, corral E, the 1 + nutrient addition, had 3 species, while the nutrient addition corral B had four species and corral D 5 species (Fig. 29).

Benthic components

Sediment chemistry. The nutrient addition treatments greatly altered the chemistry of the sediments. Oxygen consumption by sediments (Fig. 30) and ammonia release from sediments (Fig. 31) was much higher in fertilized corrals. Rates of these processes are indications of rates at which the benthic microbial community is remineralizing the sediments. Oxygen consumption increased in fertilized corrals, similar to the increase in sedimentation of carbon, by about a factor of five, and NH_4 release rates increased by about an order of magnitude over control corrals. Measurements in reference corrals compared very well with measurements in Toolik Lake and are almost in balance with nitrogen sedimentation rates (Table 8). The release rates of nitrogen from sediments in the fertilized corrals are slightly less than sedimentation rates of particulate nitrogen (Table 8).

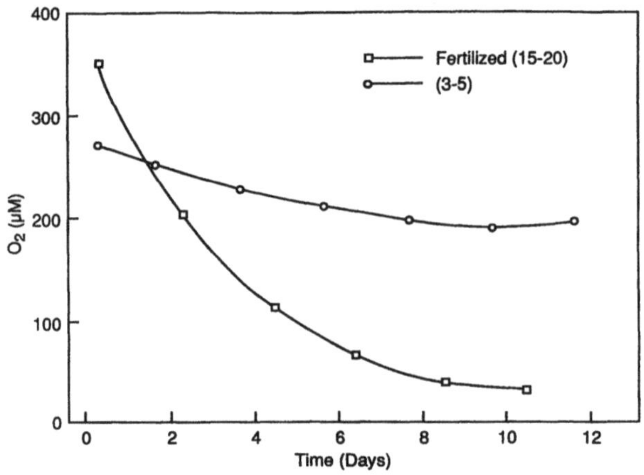

Fig. 30. Oxygen consumption rates by sediments in fertilized (open squares) and control (open circles) corrals using benthic chambers. Rates in parentheses are millimoles 02 m^{-2} dy.

Benthic microbes and decomposition. In 1984 there was no significant difference in *Carex* decomposition between the fertilized corrals and the controls. But decomposition in all corrals was significantly greater than decomposition in the bay. Acetate uptake into microbes dwelling on *Carex* litter was also not different among corral treatments.

In 1985, *Carex* samples were suspended at 1 m in the water column to see if there would be a greater weight loss than when placed on the sediment (Federle & Vestal, 1980) or if this would

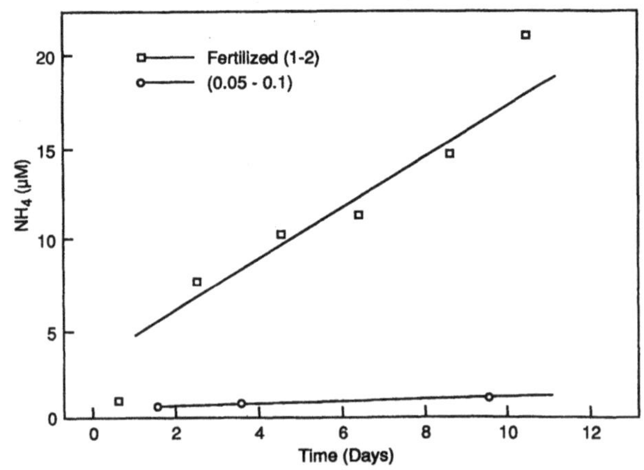

Fig. 31. Ammonia fluxes from sediments in fertilized (open squares) and control (open circles) corrals using benthic chambers Rates in parentheses are millimoles N-m^{-2} dy.

produce an effect of the nutrients on the litter decomposition. There was no difference in weight loss. There may be a difference in heterotrophic bacterial activity, but this is not too apparent and does not show a stimulation in response to nutrients, especially in corral D. The photosynthetic organisms on the litter suspended at 1 m did, however, show a stimulation above references in corrals B and D but litter in corral C, an unfertilized corral, was equal in photosynthetic activity to corral D litter.

When sediment samples were collected from the corrals during 1985,there was a decided increase in sediment bacterial activity in the nutrient enriched corrals (Fig. 32). High activity in corral A was probably an artifact caused by disturbance of the sediments during the spring 1985 and may not represent a true nutrient stimulation.

Benthic algae. The total chlorophyll quantities in 1983 for the four limnocorrals and the bay is given in Fig. 33. Two aspects of these data are striking. First, most of the chlorophyll *a* in the bay and reference limnocorrals is epibenthic and secondly, the nutrient additions had little effect on the epibenthic algal biomass. What differences there are between corrals B and D and the nutrient control corrals are almost entirely due to phytoplankton chlorophyll. It is also surprising how little of the total chlorophyll is attached to the sides of the limnocorrals. Attached algae are more

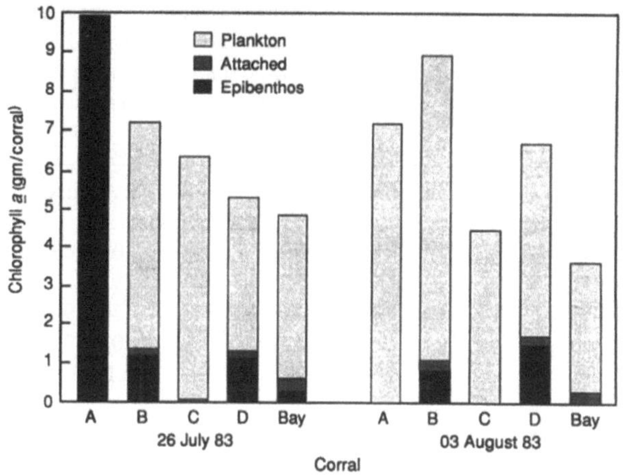

Fig. 33. Epibenthic, attached and planktonic chlorophyll densities in the corrals in 1983.

dense in the nutrient treated corrals but still make up less than 5% of the total chlorophyll in these corrals. These same trends were accentuated in the years 1984 and 1985.

The primary productivity of epibenthic algae shows hardly any treatment effect between the fertilized limnocorrals (Fig. 34) confirming the lack of impact of nutrient addition on epibenthic algae.

Benthic invertebrates. In 1983, there was no evidence that the different limnocorral treatments influenced the benthic community. At the end of the summer there was no difference among treatments for densities of chironomids and clams

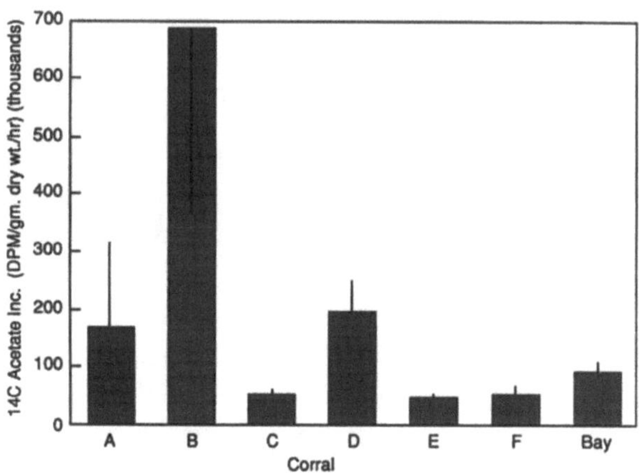

Fig. 32. Sediment acetate incorporation on 23 July 1985.

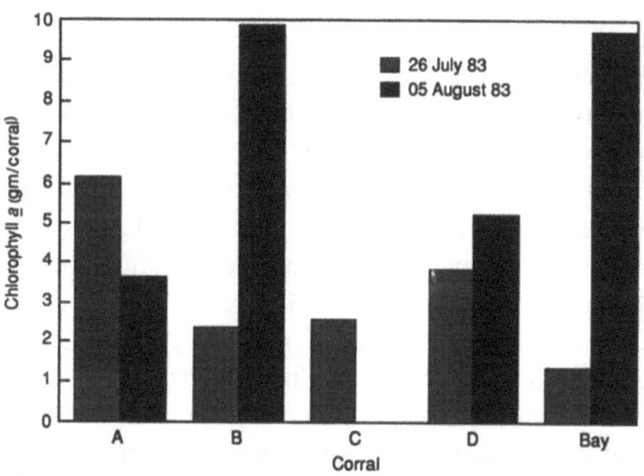

Fig. 34. Epibenthic algal production in the corrals in 1983.

Table 10. Density of major benthic taxa in limnocorrals and bay station prior to initiation of limnocorral treatments (7 July '83) and approximately 1 month after treatments were initiated (5 Aug. '83). Table entries are mean I.S.E. densities/m^2 from 3 samples.

	7 July	5 August
Limnocorral	Chironomidae	
A	6366 (956)	2075 (637)
B	7003 (555)	1910 (573)
C	6069 (1426)	2763 (942)
D	6875 (1420)	2037 (293)
Bay station	5942 (473)	A1273 (318)
	7 July	5 August
Limnocorral	Sphaeriidae	
A	849 (170)	1273 (76)
B	2292 (447)	293 (38)
C	976 (258)	853 (344)
D	1188 (42)	726 (255)
Bay station	1273 (409)	216 (115)
	7 July	5 August
Limnocorral	*Valvata*	
A	0	255 (76)
B	721 (306)	127 (–)
C	85 (85)	89 (89)
D	42 (42)	38 (38)
Bay station	340 (42)	38 (38)

Table 11. Densities (#/m^2) of Chironomidae in limnocorrals and bay station on 5 August 1983. Table entries are means of three points.

Taxon	Bay	A	B	C	D
Heterotrissocladius maeari	212	647	721	1061	467
Zalutshia sp. 1	–	42	–	85	–
Parakiefferiella triquetra	–	170	340	212	255
Psectrocladius sp. 1	42	–	–	85	85
Abiskomyia virgo	–	42	42	127	–
Cricotopus spp.	–	170	–	42	–
Stictochironomus rosenschoeldi	212	85	212	424	85
Polypedilum sp. 1	–	297	127	127	42
Tanytarsus spp.	85	85	42	–	212
Paratanytarsus spp.	–	255	–	42	85
Stempellinella minor	42	42	170	212	42
Constempellina brevicosta	–	–	127	–	–
Micropsectra sp. 1	85	–	–	–	–
Procladius nr. *vesus*	170	127	42	170	127
Ablabesmyia sp.	–	170	–	127	–
Monodiamesa bathyphila	–	–	42	42	127
Protanypus saetheri	42	– 721	42	85	170

when corrected for initial density (Table 10). Corrected values for snails did differ among treatments. However, this difference reflected differences between limnocorrals A and C and between B and D. Since these pairs of corrals were treated alike with respect to nutrients, the differences could not be due to nutrient addition.

Chironomid species composition was similar in all limnocorrals and the bay station at the initial 1983 sampling (Table 11). The dominant chironomid did not show a growth response to the nutrient addition since the mean size of *Heterotrissocladius maeaeri* was similar in all limnocorrals at the end of the year (Fig. 35).

By 1985, there were differences in chironomid density between limnocorrals and the bay station. The bay control had significantly more chironomids than any of the limnocorrals (Fig. 36). Comparisons among limnocorrals did suggest that there were significant density differences related to nutrient treatment. The limnocorral with

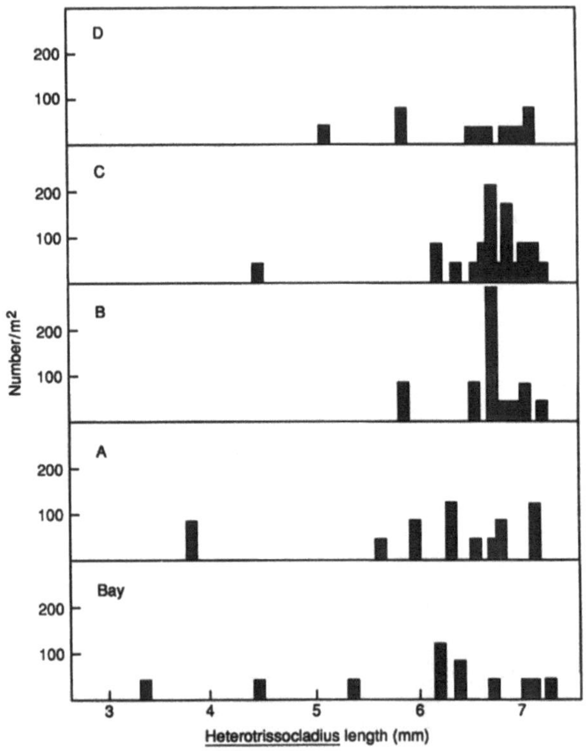

Fig. 35. Length frequency distribution of *Heterotrissocladius maeaeri* in the corrals and bay on day 30 in 1983.

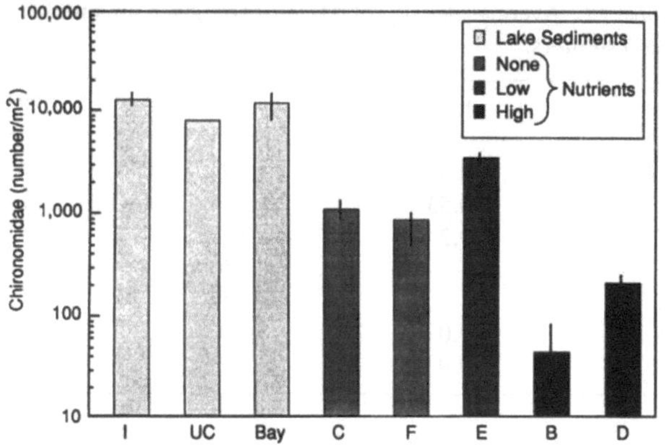

Fig. 36. Density of larval Chironomids in the corrals and bay at the onset of nutrient addition in 1984.

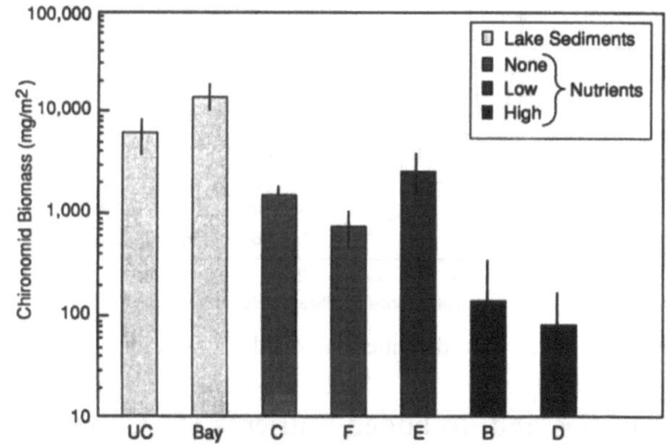

Fig. 37. Biomass of larval Chironomids in the corrals and bay at the onset of nutrient addition in 1984.

the low level of enrichment (E) had the highest chironomid density. The high nutrient limnocorrals were similar in chironomid density and had significantly fewer chironomids than either limnocorral E or the no-nutrient addition limnocorrals. The no-nutrient limnocorrals were also similar in chironomid density and also had significantly fewer chironomids than the low enrichment corral. Chironomid biomass showed the same trend as density (Fig. 37), but comparisons between limnocorrals were not significant.

There was no evidence of a chironomid growth response in 1985, based on the length frequency distributions of the dominant species in that year, *Cricotopus sylvestris* (Fig. 38). Length distributions were similar for both types of reference samples

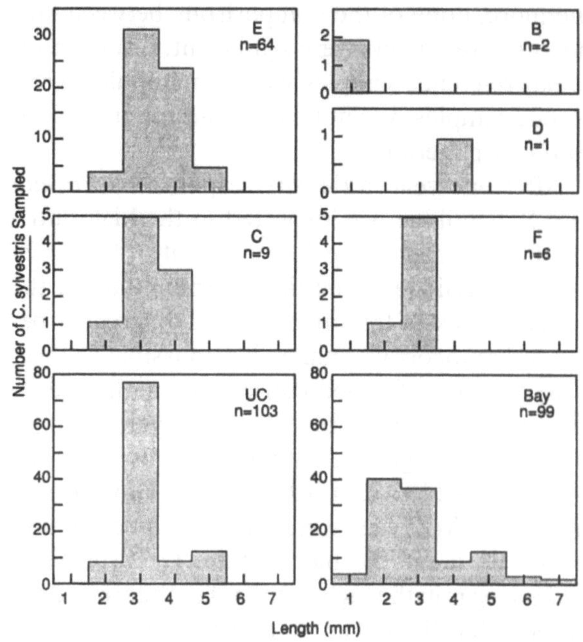

Fig. 38. Length frequency distribution of *Cricotopus sylvestris* in the limnocorrals and bay after one year of nutrient addition.

taken outside of the limnocorrals, the no-nutrient corrals, and the low nutrient limnocorral. The high nutrient limnocorrals had so few *C. sylvestris*, that there was no basis for comparison with other limnocorrals.

There was no evidence of a density response by clams or snails to the nutrient treatments (Table 12). For both snails (*Valvata*) and clams (Sphaeriidae), there were significantly higher densities outside the limnocorrals than inside. Fur-

Table 12. Density of Sphaeriidae and *Valvata lewisi* (mean no./m^2 $\pm$ SE) in limnocorral and control samples in 1985.

Treatment	*V. lewisi*	Sphaeriidae
B	127 ± 74	169 ± 112
C	43 ± 42	722 ± 279
D	424 ± 224	382 ± 127
E	169 ± 169	422 ± 297
F	42 ± 42	255 ± 195
Bay control	340 ± 85	1019 ± 195
Unmanipulated	1231 ± 449	424 ± 112
Initial	1496 ± 141	796 ± 313

thermore, none of the comparisons between limnocorral treatments were significant. Comparison of bay reference samples with unmanipulated reference samples was not significant for either *Valvata* or Sphaeriidae.

After one year, a density response by *Valvata* and Sphaeriidae was expected in the high nutrient limnocorrals. The lower density of *Valvata* and Sphaeriidae in the limnocorrals than in the bay, and lack of treatment related differences between limnocorrals, makes it impossible to evaluate the effect of added nutrients on mollusc density and growth (Table 12). However, samples from the two high nutrient limnocorrals, B and D, had 1103 ± 1103 and 1952 ± 1001 *Valvata* eggs m^2. No *Valvata* eggs were found in the other limnocorrals or at the bay station. Occurrence of *Valvata* eggs in the high nutrient limnocorrals does suggest that there was no toxicity effects on *Valvata*, and suggests that nutrient enrichment may have facilitated snail reproduction. However, since there were no eggs found in the low enrichment limnocorral, the conclusion is tentative.

Table 13. Average density of Heterocope given in numbers per 100 liters.

	1983	1984	1985
Bay	2.5	3.2	3.9
A	0.53	2.5	2.4
C	0.2	7.1	2.5
F	–	13.5	9.5
E	–	22.5	3.3
B	0.0	24.9	0.0–
D	0.9	7.7	0.17

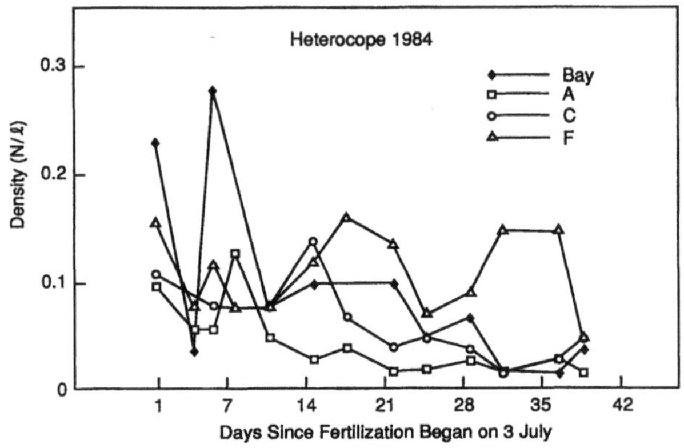

Fig. 39. Heterocope densities in corrals F, A and C in 1984.

Limnocorral response to fish treatment

As physical and chemical factors

There is no indication that the fish treatments physically or chemically altered the limnocorrals to which they were applied.

Planktonic component

In 1983, when the grayling were unrestrained in the limnocorrals, almost all the *Heterocope* were eliminated from all four limnocorrals early in the summer (Table 13) although they disappeared from the low fish treatments a week later than from the high fish treatments Despite the virtual elimination of this predator, there was no dramatic increase of herbivorous zooplankton densities over those in the bay where *Heterocope* was continually present.

In 1984, with the fish densities reduced and the fish restrained in cages, the *Heterocope* densities more closely matched those observed in the lake. In fact, in the three limnocorrals with variable fish densities and no added nutrients (F, C, and A), the *Heterocope* densities were inversely related to the fish densities by late July and early August (Fig. 39, Table 13). Furthermore, the low fish treatment had a density generally similar to the bay. Thus, in the latter half of 1984, there was a gradient of *Heterocope* density and *Heterocope* predation pressure. However, no herbivorous zooplankton species showed any major change along this gradient of *Heterocope* density. In 1985, C, E and F had the greatest *Heterocope* densities and these were all low or no fish treatments in 1984. Corrals B and D were essentially devoid of *Heterocope* in 1985 but corral F, which never had fish, had the highest density.

There was a definite fish-nutrient interaction in the case of the daphnid response in the high nutrient corrals B and D. In corral B a low fish treatment *D. middendorffiana*, a large bodied daphnid, responded to reach very high densities (Fig. 29). However, in corral D a high fish treatment *D. longiremis*, a small bodied daphnid, responded to reach high densities and *D. middendorffiana* was almost totally absent (Fig. 29).

Benthic components

Added grayling could only have affected the benthic invertebrates in 1983 since in all subsequent years the fish were restrained in cages. In 1983 however, there is no indication that even the high treatment of fish altered benthic invertebrate densities. Unfortunately we do not have data on the abundance of sculpin or burbot in any of the limnocorrals. However, since grayling are predominantly planktonic feeders in lakes, no effect of fish treatment on benthic invertebrates was expected. Not only did we not expect grayling to have an impact on benthic invertebrate communities, such an effect has been slight even where expected. For example, the influence of bottom-feeding fish on soft bottom communities in the MERL tanks was much less than had been anticipated (Pilson & Nixon, 1980).

Fish

The fish were not measured prior to placing them in the corrals so it is unknown whether there was any effect of nutrient addition on fish growth. The grayling added to the corrals seemed to fare well during all three summers with the exception of fish placed in corral B during the latter part of 1985. Several fish developed cloudy eyes and fin rot and one fish died while in corral B. The fish that were added to the corrals in 1983 were left in the corrals throughout the winter but were not seen the following summer thus presumably did not survive the winter in the corrals. All fish but grayling were intended to be excluded from the limnocorrals. However, there were doubtless some sculpin enclosed and in 1985 a group of small burbot were sighted in corral E.

Discussion

The overall impact of the two general treatments applied to the limnocorrals varied dramatically in magnitude. The nutrient treatment had strong effects on planktonic biology and chemistry as well as on benthic chemistry. The fish treatment seems to have only affected the population densities of large bodied zooplankton, *Heterocope* and *D. middendorffiana*.

Limnocorral response to nutrient treatments

Physical factors

The major way in which the nutrient treatments affected the physical and chemical processes in the corrals was through greatly stimulated photosynthesis of the phytoplankton. There was a very distinct decrease in Secchi disk depth with increasing phytoplankton density as estimated by chlorophyll concentration. The Secchi disk depth of Toolik Lake is near its maximum of 6 m at chlorophyll densities below 5 $\mu g\ l^{-1}$. In the high nutrient addition corrals Secchi depths decreased to nearly 1 m at chlorophyll densities greater than 80 $\mu g\ l^{-1}$. Light extinction values show a similar although less distinct pattern.

Chemical factors

The algal growth and resulting photosynthesis in the high nutrient addition corrals also altered the chemical environment. These corrals always had supersaturated oxygen concentrations and even more importantly developed high pH values toward the end of each summer. The pH response advanced stepwise, each year higher pH values corresponding to higher and higher levels of primary productivity and algal biomass. By the end of the summer of 1985, the pH in corral B, and to a lesser extent in corral D, reached levels which might be harmful to resident organisms. The fish put into corral B did have some problems, as noted above. It has been known for some time that photosynthetically elevated pH may cause mortality in zooplankton and O'Brien and deNoyelles (1972) demonstrated that pH just above 10.6 may be the critical point. It is also known that elevated pH may reduce *Daphnia* filtering rates (Kring & O'Brien, 1976). Malley *et al.* (1977) suggest much lower pH values may have adversely affected zooplankton in a nutrient treated lake (227) at ELA. However, both corrals B and D in 1985 had healthy populations of cladocerans although *Heterocope* disappeared from corral B and was very low in corral D.

The chemical behavior of the added ammonia and phosphorus in the high nutrient treated corrals was quite interesting. In 1983, the amount of dissolved inorganic nitrogen and phosphorus increased over the summer but at a lower rate than the daily addition. However, even the amount missing from the open water can not be accounted for by phytoplankton growth. Assuming a ratio of 1 μg of phosphorus to every microgram of Chlorophyll *a*, we would expect to see 110 mg Chl *a* m^3 in the high nutrient corrals which is about five times what actually developed. In subsequent years, we did observe more algal biomass as chlorophyll *a* but little of this seems to be due to recycling of the added nutrients of previous years. Under ice measurements in May 1985 showed there was only 1.5–7.0 moles of N (in the form of nitrate) in corrals B and D. This is only 7–10 days worth added nutrient, so recycling is supplying little N relative to the experimental nutrient additions, but of course, quite a lot relative to Toolik Lake nutrient loading.

In a nutrient budget developed for corrals B and D for 1984 (Table 9) only about one half of the added nutrients can be accounted for. Addition of sodium chloride to the corrals in 1983 and 1984 and subsequent measurement of sodium concentrations after equilibrium mixing indicate leakage of the corrals was not a problem. The most likely source of error in the nutrient budget is of the estimate of the abundance of algae attached to the sides of the corrals and of the rate of sedimentation of phytoplankton algae. Despite the possible errors, it is striking that 50-60% of the N in the fertilized corrals remains as particulate matter suspended in the water column. Another possibility is that the estimate of sedimentation could be wrong. This is the only number in Table 9 calculated from a rate so it is an excellent candidate for error. However, a reason for the error is not obvious and we would expect that sediment traps would work well in the relatively stable environment of a limnocorral. In fact, another type of measurement, ammonium release from the sediments, does agree well with the sedimentation rate for nitrogen so there is some evidence for the correctness of these sedimentation rates. Denitrification seems unlikely because we believe the sediments, even in the fertilized corrals, never became anoxic.

In 1984, the ammonia levels in the nutrient enriched corrals began declining after day 10 and declined by day 25 to nearly undetectable levels. The phosphorus levels, however, continued to increase, suggesting that a nitrogen limited system had developed. Chlorophyll increase in corral B did level off as ammonia levels declined (Fig. 6) suggesting a limiting condition. However, nothing as dramatic as a nitrogen fixing blue green algal bloom developed to confirm the ideas of a nitrogen limited system. The absence of heterocyst bluegreen algae may have been due to the high oxygen levels since both corrals B and D had supersaturated oxygen concentrations throughout most of the summer.

In August of 1984, the phytoplankton in corrals B and D appeared to be taking up nitrogen at a rate very close to that which nitrogen fertilizer was being added. Production was occurring at rates approximately 1.25 g C m^{-2} day^{-1} (Fig. 26). If the algae had a C/N ratio of 6:1, then the nitrogen requirement of the algae at this time would be about 0.2 g N m^{-2} day^{-1} while the treated corrals were receiving nitrogen at a rate of 0.16 g N m^{-2} day^{-1}. Thus, it appears that the algae were being sustained by added nutrients rather than recycled nitrogen from the water column or sediments as is the case in Toolik Lake. Measured fluxes of nitrogen from the sediments at this time were around 0.02 g N m^{-2} day^{-1}. When this flux is added to the fertilizer addition, we have a very good agreement with the estimated algal nitrogen requirement.

In 1985, both the ammonia and phosphorus levels in the nutrient enriched corrals declined to low levels. Phytoplankton densities in these corrals were higher than in either 1983 or 1984, and presumably the algal community was limited by nitrogen or phosphorus. Given the high pH values observed in 1985, it is possible that algae in highly fertilized corrals were experiencing an inorganic carbon shortage. The question remains, however, what was limiting the algae response in 1983 when both nitrogen and phosphorous were abundant and the pH low?

Planktonic components

Planktonic microbes. The bacterial community enclosed within the treated corrals responded to added nutrients by increasing about 8 times over reference levels (Fig. 9). However, in both 1983 and 1984, bacteria numbers soon declined to near reference levels. It seems clear from the observations of microheterotroph abundance made in 1984, that this decline was due to predation on bacteria by the microheterotrophs. In corral B in 1984, there was a strikingly straightforward prey-predator cycle in which bacteria numbers rose to 8–10 times those in reference corrals and then microheterotrophs increased and presumably grazed down bacteria numbers to near reference levels (Fig. 10). This has been observed in controlled laboratory situations but to our knowledge, there is only one other demonstration of this dramatic effect in larger systems (Andersen & Fenchel, 1985). Thus, while phytoplankton populations seem not to be regulated through top-down control in Toolik Lake, bacteria populations are.

Bacterial production was measured in two ways, as incorporation of thymidine and as incorporation of acetate into lipids. Happily, these two measures agreed very well in a relative sense which lends credence to both techniques. These production measurements indicate that the highest bacterial production occurs when the bacteria are under severe predation by microheterotrophs.

With regard to the interaction of planktonic algae and bacteria, bacterial activity generally followed the nutrient enrichment of algae in corrals B and D. However, when algae were experimentally removed by filtration the bacterial activity was reduced to the levels of the unfertilized corrals. This indicates that the most active bacteria were attached to particles in the nutrient enriched corrals. This also indicates that the free living bacteria had the same general activity in the fertilized and unfertilized corrals. The close coupling between the bacteria and the algae may indicate that bacteria are using algal exudates as a source of organic carbon. These data contrast with data from similar differential filtration experiments in nearby lake N-2 (Hullar *et al.*, 1986) which was divided and one side of the lake enriched with N and P. In those experiments, the algae were enriched in the nutrient addition part of the lake. However, the bacteria which were also stimulated, were not filtered out with the GFB filter indicating that they were free living (not attached to particles), but were still carbon limited and growing on algal organic sources produced by excretion.

Microbial biomass greater than 1 μm in size, mostly algae and attached bacteria, also increased in the fertilized corrals (corral D) as measured by lipid phosphate (Fig. 13). This indicates that viable biomass determined by chlorophyll *a* content (Fig. 21) and lipid phosphate (Fig. 13) are similar and can be used to measure the same planktonic components.

The decomposition of litter as shown by weight loss and microbial activity was not affected by the nutrient addition to the corrals whether the litter was incubated on the sediment surface or at 1 m of depth in the water column. These results essentially confirm the results of Federle & Vestal (1980, 1982) and Federle *et al.* (1982) that added nutrients do not stimulate *Carex* decomposition in Toolik Lake.

However, there was a distinct effect of nutrients on the heterotrophic microbiota in the top 1 cm of sediment material in corrals B and D. This response, observed in 1985, reflects a response to one year of nutrient addition, since these sediments were placed in the limnocorrals from July 1984 to July 1985.

The litter and sediment microbiota results are consistent with observations regarding nitrogen release and oxygen consumption by the sediments (Figs. 30 and 31). Measurements of C and N in the sediments of corrals B and D (not reported here) show that increased algal sedimentation resulted in only a very thin layer of new material at the sediment-water interface during 1983 and 1984. Iron and Mn measurements made concurrently with the nitrogen and oxygen measurements (Kipphut, 1984), indicate that sediment remineralization is largely confined to the thin layer of freshly sedimented algal material. Remineralization of the underlying sediments (which are similar in nature to the material used in the litter bags) apparently was not stimulated. Likewise, it apparently was not until 1985 before enough fresh algal material had accumulated so that increases in heterotrophic activity could be directly measured in the sediments. The sediments of Toolik Lake are responding much more slowly than the overlying waters to the nutrient enrichments.

Phytoplankton. The phytoplankton were clearly stimulated by the addition of nitrogen and phosphorus. In fact, each year the overall density of phytoplankton in the high nutrient addition corrals increased to levels greater than the year before. This was true in 1985 as well, even though the nutrient treatment was decreased from a 10 × to a 5 × level in 1985 (Fig. 23).

The development of chlorophyll concentrations in Toolik Lake and in the limnocorrals does not conform too well to the phosphorus loading relationships developed by Ahl (1975) for some large temperate lakes and applied by Persson *et al.* (1975) to some subarctic Swedish lakes, two of which were experimentally fertilized with phosphorus and nitrogen plus phosphorus (Fig. 24). Given the phosphorus loading estimate for Toolik Lake by Whalen & Cornwell (1985) of around $0.1\ g\ m^{-2}\ yr^{-1}$, the chlorophyll that developed in the lake and nutrient control corrals is generally below the level predicted by Ahl (1975), sometimes considerably below. However, Hobbie *et al.* (1983) showed that much of the inflow and thus nutrient loading of Toolik Lake comes with first melt and has a very low residence time in the lake. Whalen & Cornwell (1985) show that retention of phosphorus is quite low, only 30%. If we then use $0.03\ g\ m^{-2}\ yr^{-1}$ as the best estimate of the effective phosphorus loading of the lake, the values for the lake and nutrient control corrals nicely cluster around the line given in Ahl (1975).

However, all the high nutrient addition corrals have chlorophyll values greater than that predicted by Ahl (1975) although only the 1985 chlorophyll values for corrals B and D are seriously greater than predicted. One contributing factor to these high chlorophyll values is that none of the limnocorrals were lifted between the years of 1984 and 1985 and thus did not flush. Toolik Lake and the lakes reported by Ahl (1975) and Persson *et al.* (1975) all have some degree of annual flushing. This lack of flushing reduces the phosphorus loading of the nutrient control corrals but may greatly increase the loading of the treated corrals relative to lakes. However, even if one considers the total cumulative phosphorous loading of the treated corrals over the three years of nutrient addition as the appropriate abscissa value for Fig. 24, the chlorophyll values are still greater than would be predicted by this relationship. A further contributing factor may be that the treated corrals are not just loaded with phosphorus but simultaneously with nitrogen as well. Persson *et al.* (1975) treated Lake Hymenjaure from 1972 to 1974 with just phosphorus and the chlorophyll levels in the lake hardly tripled and were always below that predicted by the relationship found by Ahl (1975). However in 1975 Persson *et al.* (1975) treated the lake with phosphorus and nitrogen and the chlorophyll increased by an order of magnitude over what it had been in 1974. By 1975, the chlorophyll concentration was considerably higher than that predicted. Thus, the response of the treated corrals is consistent with this result.

Another possible explanation for the high values of chlorophyll concerns the total phosphate (the abscissa in Fig. 24). Much of the total phosphorus measured in lakes is organically bound phosphate and would likely be much less available to algae than the inorganic phosphorus added to the corrals. This would be true for nitrogen as well.

In comparing these results with other aquatic ecosystem nutrient enrichment studies it is important to keep in mind the low grazer densities in Toolik Lake and the limnocorrals. Hrbacek *et al.* (1978) showed that density of planktivous fish, presumably altering the density of grazers, altered the phosphorus to chlorophyll *a* relationship. Thus, very low grazing could be an explanation for the high chlorophyll *a* values observed in the limnocorrals.

There are several other interesting alternatives to nitrogen and phosphorus as the main regulating factor for phytoplankton. Throughout the summer of 1983 there were very high levels of nitrogen and phosphorus in solution in the treated corrals and it appeared that phytoplankton growth and nutrient uptake did not proceed nearly as rapidly as nutrients were being experimentally added. This contrasts with single pulse experiments with added nutrients to lakes in which phosphorus rapidly declines to pretreatment levels in a matter of days if not hours (Lean, 1973). Thus, with plenty of soluble nitrogen and phosphorus available, the question remains as to why the phytoplankton did not increase still further. The possibility of light limitation is obvious but in 1985 almost 3 times the chlorophyll density was produced in corral B than in either corral in 1983. Furthermore, experiments in 1983 in which phytoplankton from the high nutrient corrals were incubated at constant, high light showed no higher primary productivity in these experiments than in phytoplankton experiments incubated at depth in the corrals. On August 8, the only date we have simultaneous measurements, the in situ production was 800 and 650 mg C m^{-2} day^{-1} in B and D while at constant light the production was 840 and 600 mg Cm^{-2} day^{-1}, respectively.

Zooplankton. The zooplankton response to the high phytoplankton densities created through plant nutrient addition was slow developing. Most species either did not respond or responded in an erratic fashion. In both corrals B and D in 1983 and 1984, temporal replicates, neither the abundance nor species present changed much from that observed in the bay (Fig. 29). However, *D. longiremis* showed a small but consistent response in 1983 and 1984 and a pronounced response by 1985. It may well be that this species needs several years in which to build up a response to increased phytoplankton. A much more dramatic response occurred in corral B in 1985 where *D. middendorffiana* grew to very high densities at the very end of the summer.

The overall modest response of zooplankton population densities to increased phytoplankton biomass is not entirely unprecedented. In the whole lake fertilization experiments done at ELA in Canada, few have shown any increase in zooplankton (Malley *et al.*, 1977). Bare Lake in southern Alaska was fertilized and this stimulated phytoplankton growth but there was no increase in zooplankton density. In the longterm, large lake fertilization experiment in Sweden, phytoplankton was definitely increased but zooplankton only increased after two years. In the present study, we only had one year for such a carryover to be expressed since the corrals were lifted and started again in 1984.

Furthermore, the fertilized limnocorrals may have provided an unfavorable overwintering habitat for *Heterocope* which in 1985 was absent or very low in the high nutrient corrals but present in all others. It is not known what was responsible for this disappearance, but this species does release resting eggs which sink to the bottom and hatch the following spring. It would seem possible that the high sedimentation in these corrals may have created an environment not very suitable for such eggs although *Heterocope* does live in many arctic ponds which have thick layers of flocculent sediments. It is likely that corrals B and D had anoxic sediments during the spring of 1985. Oxygen measurements under the ice taken in May 1985 showed levels in the water column less than 2 mg O_2 l^{-1}. As a result of sewage eutrophication of Meretta Lake in the Canadian Arctic, a large copepod, *Limnocalanus* disappeared apparently due to egg mortality (Schindler *et al.*, 1974a.b; Kalff & Welch, 1974; Kalff *et al.*, 1975).

Benthic components

Benthic invertebrates. We did not expect any effect of nutrient addition on chironomid density for the short term 1983 experiment because density differences could only occur by differential recruitment success among limnocorrals. Most species emerge during July and August (Hershey, 1985b), and we did not see evidence of the new generation during the 1983 season. Scraping of the bottom sediments during the ice-out movement of the limnocorrals precluded measurement of recruitment success the following spring.

Lack of an observed growth effect in 1983 on the dominant chironomid *H. maeaeri* in the nutrient enriched limnocorrals may be attributed to either of two factors: 1) There may have been insufficient time for a growth response to have occurred; or 2) the chironomids may not have been resource limited. Growth of chironomids has been correlated with planktonic primary production in other systems (Jonasson, 1972), but such stimulation probably results from the settling to the bottom of an algal bloom which may not have occurred soon enough or with sufficient intensity to stimulate *H. maeaeri* growth. Alternatively, *H. maeaeri* may not have been resource limited. We know from a previous study that chironomid density in the bare sediments (not covered by macrophytes) is controlled by sculpin predation (Hershey, 1985a).

The lack of an observed growth effect in 1985 on *C. sylvestris* supports the conclusion that chironomids were not resource limited. Length distributions were similar for both types of control samples taken outside of the limnocorrals, the no nutrient corrals, and the low nutrient limnocorral (Fig. 38). The high nutrient limnocorrals had so few *C. sylvestris* that there was no basis for comparison with other limnocorrals.

Differences in the density of chironomids between limnocorrals appear to be inversely correlated to nutrient treatment regardless of the poor recruitment success of the larvae. There are at least two interpretations for the lower density of chironomids in the two high nutrient limnocorrals: 1) the high nutrient concentration could have been toxic to larvae adapted to oligotrophic conditions, either due to high ammonia concentrations or low O_2 concentrations; or 2) the high nutrient concentration could have facilitated larval development rate,such that larvae in the high nutrient limnocorrals emerged sooner, leaving fewer larvae in samples to be collected. Toxicity due to ammonia or other measured water quality parameters is not known for arctic chironomids, but ammonia concentrations were higher in limnocorrals B and D. Lower oxygen concentrations did not occur in the water near the sediments during the summer, but increased benthic O_2 consumption did occur in limnocorrals B and D; which may have affected O_2 availability to chironomids within the sediments.

The second alternative, faster development rate, is not supported by other data. If development rate were faster, an intermediate density of chironomids would be expected in limnocorral E, rather than the higher density observed. Furthermore, a faster growth would be supporting evidence for a faster development rate, and length data for *C. sylvestris* (Fig. 38) in 1985, and for *H. maeaeri* (Fig. 35) in 1983 do not indicate that growth was facilitated in B and D.

Since limnocorral E had the highest density of chironomids, it is clear that the low level of nutrient addition did not cause toxicity. Because of this higher density of chironomids in limnocorral E, it is tempting to conclude that low level of nutrients stimulated chironomid production, but high level of nutrients was detrimental. But, with careful consideration of the data, this interpretation is not supported. If stimulation occurred we would expect to also detect higher biomass in limnocorral E than in C and F where no nutrients were added. There appeared to be a trend in this direction, but not enough to support this idea. Since chironomid weight increases exponentially with length (e.g., Smock, 1980), the difference in biomass should be proportionally greater than the difference in density if nutrient addition stimulated production; however, the difference in biomass was less than the difference in density. Furthermore, *C. sylvestris*, the dominant species in 1984, did not have a higher growth rate than in limnocorrals C and F, again suggesting no stimulation due to nutrient addition. The higher density (but not higher biomass) can best be explained by more small chironomids in E, which is an indication of higher recruitment success in E compared to the other limnocorrals but still low success compared to the bay. If this is correct, then the conclusion that follows is that chironomids are not food limited in Toolik Lake.

Limnocorral response to fish treatments

Only large-bodied zooplankton species seem to have been affected by the fish treatment. *Heterocope* was virtually eliminated from all four corrals in 1983 and *Heterocope* densities were considerably reduced in the high fish treatments in 1984 and greater densities in the no fish treatment (corral F) in 1984 and 1985 (Table 13).

The lack of increase in herbivorous zooplankton species populations with the near complete elimination of *Heterocope* from the corrals in 1983 and reduction of *Heterocope* densities in high fish corrals in 1984 and 1985 is puzzling. There seems little doubt that *Heterocope* can be a voracious predator on those species (Luecke & O'Brien, 1983; Dodson, 1984) and that it may be responsible for declines of these populations in Toolik Lake. However, it must be remembered that the fish treatment to manipulate *Heterocope* only decreased the density of *Heterocope* but never dramatically increased its density. Thus, if the herbivorous populations of zooplankton in Toolik Lake are food limited, they might well not increase much in density in the absence of *Heterocope* but still decline when it was abundant.

The fish treatment also seems to have determined which daphnid species responded to the increased phytoplankton in the nutrient enriched corrals B and D in 1985. In corral B, a low fish treatment, the large bodied *Daphnia middendorffiana* dramatically increased in abundance. However, in corral D, a high fish treatment, the small bodied *D. longiremis* dramatically increased in abundance.

General discussion

The use of enclosures to answer ecosystem level questions is quite new. In fact, even the bottom up-top down question of ecosystem control is newly recognized. Here we address how our experiment faired in these topical areas.

Limnocorral as experimental units

The use of limnocorrals and other such enclosures is now common in both freshwater and marine systems, where they are termed mesocosms (Grice & Reeve, 1982). Their use is surprisingly recent where the first was in 1957 in a Czechoslovakian reservoir to study the effects of algicides on phytoplankton (Stepanek & Zelinka, 1961). However, this paper was largely overlooked and the papers which most stimulated future research using limnocorrals stem from Thomas (1958) and Goldman (1962). Goldman actually constructed a limnocorral much like the ones used here but of smaller diameter (0.6–1.2 m).

The size and time scale of our limnocorral experiment is within the guidelines suggested by Parsons (1982) where when small fish are included in the experimental design he suggests volumes that range from 10 to 1000 m^3 and time scales from 2 weeks to a year. Volumes must be sufficient to withstand destructive sampling and sufficient surface to volume ratio to minimize edge effects (Sanders, 1985). Large sized limnocorrals such as those used by Lund & Reynolds (1982) have maintained the natural phytoplankton and zooplankton communities for at least 2.5 yrs. There was, however, a progressive oligotrophication which resulted in reduced phytoplankton biomass. Increased time is necessary to allow for various time delays. For example, Lean & Charlton (1976) observed a 55 day lag time between nutrient enrichment and response of phytoplankton biomass due to sediments acting as a sink for phosphorus. Even larger enclosures may have problems with the duration of the experiment. The CEPEX marine enclosures caused a rapid sedimentation of large diatoms soon after deployment (Thomas & Siebert, 1977). However, in the present experiment, observed nutrient dynamics in the nutrient addition limnocorrals behaved differently in each 3 summers of the experiment demonstrating both the problem and importance of long term experiments.

Limnocorrals have been used for a variety of research purposes. They have been used to study eutrophication processes (Lack & Lund, 1974; Lund, 1978; Leah *et al.*, 1978) where the focus is primarily on studying phytoplankton-nutrient interactions. More recently, they have been used to study the effects of specific pollutants such as pesticides (Solomon *et al.*, 1980; Solomon *et al.*, 1982; Solomon *et al.*, 1985; Kaushik *et al.*, 1985). Limnocorrals treated similarly have been found to show low between corral variability (Uehlinger *et al.*, 1984). However, few have really been treated as separate, isolated ecosystems with most of the populations and ecosystem processes studied. Furthermore, most studies have been short-term. However, their use in manipulations of trophic structure of natural ecosystems is also reported (Andersson *et al.*, 1978; Borsheim *et al.*, 1982). Still, limnocorrals cannot be thought of as truly replicatable. For example, the control MERL tanks showed tremendous temporal variability in phytoplankton biomass throughout 1976–1977 (Pilson & Nixon, 1980).

Even the largest limnocorral may cause abnormalities through shading and reduced turbulence. Bloesch *et al.* (1988) noted a ten-fold reduction in eddy diffusion and zooplankton migration was reduced due to shading. In the present case the limnocorrals were constructed of semitransparent material that absorbed most wavelengths of light, about 50%. However, Aufwuchs and degradation through time undoubtably reduced this transmission. In a 10 m diameter and 20 m deep enclosure at temperate latitudes, at least 50% of the enclosed water volume will be shaded and will be exposed to only diffuse light during sunny days (Takahashi & Whitney, 1977). Several enclosure experiments have used non-transparent material (the MERL tanks, the Lund tubes, MELIMEX and others). Such a choice was generally made to obtain a stronger material and does have the advantage that the effect is constant and not subject to change due to Aufwuchs growth. Suspending strips of sidewall material in a vertical series will allow quantification of the effect of Aufwuchs buildup (Sanders, 1985). We did this for individual years but not over the full extent of the experiment and thus could not document the effect of ice scour and temporal degradation of Aufwuchs material. During the arctic summer, however, light limitation seems unlikely. Miller (personal communication) has been unable to demonstrate light limitation in phytoplankton in the present limnocorral experiment even from highly enriched limnocorrals which developed dense algal blooms. However, Uehlinger (1982), using semi-transparent limnocorrals of similar size to those used here (3 m diameter and 11 m deep), found appreciable shading on sunny days which affected the distribution of zooplankton and phytoplankton production. On overcast days they found no observable effects.

One of the least understood effects in lakes and oceans is turbulence and unfortunately it is an effect which limnocorrals are most likely to perturb (Lundgren, 1985). The dramatic loss of large diatoms in the CEPEX enclosures is obviously due to reduced turbulence (Grice *et al.*, 1977). However, diatoms generally do not make up a major portion of the Toolik Lake phytoplankton community and this was not a significant problem. The reduction of wind shear stress in 12 m diameter enclosures resulted in more shallow thermocline than found in the adjacent lakes (Imboden *et al.*, 1979). However, in the present case the bottom depth of the limnocorrals was situated near the bottom of the thermocline for Toolik Lake.

Overall the response of the contained ecosystems in the limnocorrals was heartening in that unmanipulated corrals responded similarly to the lake. Furthermore the manipulated corrals responded as expected although there were some surprises. The bottom up response was considerably more pronounced than the top down response.

Bottom-up control

The phytoplankton clearly and dramatically responded to the added nutrients of inorganic nitrogen and phosphorus. While the phytoplankton did respond in the first year of the experiment the phytoplankton did not take up all the available nitrogen or phosphorus leaving open the possibility of some other control mechanism. However, since the zooplankton did not increase in abundance in 1983 it does not seem that the phytoplankton population density was limited by zooplankton grazing. In 1984 and especially in 1985 when both inorganic nitrogen and phosphorus were reduced to very low levels, it seems much more certain that phytoplankton population densities were controlled by the level of available nutrients experimentally provided. Thus by inference the phytoplankton populations of Toolik Lake are also likely limited by available nutrients.

It also appears that open water bacterial populations are also, if not controlled, certainly stimulated from the bottom up. That is bacterial populations only increased in the nutrient treated corrals and then only after large populations of phytoplankton had developed. However, it seems that the ultimate control of these populations is from the top down by microheterotrophs since once the bacteria population densities increased they appeared to be quickly grazed down by microheterotrophs.

Zooplankton populations, especially daphnid species populations of *D. longiremis* and *D. middendorffiana* also responded to added nutrients and the resultant phytoplankton increase. However, the response was not observed in dramatic fashion until 1985 with two years of nutrient addition and the possibility of a carryover effect.

The reasons for the delay in response are not clear but appear to be quite real since a similar effect has been observed in a small arctic lake which has been divided and one section treated in a manner similar to the high nutrient corrals (O'Brien, personal communication).

Rates of sediment metabolism also appear to respond to nutrient addition. This response is likely due to the increased sedimentation of phytoplankton in the nutrient enriched corrals due to the more labile nature of this material than normal Toolik Lake organic sediment.

However, several components of the Toolik Lake ecosystem did not respond to added nutrients and thus appear not to be regulated from the bottom up. While the actual sediments in the nutrient addition corrals had much higher respiration than did the nutrient control corrals the added nutrients did not stimulate the decomposition of such components as *Carex* leaves. Nor did added nutrients and the resulting increased sedimentation of labile detritus stimulate benthic invertebrate population increases. However, as outlined previously the limnocorrals did not provide a strong test of this possibility since chironomids failed to recruit into them and corral movement destroyed a good means of observing multi-year effects. The grayling added for the fish treatment of the experiment were neither weighed nor measured for length prior to stocking in the corrals and thus any impact of added nutrients could not be determined. However, due to the small number of fish involved such a result, whether positive or negative, would be suspect.

Top-down control

The impact of the added grayling was much less pronounced than the effect of the added nutrients. The only direct effect was a decided impact on large bodied zooplankton. This was clearly and dramatically shown in the first year of the experiment where added grayling almost completely eliminated the large bodied zooplankter *H. septentrionalis* from all four corrals. In subsequent years with a modified fish addition the effect was less pronounced but generally evident. The added fish also affected the response of zooplankton to added nutrients. The large-bodied zooplankton *D. middendorffiana* responded to high nutrient addition in a low fish treatment but was likely prevented from responding in the high fish treatment. In this treatment the small-bodied zooplankter, *D. longiremis* responded to added nutrients. However, there is no indication that small-bodied zooplankton species were affected by the fish addition. This is not unexpected because arctic grayling are only facultative planktivorous on small zooplankton when the fish are small. Even at small size they are reluctant planktivores on small-bodied zooplankton (Evans, 1986).

The only other component of the ecosystem which would likely be vulnerable to increased fish predation was the benthic invertebrate populations and in 1983 there was no observable effect of high fish densities on benthic invertebrate densities. However, grayling are primarily planktivorous, thus the grayling manipulation does not constitute a test of top-down control of benthic invertebrates. In subsequent years the fish were restrained in large mesh cages and thus could not physically feed on bottom dwelling organisms.

Interestingly, there was no increase in phytoplankton with the fish addition as suggested by Reinerten *et al.* (1986), Vanni & Findlay (1990) and Threlkeld (1988). The high fish addition in 1983 was in the range of 3–5 g m^3. Vanni & Findlay (1990) stocked several limnocorrals at about 10 g m^3 and observed phytoplankton stimulation that they took to be caused by the fish addition. However, they postulate the fish effect is partially due to zooplankton recycling nutrients more rapidly when small bodied. In the present situation the zooplankton are probably too sparse to have such an effect. We do not know if the fish died during this experiment but at least some of the fish were observed throughout the summer. In the subsequent summers smaller fish were used and were present in one-half the fish days leading to a stocking rate closer to 1 g m^3.

References

Aagaard, K., 1978. The chironomids of Lake Malsjoen. A phenological, diversity, and production study. Norw. J. Ent. 25: 21–37.

Ahl, T., 1975. Effects of man-induced and natural loading of phosphorus and nitrogen on the large Swedish lakes. Verh. int. Ver. Limnol. 19: 1125–1132.

Anderson, R. 0., 1959. A modified flotation technique for sorting bottom fauna samples. Limnol. Oceanogr. 4: 223–225.

Anderson P. & T. Fenchel, 1985. Bacterivory by microheterotrophic flagellates in seawater samples. Limnol. Oceanogr. 30: 198–202.

Andersson, G., H. Berggren, G. Cronberg & C. Gelin, 1978. Effects of planktivorous and benthivorous fish on organisms and water chemistry in eutrophic lakes. Hydrobiologia 59: 9–15.

Ball, R. C. & I. D. W. Hayne, 1952. Effects of the removal of the fish populations on the fish-food organisms of a lake. Ecology 33: 41–48.

Benndorf, J., H. Kneschke, K. Kossatz & E. Penz, 1984. Manipulation of the pelagic food web by stocking with predacious fishes. Int. Revue ges. Hydrobiol. 69: 407–428.

Bligh, E. G. & W. J. Dyer, 1959. A rapid method of total lipid extraction and purification. Can. J. Biochem. Physiol. 37: 911–917.

Bloesch, J., P. Bossard, H. Bührer, H. R. Bührer & Uehlinger, 1988. Can results from limnocorral experiments be transferred to *in situ* conditions? (Biomanipulation in limnocorrals Vl) Verh. int. ver. Limnol. 23: 762–763.

Borsheim, Y., J. I. Koksvik, A. Langeland, B. Ness, Y. Olsen & H. Reinertsen, 1982. Karbon-og fosforbudsjett for planktonsamfunn i plastinnhegninger, med spesiell vekt pa bakterienes utvikling og betydning som fode for zooplanktonet. Norges Teknisk- Naturvitenskaplige Forskningsrad. Utvalg for eutrofieringsforskning. Intern Rapp. 14/82: 1–46.

Brooks, J. L. & S. J. Dodson, 1965. Predation, body size, and the composition of the plankton. Science 150: 28–35.

Brown, K. M., 1982. Resource overlap and competition in pond snails: An experimental analysis. Ecology 64: 412–422.

Brown, K. M. & D. R. DeVries, 1985. Predation and the distribution of a pulmonate pond snail. Oecologia 66: 93–99.

Caron, D. A, 1983. Technique for enumeration of heterotrophic and phototrophic nanoplankton, using epifluorescence microscopy, and comparison with other procedures Appl. Envir. Microbiol. 46: 491–498.

Carpenter, S. R. & J. F. Kitchell, 1984. Plankton community structure and limnetic primary production. Am. Nat. 124: 159–172.

Carpenter, S. R., J. F. Kitchell & J. R. Hodgson, 1985. Cascading trophic interactions and lake productivity. Bioscience 35: 634–639.

Carpenter, S. R., J. F. Kitchell, J. R. Hodgson, P. A. Cochran, J. J. Elser, M. M. Elser, D. M. Lodge, D. Kretchmer, X. He & C. N. von Ende, 1987. Regulation of lake primary productivity by food web structure. Ecology 68: 1863–1876.

Chanton, J. P., C. S. Martens & G. W. Kipphut, 1983. Lead-210 geochronology in a changing environment. Geochim. Cosmochim. Acta 47: 1791–1804.

Crowder, L. B. & W. E. Cooper, 1982. Habitat structural complexity and the interaction between bluegills and their prey. Ecology 63: 1802–1813.

Cuker, B. E., 1981. Control of epilithic community structure in an arctic lake by vertebrate predation and invertebrate grazing. Ph.D. thesis, North Carolina State University, Raleigh, 96 p.

Cuker, B. E., 1983. Competition and coexistence between the grazing snail Lymnea, Chironomidae, and Microcrustacea in an Arctic epilithic lacustrine community. Ecology 64: 10–15.

Dillon, P. J. & F. H. Rigler, 1974. The phosphorus-chlorophyll relationship in lakes. Limnol. Oceanogr. 19: 767–773.

Dodson, S. I., 1974. Zooplankton competition and predation: An experimental test of the size-efficiency hypothesis. Ecology 55: 605–613.

Dodson, S. I., 1979. Body size patterns in arctic and temperate zooplankton. Limnol. Oceanogr. 24: 940–949.

Dodson, S. I., 1984. Predation of *Heterocope septentrionalis* on two species of *Daphnia*: Morphological defenses and their cost. Ecology 65: 1249–1257.

Drenner, R. W., S. T. Threlkeld, J. D. Smith, J. R. Mummert & P. A. Cantrel, 1989. Interdependence of phosphorus, fish, and site effects on phytoplankton biomass and zooplankton. Limnol. Oceanogr. 34: 1315–1321.

Eisenberg, R. M., 1966. The role of food in the regulation of the pond snail, *Limnaea elodes*. Ecology 47: 889–906.

Evans, B. I., 1986. Strategies and tactics of search behavior in salmonid and centrarchid planktivorous fish. Ph.D. Thesis. University of Kansas. 83 pp.

Federle, T. W. & J. R. Vestal, 1980. Lignocellulose mineralization by arctic lake sediments in response to nutrient manipulation. Appl. Envir. Microbiol. 40: 32–37.

Federle, T. W. & J. R. Vestal, 1982. Evidence of microbial succession on decaying leaf litter in an arctic lake. Can. J. Microbiol. 28: 686–695.

Federle, T. W., V. L. McKinley & J. R. Vestal, 1982. Effects of nutrient enrichment on the colonization and decomposition of plant detritus by the microbiota of an arctic lake. Can. J. Microbiol 28: 1199–1205.

Fogged, N., 1981. Diatoms of Alaska. Bibliotheca Phycologica 53: 1–317. J. Cramer, Germany.

Fuhrman, J. A. & F. Azam, 1982. Thymidine incorporation as a means of heterotrophic bacterioplankton production in marine surface waters: evaluation and field results. Mar. Biol. 66: 109–120.

Gilinsky, E., 1984. The role of fish predation and spatial heterogeneity in determining benthic community structure. Ecology 65: 455–468.

Goldman, C. R., 1962. A method of studying nutrient limiting factors *in situ* in water columns isolated by polyethylene film. Limnol. Oceanogr. 7: 99–101.

Grice, G. D., M. R. Reeve, P. Koeller & D. W. Menze, 1977. The use of large volume, transparent, enclosed sea-surface water columns in the study of stress on plankton ecosystems. Helgol. Wiss. Meeresunters. 30: 118–133.

Grice, G. D. & M. R. Reeve (eds), 1982. Marine Mesocosms. Springer-Verlag, New York.

Hairston, N. G., F. E. Smith & L. B. Slobodkin, 1960. Community structure, population control, and competition. Am. Nat. 94: 421–425.

Hall, D. J., W. E. Cooper & E. E. Werne, 1970. An experimental approach to the production dynamics and structure of freshwater animal communities. Limnol. Oceanogr. 15: 839–928.

Hall, D. J., S. T. Threlkeld, C. W. Burns & P. H. Crowley, 1976. The size efficiency hypothesis and the size structure of zooplankton communities. Ann. Rev. Ecol. Syst. 7: 177–208.

Hayne, D. W. & R. C. Ball, 1956. Benthic productivity as influenced by fish predation. Limnol. Oceanogr. 1: 162–175.

Hershey, A. E., 1985a. Effects of predatory sculpin on the chironomid community in an arctic lake. Ecology 66: 1131–1138.

Hershey, A. E., 1985b. Littoral chironomid communities in an arctic Alaskan lake. Holarctic Ecology 8: 39–48.

Hershey, A. E., 1990. Snail populations in arctic lakes: competition mediated by predation. Oecologia 82: 26–32.

Hershey, A. E. & M. E. McDonald, 1985. Diet and digestion rates of the slimy sculpin, *Cottus cognatus*, in an Alaskan arctic lake. Can. J. Fish. aquat. Sci. 42: 483–487.

Hillbricht-Ilkowska, A. & T. Weglenska, 1970. Some relations between production and zooplankton structure of two lakes of a varying trophy. Pol. Arch. Hydrobiol. 17: 233–240.

Hobbie, J. E. & J. J. Cole, 1984. Response of a detrital food web to eutrophication. Bull. mar. Sci. 35: 357–363.

Hobbie, J. E., R. J. Daley & S. Jasper, 1977. Use of Nucleopore filters for counting bacteria by fluorescence microscopy. App. envir. Microbiol. 33: 1225–1228.

Hobbie, J. E., T. L. Corliss & B. J. Peterson, 1983. Seasonal patterns of bacterial abundance in an arctic lake. Arct. Alp. Res. 15: 253–259.

Holmgren, S. K., 1984. Experimental lake fertilization in the Kuokkel area, Northern Sweden. Phytoplankton biomass and algal composition in natural and fertilized subarctic lakes. Int. Revue ges. Hydrobiol. 69: 781–817.

Hrbacek, J., 1962. Species composition and the amount of zooplankton in relation to the fish stock. Rozpr. Cesk. Akad. Ved. Rada Mat. Prir. Ved. 72: 1–114.

Hrbacek, J., B. Desortova & J. Popovsky, 1978. Influence of the fish stock on the phosphorus-chlorophyll ratio. Verh. int. Ver. Limnol. 20: 1624–1628.

Huber-Pestilozzzi, G., 1961. Das Phytoplankton des Sussawassers: Chlorophyceae (Grunalgen) Ordnung: Volvocales. Die Binnengewasser 5. Stuttgart, Germany, 728 p + 158 Plates.

Huber-Pestilozzi, G., 1968. Das Pytoplankton des Susswassers: Cryptopyceae, Chloromonadophyceae, Dinophyceae. Die Binnengewasser 16(3) Auflage 2. Stuttgart, Germany. 322 p.

Huber-Pestilozzi, G., 1976. Das Phytoplankton des Susswassers: Chrysophyceen. Farblose Flagellaten, Heterokonten. Die Binnengewasser 16(2) Halfte 1. Stuttgart, Germany 365 p.

Huffaker, C. B., 1958. Experimental studies on predation: dispersion factors and predator–prey oscillations. Hilgardia 27: 343–383.

Hullar, M. A. J., M. J. Coffman & J. R. Vestal, 1986. The effects of nutrient enrichment on the distribution of microbial heterotrophic activity in arctic lakes. Perspectives in Microbial Ecology, Proceedings of the 4th International Symposium of Microbial Ecology: 207–212.

Hurlbert, S. H., J. Zedler & D. Fairbanks, 1972. Ecosystem alteration by mosquito fish (*Gambusia athinis*). Science 175: 639–641.

Hyatt, K. D. & J. G. Stockner, 1985. Responses of sockeye salmon (*Oncorhynchus nerka*) to fertilization of British Columbia coastal lakes. Can. J. Fish. aquat. Sci. 42: 320–331.

Imboden, D. M., B. S. F. Eid, T. Joller, M. Schurter & J. Wetzel, 1979. MELIMEX, and Experimental Heavy Metal Pollution Study: Vertical Mixing in a Large Limno-Corral. Schweiz. Z. Hydrol. 41: 177–189.

Jansson, M., 1978. Experimental lake fertilization in the Kuokkel area, northern Sweden: Budget calculations and the fate of nutrients. Verh. int. Ver. Limnol. 20: 857–862.

Johnson, L., 1972. Kellery Lake: Characteristics of a culturally unstressed salmonid community. J. Fish Res. Bd Can. 29: 731–740.

Jonasson, P. M., 1972. Ecology and production of the profundal benthos in relation to phytoplankton in Lake Esrom. Oikos Supplementum 14: 1–148.

Kalff, J. & H. E. Welch, 1974. Phytoplankton production in Char Lake, a natural polar lake, and in Meretta Lake, a polluted polar lake, Corwallis Island, Northwest Territories. J. Fish. Res. Bd Can. 31: 621–636.

Kalff, J., H. J. Kling, S. H. Holmgren & H. E. Welch, 1975. Phytoplankton, phytoplankton growth and biomass cycles in an unpolluted and in a polluted polar lake. Verh. int. Ver. Limnol. 19: 487–495.

Kaushik, N. K., G. L. Stephenson, K. R. Solomon & K. E. Day, 1985. Impact of permethrin on zooplankton communities in limnocorrals. Can. J. Fish. aquat. Sci. 42: 77–85.

Keller, S. A., R. H. Morris & D. L. Detterman, 1961. Geology of the Shaviovik and Sagavanirktok Rivers region, Alaska. U. S. Geol. Surv. Prof. Pap. 303D: 169–222.

Kelso, J. R. M., 1985. Standing stock and production of fish in a cascading lake system in the Canadian Shield. Can. J. Fish. aquat. Sci. 42: 1315–1320.

Kerfoot, W. C., 1977. Implications of copepod predation. Limnol. Oceanogr. 22: 316–325.

Kipphut, G. W., 1984. Sediment-water chemical exchange in an arctic lake. EOS 65: 897.

Kitchell, J. F. & S. R. Carpenter, 1987. Piscivores, planktivores, fossils, and phorbins. Pages 132–145. In W. C. Kerfoot & A. Sih, (eds) Predation: direct and indirect impacts on aquatic communities. University Press of New England, Hanover, New Hampshire, USA.

Kring, R. L. & W. J. O'Brien, 1976. Accommodation of *Daphnia pulex* to altered pH conditions as measured by feeding rate. Limnol. Oceanogr. 21: 313–315.

Lack, T. J. & W. G. Lund, 1974. Observations and experiments on the phytoplankton of Blelham Tarn, English Lake District. Freshwat. Biol. 4: 399–415.

Leah, R. T., B. Moss & D. E. Forrest, 1978. Experiments with large enclosures in a fertile, shallow, back lake, Hickling Broad, Norfolk, United Kingdom. Int. Rev. Gesameter Hydrobiol. 63: 291–310.

Lean, D. R. S., 1973. Phosphorus dynamics in lake water. Science 179: 678–680.

Lean, D. R. S. & M. N. Charlton, 1976. A study of phosphorus kinetics in a lake ecosystem, in J. O. Nriagu (ed.), Environmental Biogeochemistry, Vol. 1, Carbon, Nitrogen, Phosphorus, Sulfurand Selenium Cycles. Ann Arbor Science Publ., Inc., Ann Arbor, Michigan: 283–294.

LeBrasseur, R. J. & O. D. Kennedy, 1972. The fertilization of Great Central Lake II. Zooplankton standing stock. Fish. Bull. 70: 25–36.

Liebig, T., 1874. Chemistry and its application to agriculture and physiology. Taylor and Walton, London.

Losos, B. & J. Hetesa, 1973. The effect of mineral fertilization and carp fry on the composition and dynamics of plankton. Hydrobiol. Stud. 3: 173–217.

Luecke, C., 1990. Changes in abundance and distribution of benthic macroinvertebrates after introduction of cutthroat trout into a previously fishless lake. Trans. am. Fish. Soc. 119: 1010–1021.

Luecke, C. & W. J. O'Brien, 1983. The effects of *Heterocope* predation on zooplankton communities in arctic ponds. Limnol. Oceanogr. 28: 367–377.

Lund, J. W. G., 1978. Experiments with lake phytoplankton in large enclosures. Rep. Freshwat. Biol. Ass. 46: 32–39.

Lund, J. W. G. & C. S. Reynolds, 1982. The development and operation of large limnetic enclosures in Blelham Tarn, English Lake District, and their contribution to phytoplankton ecology. Progress in Phycological Res. 1: 1–65.

Lundgren, A., 1985. Model ecosystems as a tool in freshwater and marine research. Arch. Hydrobiol. Suppl. 70: 157–196.

Lunte, C. C. & C. Luecke, 1990. Trophic interactions of *Leptodora* in Lake Mendota. Limnol. Oceanogr. 35: 1091–1100.

Lynch, M. & J. Shapiro, 1981. Predation, enrichment and phytoplankton community structure. Limnol. Oceanogr. 26: 86–102.

Macan, T. T., 1966. Predation by *Salmo trutta* in a moorland fishpond. Verh. int. Ver. Limnol. 15: 1081–1087.

Malley, D. F., P. S. S. Chang & D. W. Schindler, 1977. Decline of zooplankton populations following eutrophication of Lake 227, Experimental Lakes Area, Ontario: 1969–1974. Department of Fisheries and Oceans, Freshwater Institute, Winnipeg, Manitoba.

McDonald, M. E., B. E. Cuker & S. C. Mozley, 1982. Distribution, production, and age structure of slimy sculpin in an arctic lake. Envir. Biol. Fish. 7: 171–176.

McDonald, M. E. & A. E. Hershey, 1989. Size structure of a lake trout (*Salvelinus namaycush*) population in an arctic lake: influence of angling and implication for fish community structure. Can. J. Fish. aquat. Sci. 46: 2153–2156.

McKinley, V. L., T. W. Federle & J. R. Vestal, 1982. Effects of petroleum hydrocarbons on plant litter microbiota in an arctic lake. Appl. Envir. Microbiol. 43: 129–135.

McQueen, D. J., J. R. Post & E. L. Mills, 1986. Trophic relationships in freshwater pelagic ecosystems. Can. J. Fish. aquat. Sci. 43: 1571–1581.

McQueen, D. J. & J. R. Post, 1988. Limnocorral studies of cascading trophic interactions. Verh. int. Ver. Limnol. 23: 739–747.

Merrick, C. W., A. E. Hershey & M. E. McDonald, 1991. Lake Trout (*Salvelinus namaycush*) control of snail density and size distribution in an Arctic lake. Can. J. Fish. aquat. Sci. 48: 498–502.

Miller, M., P. Spatt, P. Westlake, D. Yeakel & G. Hater, 1986. Primary production and its control in Toolik Lake. Arch. Hydrobiol 74: 97–131.

Mills, K. H., 1985. Responses of lake whitefish (*Coregonus clupeaformis*) to fertilization of Lake 226, the Experimental Lakes area. Can. J. Fish. aquat. Sci. 42: 129–138.

Moore, J. W., 1978. Some factors influencing the diversity and species composition of benthic invertebrates in twenty arctic and subarctic lakes. Int. Revue ges. Hydrobiol. 63: 757–771.

National Academy of Science, 1969. Eutrophication: Causes, Consequences, Correctives. Washington, D.C., National Academy of Sciences. 661 pp.

Neill, W. E. & A. Peacock, 1980. Breaking the bottleneck: Interactions of invertebrate predators and nutrients in oligotrophic lakes. In W. C. Kerfoot (ed.), Evolution and Ecology of Zooplankton Communities. University Press of New England, Hanover, New Hampshire: 715–724.

Nelson, P. R. & W. T. Edmondson, 1955. Limnological effects of fertilizing Bare Lake, Alaska. U. S. Fish Wildl. Serv., Fish. Bull. 56: 414–436.

Nilsson, N. A. & B. Pejler, 1973. On the relationship between fish fauna and zooplankton composition in north Swedish lakes. Rep. Instit. Freshwat Res. Droltninghom 53: 51–77.

Nilsson, N. A., 1972. Effects of introductions of salmonids into barren lakes. J. Fish. Res. Bd Can. 29: 693–697.

Northcote, T. G., 1988. Fish in the structure and function of freshwater ecosystems: A 'topdown' view. Can. J. Fish. aquat. Sci. 45: 361–379.

O'Brien, W. J., 1975. Some aspects of the limnology of the ponds and lakes of the Noatak drainage basin, Alaska. Verh. int. Ver. Limnol. 19: 472–479.

O'Brien, W. J. & F. deNoyelles, 1972. Photosynthetically elevated pH as a factor in zooplankton mortality in nutrient enriched ponds. Ecology 53: 605–614.

O'Brien, W. J. & F. deNoyelles, Jr., 1974. Relationship between nutrient concentrations of phytoplankton density, and zooplankton density in nutrient enriched experimental ponds. Hydrobiologia 44: 105–125.

O'Brien, W. J., C. Buchanan & J. F. Haney, 1979a. Arctic zooplankton community structure: Exceptions to some general rules. Arctic 32: 237–247.

O'Brien, W. J., D. Kettle & H. P. Riessen, 1979b. Helmets and invisible armor: Structures reducing predation from tactile and visual planktivores. Ecology 60: 287–294.

O'Brien, W. J. & C. Luecke, 1988. The coexistence of a predaceous copepod and a daphniid: Weeding and gardening in the Arctic. Verh. int. Ver. Limnol. 23: 2069–2074.

Paine, R. T., 1966. Food web complexity and species diversity. Am. Nat. 100: 65–75.

Parsons, T. R., 1982. The future of controlled ecosystem enclosure experiments. In Grice, D. G. & M. R. Reeve (eds). Marine mesocosms. Biological and chemical research in experimental ecosystems. Springer, New York: 411–418.

Persson, G., S. K. Holkgren, M. Jansson, A. Lundgren, B. Nyman, D. Solander & C. Anell, 1975. Phosphorus and nitrogen and the regulation of lake ecosystems: Experimental approaches in subarctic Sweden. Section III. Freshwater Ecology, National Research Council, Canada.

Pilson, M. E. Q. & S. W. Nixon, 1980. Marine Microcosms in Ecological Research, in J. P. Giesy, Jr. (ed.), Microcosms in Ecological Research, Technical Information Center, U.S. Department of Energy, Springfield, Virginia, 724–741.

Porter, K., 1976. Enhancement of algal growth and productivity by grazing zooplankton. Science 192: 1332–1334.

Porter, K. & Y. S. Feig, 1980. The use of DAPI for identifying and counting the aquatic microflora. Limnol. Oceanogr. 25: 943–948.

Post, J. R. & D. J. McQueen, 1987. The impact of planktivorous fish on the structure of a plankton community. Freshwat. Biol. 17: 79–89.

Prescott, G. W., 1962. Algae of the Western Great Lakes. Wm. C. Brown, Co. Dubuque, Iowa. 977 p.

Reinerten, H., A. Jensen, A. Langeland & Y. Olsen, 1986. Algal competition for phosphorus: The influence of zooplankton and fish. Can. J. Fish aquat. Sci. 43: 1135–1141.

Rublee, P. A., 1992. Community structure and bottom-up regulation of heterotrophic microplankton in arctic LTER lakes. Hydrobiologia 240: 133–141.

Rudd, J. W. M., M. A. Turner, B. E. Townsend, A. Swick & A. Furutani, 1980. Dynamics of selenium in mercury-contaminated experimental freshwater enclosures. Can. J. Fish. aquat. Sci. 37: 848–857.

Sanders, F. S., 1985. Use of large enclosures for perturbation experiments in lentic ecosystems: A review. Envir. Monit. Assess. 5: 55–99.

Schindler, D. W., 1974. Eutrophication and recovery in experimental lakes: Implications for lake management. Science 184: 897–899.

Schindler, D. W., J. Kalff, H. E. Welch, G. J. Brunskill, H. Kling & N. Kritsch, 1974. Eutrophication in the high arctic – Meretta Lake, Cornwallis Island (75'N lat.). J. Fish. Res. Bd Can. 31: 647–662.

Schindler, D. W., H. E. Welch, J. Kalff, G. J. Brunskill N. Kritsch, 1974. Physical and chemical limnology of Char Lake, Cornwallis Island (75°N lat.). J. Fish. Res. Bd Can. 31: 585–607.

Shapiro, J. & D. I. Wright, 1984. Lake restoration by biomanipulation. Freshwat Biol. 14: 371–383.

Shapiro, J., 1990. Biomanipulation: The next phase making it stable. Hydrobiologia 200/201: 13–27.

Skuja, H. 1958. Taxonomie Des Phytoplanktons Einiger Seen In Uppland, Schweden. Symbolae Botanicae Upsasaliensis IX. Sweden.

Smith, M. W., 1969. Fertilization and predator control to increase growth rate and yield of trout in a natural lake. J. Fish. Res. Bd Can. 25: 2011–2036.

Smith, S. H., 1972. Factors in ecological succession in oligotrophic fish communities of the Laurentian Great Lakes. J. Fish. Res. Bd Can. 29: 717–730.

Smock, L. A., 1980. Relationships between body size and biomass of aquatic insects. Freshwat. Biol. 10: 375–383.

Solarzano, L., 1969. Determination of ammonia in natural waters by the phenolhypochlorite method. Limnol. Oceanogr. 14: 799–801.

Solomon, K. R., K. E. Smith, G. Guest, J. Yoo & K. Kaushik, 1980. The use of limnocorrals in studying the effects of pesticides in aquatic ecosystems. Can. Tech. Rep. Fish. aquat. Sci. 975: 1–9.

Solomon, K. R., K. E. Smith & G. L. Stephenson, 1982. Depth integrating samples for use in limnocorrals. Hydrobiologia 94: 71–75.

Solomon, K. R., J. Y. Yoo, D. Lean, N. K. Kaushik, K. E. Day & G. L. Stephenson, 1985. Dissipation of permethrin in limnocorrals. Can. J. Fish. aquat. Sci. 42: 70–76.

Stainton, M. P., M. J. Capel & F. A. J. Armstrong, 1977. The chemical analysis of freshwater, 2nd Edition. Can. J. Fish. mar. Serv. Misc. Spec. Publ. 25: 180 p.

Stein, R. A., J. F. Kitchell & B. Knezevic, 1975. Selective predation by carp (*Cyprinus carpio* L.) on benthic molluscs in Skadar Lake, Yugoslavia. J. Fish. Biol. 391–399.

Stein, R. A., C. G. Goodman & A. Marschall, 1984. Using time and energetic measures of cost in estimating prey value for fish predators. Ecology 65: 702–715.

Stenson, J. A. E., T. Bohlin, L. Henrikson, B. I. Nilsson, H. G. Nyman, H. G. Oscarson & P. Larsson, 1978. Effects of fish removal from a small lake. Verh. int. Ver. Limnol. 20: 794–801.

Stepanek, M. & M. D. Zelinka, 1961. Limnological study of the reservoir Sedlice near Zeliv. XVIII. The development of phytonanoplankton in silon bags. (In Czechoslovakian.) Sbornik Vys. Skoly Chem. Technol. Praze. Oddil Technol. vody (Sci. Pap. Inst. Chem. Technol. Prague, Water Technol.) 5: 275–323.

Strickland, J. D. H. & T. R. Parsons, 1972. A practical handbook of seawater analysis. Bull. 167 Fish. Res. Bd Can. Ottawa. 310 pp.

Takahashi, M. & F. A. Whitney, 1977. Temperature, salinity, and light penetration structures: controlled ecosystem pollution experiment. Bull. mar. Sci. 27: 8–16.

Thienemann, A., 1928. Der sauerstoff im eutrophen und oligotrophen See. einbeitrag zur seetypenlehre. Die Binnengewasser 4: 175 pp.

Thomas, E. A., 1958. Das plankton-test-lot, ein Gerät zum studium des verhaltens von planktonorganismen in sec. Monatsbull. Schweiz. ver, gas-u. Wasserfachm., Nr 1, 85.

Thomas, W. H. & D. L. R. Siebert, 1977. Effects of copper on the dominance and diversity of algae: Controlled ecosystem pollution experiment. Bull. Mar. Sci. 27: 23–33.

Thorpe, J. H. & E. A. Bergey, 1981. Field experiments on responses of a freshwater benthic macroinvertebrate community to vertebrate predators. Ecology 62: 365–375.

Threlkeld, S. T., 1987. Experimental evaluation of trophic-cascade and nutrient mediated effects of planktivorous fish on plankton community structure. Pages 161–173. In W. C. Kerfoot & A. Sih, (ed), Predation: direct and indirect impacts on aquatic communities. University Press of New England, Hanover, New Hampshire, USA.

Threlkeld, S. T., 1988. Planktivory and planktivore biomass effects on zooplankton, phytoplankton, and the trophic cascade. Limnol. Oceangr. 33: 1362–1375.

Tippets, W. E. & P. B. Moyle, 1978. Epibenthic feeding by rainbow trout (*Salmo gairdneri*) in the McCould River, California. J. Ann. Ecol. 47: 549–559.

Uelinger, U., 1982. Use of limnocorrals in a mesotrophic lake: Experiences and preliminary results. SIL in Trondheim, Norway.

Uehlinger, U., P. Bossard, J. Bloesch, H. R. Bürgi & H. Bühre, 1984. Ecological experiments in limnocorrals: Methodological problems and quantification of the epilimnetic phosphorus and carbon cycles. Verh. int. Ver. Limnol. 22: 163–171.

Vanni, M. J. & D. L. Findlay, 1990. Trophic cascades and phytoplankton community structure. Ecology 71: 924–937.

Vermeij, G. J. & A. P. Covich, 1978. Coevolution of freshwater gastropods and their predators. Am . Nat. 112: 833–843 .

Ware, D. M, 1972. Predation by rainbow trout (*Salmo gairdneri*): the influence of hunger, prey density, and prey size. J. Fish. Res. Bd Can. 29: 1193–1201.

Welch, H. E., J. A. Legault & H. J. Kling, 1989. Phytoplankton, nutrients, and primary production at Saquaqjuac, N. W. T. Can. J. Fish. aquat. Sci. 46: 90–107.

Werner, E. E., J. F. Gilliam, D. J. Hall & G. G. Middlebach, 1983. An experimental test of the effects of predation risk on habitat use in fish. Ecology 64: 1540–1548.

Wetzel, R. G. & G. E. Likens, 1979. Limnological analyses. W. B. Saunders, Philadelphia, PA. 357 pp.

Whalen, S. C. & J. C. Cornwell, 1985. Nitrogen, phosphorus, and organic carbon cycling in an arctic lake. Can. J. Fish. aquat. Sci. 42: 797–808.

White, D. C., R. J. Bobbie, S. J. Morrison, D. K. Oosterhof, C. W. Taylor & D. A. Meete, 1977. Determination of microbial activity of estuarine detritus by relative rates of lipid biosynthesis. Limnol. Oceanogr. 22: 1089–1099.

White, D. C., W. M. Davis, J. S. Nichels, J. D. King & R. J. Bobbie, 1979. Determination of the sedimentary microbial biomass by extractable lipidphosphate. Oecologia 40: 51–62.

Hydrobiologia **240**: 189–201, 1992.
W.J. O'Brien (ed.), Toolik Lake – Ecology of an Aquatic Ecosystem in Arctic Alaska.

A comparison of slimy sculpin (*Cottus cognatus*) populations in arctic lakes with and without piscivorous predators

Kristi L. Hanson,[1] Anne E. Hershey[1,*] & Michael E. McDonald[2]
Departments of [1] Biology; [2] Chemical Engineering; University of Minnesota, Duluth, MN 55812, USA
(author for correspondence)*

Abstract

Arctic slimy sculpin were sampled by passive trapping in lakes containing the predators lake trout and burbot (LT lakes), and lakes lacking sculpin predators (NoLT lakes). Sculpin food abundance (chironomid biomass) from the rocky littoral zone was compared with that from the deep water sediment zone. Distribution, size, growth, age, condition and relative abundance of sculpin were examined.

Spatial distribution of sculpin was different between lake types, with more and larger sculpin found over the sediment zone in NoLT lakes. There were no seasonal patterns evident in this distribution and catch per unit effort was not significantly different between LT and NoLT lakes. Biomass of chironomids, the major food of the sculpin, was higher in the sediments than on the rocks, suggesting that sediments should be the preferred environment in the absence of piscivores. Longevity of sculpins varied between IV and VIII years and was not correlated with lake type. Sculpin size frequency distributions were shifted toward slightly larger fish in NoLT lakes. Sculpin growth curves and condition estimates did not reveal a difference between lake types, but comparison of mean ototlith interannular distances between lake types showed a trend, significant in year 4, toward more growth in no lake trout lakes. These results suggest that the presence of piscivores is an important factor limiting arctic slimy sculpin distribution and may act in concert with food supply to impact sculpin growth.

Introduction

Fish predation on benthos has implications for benthic community structure (Gilinsky, 1984) and can regulate abundance or biomass of benthic invertebrates (Crowder & Cooper, 1982; Hershey, 1985). Through cascading effects on alternate trophic levels, piscivores can regulate lake productivity (Carpenter *et al.*, 1985). To minimize predation risk, prey fishes employ optimal feeding strategies (Pyke *et al.*, 1977), which can influence their habitat use (Mittelbach, 1981, 1986; Crowder & Cooper, 1982; Brandt, 1986a, b; Werner & Hall, 1988), and result in habitat selections that offer lower prey abundance than higher risk habitats (Mittelbach, 1981, 1986; Werner *et al.*, 1983a, b; Power, 1984).

Arctic lakes are simplified systems in which predation appears to an important factor controlling benthic abundances (Hershey, 1985, 1990; Merrick *et al.*, 1991). Slimy sculpin (*Cottus cognatus* Richardson), small, bottom-dwelling fish found in lakes and streams throughout northern North America (Scott & Crossman, 1973), are widely distributed in Alaska (Craig & Wells, 1976). Sculpin have been shown to be important in the trophic interactions of lakes near the Toolik Lake Long Term Ecological Research (LTER) site by controlling the densities of their main prey item, chironomid larvae (Hershey, 1985; Goyke

& Hershey, 1992). Similarly, lake trout (*Salvelinus namaycush*), present in some LTER lakes, control the abundance and size distribution of their major prey, the snail *Lyrnnaea elodes* (Hershey, 1990, Merrick *et al.*, 1991).

Lakes in this area host up to 5 fish species; two of these species are predators on sculpin: lake trout (Petrosky & Waters, 1975; Moyle, 1977; Merrick, 1989) and burbot (*Lota lota*) (Van Vliet, 1964; Godkin *et al.*, 1982). Lake trout in Toolik area lakes feed predominantly on snails, but 12% of lake trout stomachs that have been analyzed also contained sculpin (Merrick *et al.*, 1992). Growth efficiency models (Kerr, 1971) suggest that stunted lake trout may require occasional larger prey items to attain large body size (Konkle & Sprules, 1986). Burbot are voracious predators whose role in Toolik-area lakes is not well understood. In many lakes and rivers they will feed on any available prey (Becker, 1983) and burbot in North Slope lakes are known to eat slimy sculpin (K.L. Hanson, pers. observation).

Prey exposure is a major determinant of prey risk in benthic food chains (Ware, 1973; Hershey & Dodson, 1985; Hershey, 1987), and increased spatial heterogeneity may allow prey items a refuge from predation (Crowder & Cooper, 1982; Gilinsky, 1984; Hershey, 1985, Luecke, 1990). Sculpin in arctic lakes inhabit both the rocky littoral zone and the bare sediments of deeper water, and these habitats may host different food sources. Slimy sculpin may exhibit different behavior in the presence and absence of predation threat, which could have implications for sculpin growth and population structure.

In this study, slimy sculpin populations were sampled in 4 lakes containing lake trout and burbot, and 3 lakes containing no predators in order to determine whether the presence of piscivores affected distribution, size, age, growth and condition, and relative abundance of sculpin. Chironomid biomass on substrates was examined to determine if prey abundance differed among habitats used by sculpin. Sculpin were also sampled in 2 lakes which were manipulated by experimental removal and addition of lake trout to examine growth history, but distribution data from manipulated lakes were not included in the analyses.

Study area

All lakes sampled were in the vicinity of the Toolik Lake LTER site (68°N, 149°W) (Fig. 1), located 24 km north of the Brooks Mountain Range and 200 km south of Prudhoe Bay, Alaska. This is a region of rolling tundra with numerous, highly oligotrophic lakes (Miller *et al.*, 1986). The open-water season is short, usually lasting about 3 months and surface water temperatures vary seasonally from approximately 0–18 °C (M. C-. Miller, Univ. of Cincinnati, pers. comm.).

Lakes were categorized according to the presence or absence of lake trout, which are the major predator on sculpin. Our knowledge of burbot,

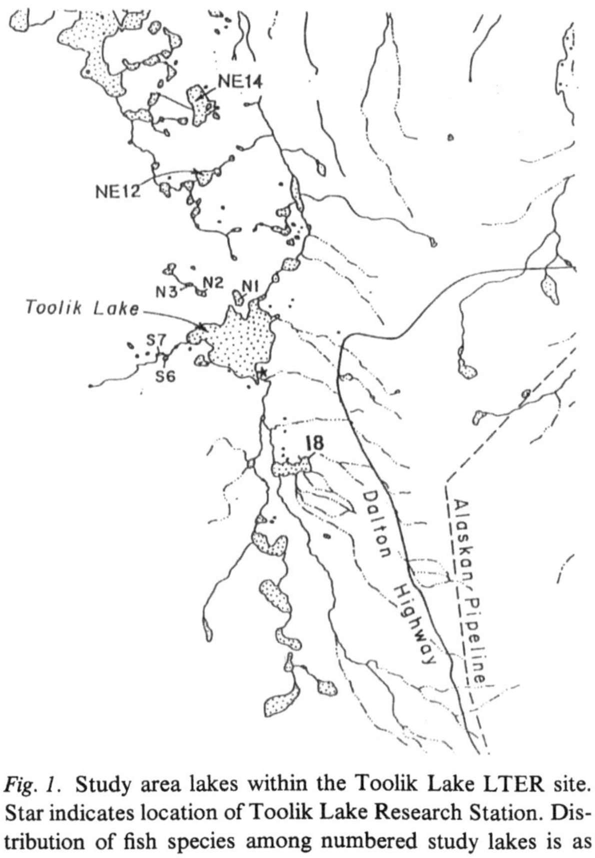

Fig. 1. Study area lakes within the Toolik Lake LTER site. Star indicates location of Toolik Lake Research Station. Distribution of fish species among numbered study lakes is as indicated in Table 1.

the other piscivorous predator on sculpin, is more limited, but our observations suggest that they usually co-occur with lake trout. Lakes N-1, NE-14, Toolik and I-8 contained sculpin, lake trout, and burbot and were categorized as 'lake trout' (LT) lakes. Lakes S-7, S-6, N-2 and N-3 hosted sculpin but no significant predators on sculpin, and were termed 'no lake trout' (NoLT) lakes. These lakes vary in size from 2–149 ha and in average depth from 5–25 m. Arctic grayling were present in all lakes, except possibly NE-14. Round whitefish also were present in I-8 and Toolik.

Methods

Sculpin were captured using unbaited traps constructed from clear glass jars and lucite funnels (McDonald *et al.*, 1982). The traps were tethered in pairs and placed on the lake bottom across four depth substrate zones designated: zone 1 (shallow/rock), zone 2 (mid-depth/rock), zone 3 (mid-depth/rock-sediment interface) and zone 4 (deep/sediments.) Jar traps were set in three replicate sites per lake; fishing effort was 24 jars/lake. All traps were within 25 m of shore.

Traps in N-2, N-3, S-7, N-l, NE-14, Toolik, and S-6 were checked by boat or raft approximately every 3 days for 5 weeks from late June through early August in 1988. Lakes S-6 and I-8 were similarly sampled in 1989. Sculpin were recovered from traps, weighed wet, and measured for total length (TL) to the nearest millimeter. Approximately half of the sculpin trapped in each lake were preserved in 95% ethanol for later aging. When measurements could not be performed in the field, preserved weights were converted to fresh weights by the linear regression formula ($Y = -0.06 + 1.27X$), calculated from 38 specimens ranging from 0.30 to 6.40 g; dessication averaged 16%. Length shrinkage was negligible.

Saggital otoliths were removed from sacrificed fish and placed in glycerine to promote clearing (Cailliet *et al.*, 1986). Otoliths were examined whole using a dissecting microscope and reflected light. The numbers of annuli were determined, and measurements from focus to each annulus were made using a camera lucida and digitizing pad. The ability to accurately distinguish annuli on whole otoliths was verified by mounting and grinding one each from 12 otolith pairs. Further aging verification was provided by Robert Muth, Colorado State University Larval Fish Laboratory.

Invertebrates were sampled in the rock zone of each lake by selecting 4 rocks which covered approximately 2/3 the bottom surface area of a 30 × 40 cm (top dimensions) dishpan, giving an estimated rock area of 363 cm^2. This sampling was conducted to confirm the assumption that rocks offer sculpin less food than sediments. Rocks were scrubbed with a stiff brush and the resulting sample was strained through a 70 μm net and preserved in 95% ethanol. One sample per lake was taken in July 1988 and in August 1989. Invertebrates were removed by sorting each sample in a clear glass dish over a light table. Insects were identified to family and measured for length using a dissecting microscope, camera lucida and digitizing pad. Lengths for chironomids and caddisflies were converted to dry biomass using the formulae: $w = 0.0051$ ($TL^{2.32}$) and $w = 0.0049$ ($TL^{2.84}$), respectively (Smock, 1980). Mean biomass of chironomids and caddisflies, and chironomids alone, was calculated for the rock zone of LT and NoLT lakes. Data were compared to chironomid biomass estimates from the sediment zone (Goyke & Hershey, this issue) using t-tests.

Sculpin distribution was enumerated by pooling replicate site counts. Zone totals were converted to percentages in each lake. Distribution by treatment was calculated as mean percent of 4 LT or 3 NoLT lakes. Lakes S-6 and NE-12 were not included in the sculpin distribution study because these lakes were the site of a lake trout introduction and removal experiments, respectively, initiated simultaneously with the present study. For the remaining lakes, data from the two rock zones were combined and averaged. Distribution between LT and NoLT treatments was analyzed by t-test for each zone.

All young-of-year (yoy) sculpin were recorded

and compiled in a separate data set. Fish 20 mmTL or larger were measured to the nearest mm. Fish smaller than approximately 20 mmTL were preserved in 95% ethanol and measured in the lab with a dissecting microscope and digitizing pad to the nearest 0.1 mm. Several yoy otoliths were examined to assist in aging older fish, but most yoy fish were categorized to age class strictly by size.

Seasonal distribution was plotted as cumulative percent over time, using the cumulative total fish per zone divided by the cumulative total captured, and recalculated for each sampling date. A seasonal change, though not expected might have occurred if breeding males had comprised a large portion of the fish sampled.

Growth curve estimates were made from age-length means (Bagenal & Tesch, 1978) and from 2 types of back-calculated means (Carlander, 1981). Since growth studies integrate historical effects on the fish rather than current effects, Lake S-6 and NE-12 data were used in growth comparisons. Sculpin from Lake I-8 were not sacrificed, thus these were not included in the growth analysis. Age-length growth curves were plotted from the average TL of each age group. Back-calculation procedures used a linear least squares equation (regression method) and a proportional equation (Fraser-Lee method). The y-intercept value in the proportional equation, ($a = 10.97$), was the mean value from the TL/otolith regressions of 7 lakes. Growth curves were statistically compared between lake types by using a t-test on the slopes of the mean TL least-squares regression lines. Interannulus distance at each age were also compared directly to avoid assumptions inherent in back calculation methods. In these comparisons, mean interannulus distance (or focus to annulus distance for year I growth) for years I–IV of growth was calculated for all fish in a given lake. These values were used as the variates representing growth in the analysis. Only ages I–IV were used because older fish were not present in all lakes. Statistical comparisons of mean growth at ages I–IV in 4 No LT lakes and 4 LT lakes were made using t-tests on $\log_e$ transformed interannular distances.

All fish from a given zone in a lake were pooled, and mean TL and weight was calculated by zone in each lake. T-tests were used to compare TL and weight of fish in each zone between LT and NoLT lakes using the mean value/lake for the respective zones as the variate. To determine if sculpin were largest in the sediment zone and smallest in the rock zone in all lakes, distributions were compared using the Page test for the ordered alternatives: sediments > interface > rocks (Hollander & Wolfe, 1973).

Fish condition was compared using the slope of the line: log weight/log TL. This method was selected over K-factors to measure fish condition, sensu Cone (1989), due to errors encountered when comparing fish of different lengths. Slopes were compared by t-test between LT and NoLT lakes.

Catch per unit effort (CPUE) was calculated by:

$$\text{Catch per day} = \frac{\text{total \# fish captured per lake}}{\text{\# sampling days}},$$

$$\text{Catch per trap} = \frac{\text{catch per day}}{\text{24 traps per lake}}.$$

CPUE was analyzed by t-test comparing 3 LT and 3 NoLT catch per trap values.

Results

Over 1000 slimy sculpin were trapped, with a size range from 8.3 mm TL (yoy) to 93 mm TL. The maximum individual weight recorded was 7.35 g (lake N-3).

The percentages of slimy sculpin captured over each of 3 zones differed significantly among lakes ($t = 3.12$, $p = 0.017$, Fig. 2). NoLT lakes showed higher percentages of fish over the sediment than in other zones. In LT lakes, sculpin catch was distributed more evenly over rock, interface and sediment.

There were no apparent seasonal shifts in distribution across 4 substrate zones during sam-

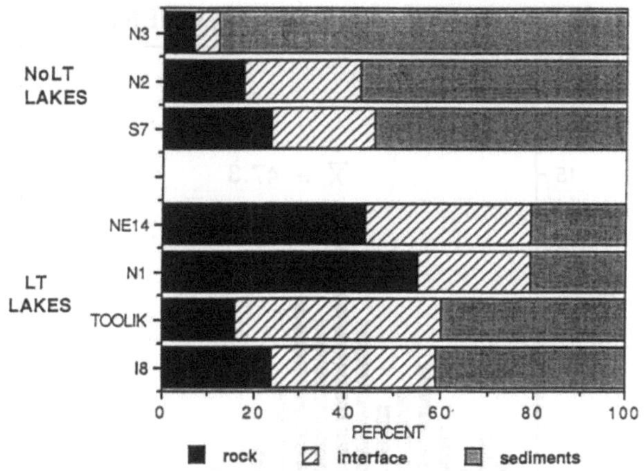

Fig. 2. Percentage distribution of sculpin trapped in rocky, interface, and soft-sediment zones of each study lake. NoLT lakes showed higher percentages of fish over the sediments than in the other zones ($T = 3.12$, $p < 0.02$).

pling from late June – early August, 1988. NoLT lakes consistently had more sculpin over sediments throughout the season than in other zones.

As expected, chironomid biomass estimates from the rocks of both LT and NoLT lakes were significantly less than the biomass estimates from sediment zones in Toolik-area LTER lakes (Table 1; sediment data from Goyke & Hershey, this issue). Goyke & Hershey (1992) found significantly less chironomid biomass in NoLT lake sediments than in the sediments of LT lakes. Chironomid biomass appeared less in LT lake rock zones than in NoLT rock zones, but the difference was not significant (Table 1). Incidence of relatively large caddisfly larvae was low; when caddisflies were included in biomass estimates, rock biomass was still significantly less than sediment biomass (Table 1).

The sculpin length frequency distributions (Fig. 3) do not give a clear indication of age classes in most lakes. LT and NoLT size distributions overlap, with NoLT distributions shifted slightly toward larger TL's.

Sculpin mean TL's and weights by lake and zone show a significant trend toward smaller sculpin in the rock zone, larger sculpin at the interface, and largest sculpin in the sediment zone in each lake regardless of the presence of piscivores (Page test for ordered alternatives, length $p < 0.05$, weight $p < 0.01$; Table 2). Between LT

Table 1. Mean (SE) sculpin food availability (chironomid and caddisfly biomass, g/m^2) in rocky and soft-sediment zones of study area lakes. Sediment data from Goyke & Hershey, 1992.

	Chiron. only			
	Sediments	Rock	t	p
LT	1.23 (0.16)	0.08 (0.04)	6.94	<0.001
NoLT	0.64 (0.07)	0.29 (0.16)	2.05	<0.05
	Chiron. + caddis			
	Sediments	Rock	t	p
LT	1.23 (0.16)	0.29 (0.14)	4.40	<0.001
NoLT	0.64 (0.07)	0.31 (0.17)	1.77	<0.05

Table 2. Mean length and weight of sculpin captured in rocky, interface, and soft-sediment zones of study area lakes. Page tests show that sediments > interface > rocks for length ($L = 104$, $n = 8$, $p = 0.05$) and weight ($L = 108$, $n = 8$, $p < 0.01$).

Lake	Mean total length (mm) by zone		
	Rock zones 1 & 2	Interface zone 3	Sediment zone 4
Toolik	43.5	53.8	53.1
N1	54.5	56.4	48.7
NE14	45.8	43.7	49.0
NE12	46.9	41.9	51.0
S7	54.6	59.9	59.4
N2	44.7	47.5	57.7
N3	56.3	55.0	80.2
S6	49.5	54.5	58.7
Lake	**Mean weight (g) by zone**		
	Rock zones 1 & 2	Interface zone 3	Sediment zone 4
Toolik	0.86	1.39	1.44
N1	1.47	1.78	1.35
NE14	0.86	0.67	1.08
NE12	0.89	0.68	1.26
S7	1.59	2.02	2.10
N2	0.73	0.92	1.62
N3	2.17	1.25	5.36
S6	1.04	1.53	1.80

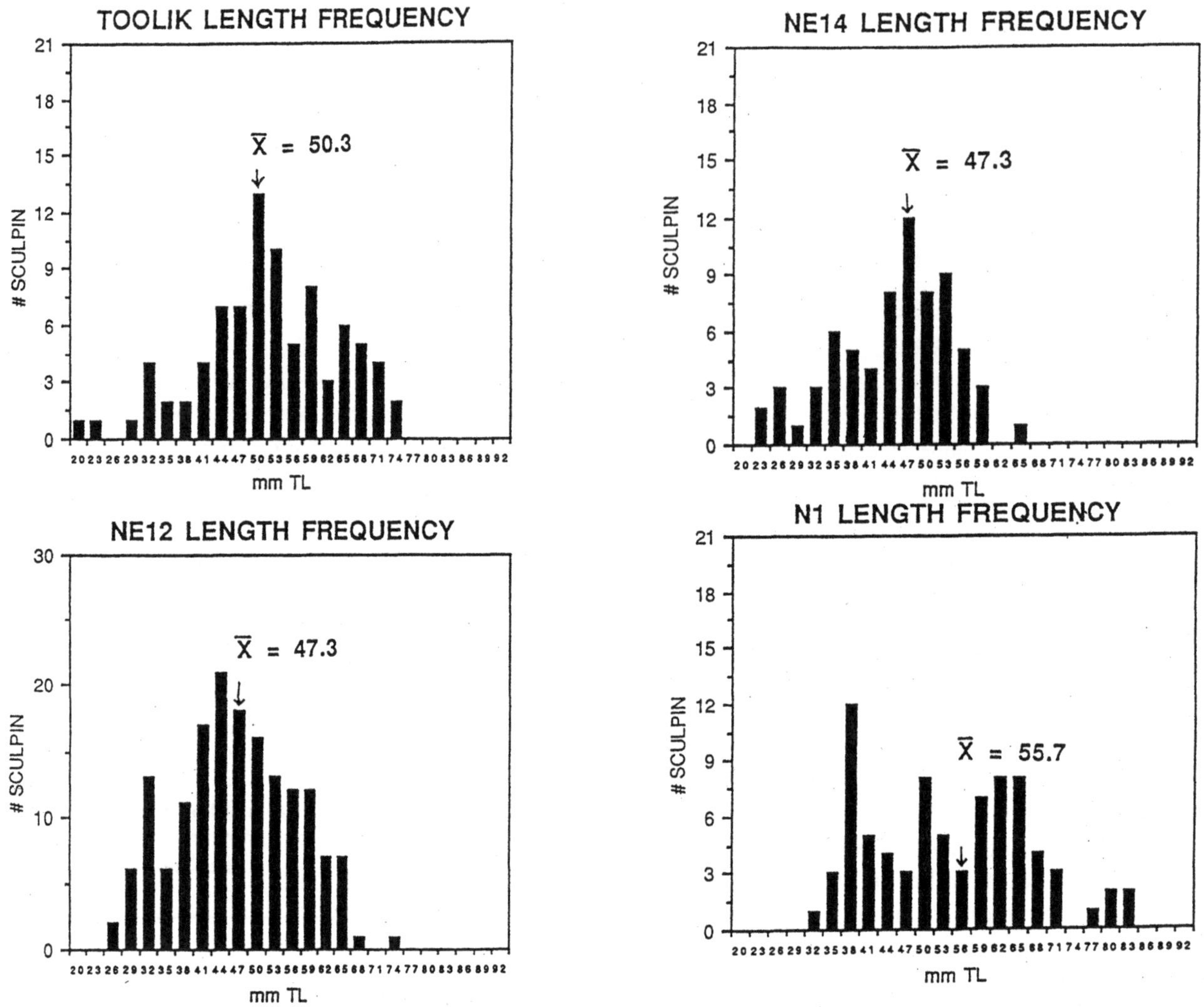

and NoLT lake-types, sculpin from the sediment zone but not other zones had significantly greater total length ($df = 6, t = 2.929, t < 0.05,$) and weight ($df = 6$, $t = 2.524$, $p < 0.05$) in NoLT lakes (Table 2).

Sculpin in LT lakes aged I–VI in Toolik Lake, I–V years in NE 14, and I–IV in N-1 and NE-12 (Fig. 4a). Sculpin captured from NoLT lakes N-3, S-6, and S-7 aged I–IV years, and sculpin from NoLT lake N-2 aged I–VIII years (Fig 4b). Mean age of sculpin from 4 NoLT lakes was 2.7 ± 0.4 and mean age of sculpin from 4 LT lakes was 2.8 ± 0.3. These means were not significantly different.

Sculpin growth curves of TL by back-calculation using a linear and a proportional equation were similar to slightly lower than population growth curves from actual age-length. The slopes of these lines vary slightly between each of the 3 analyses, but the rank of each lake remains nearly the same (Table 3). There was no significant difference between the mean of slopes from LT and NoLT growth curves regardless of technique. However, examination of interannular distances indicated a slight but consistent trend toward faster growth in NoLT lakes. Interannular distance was significantly greater for year IV growth in NoLT lakes ($t = 2.46$, $df = 6$, $p < 0.05$), but not

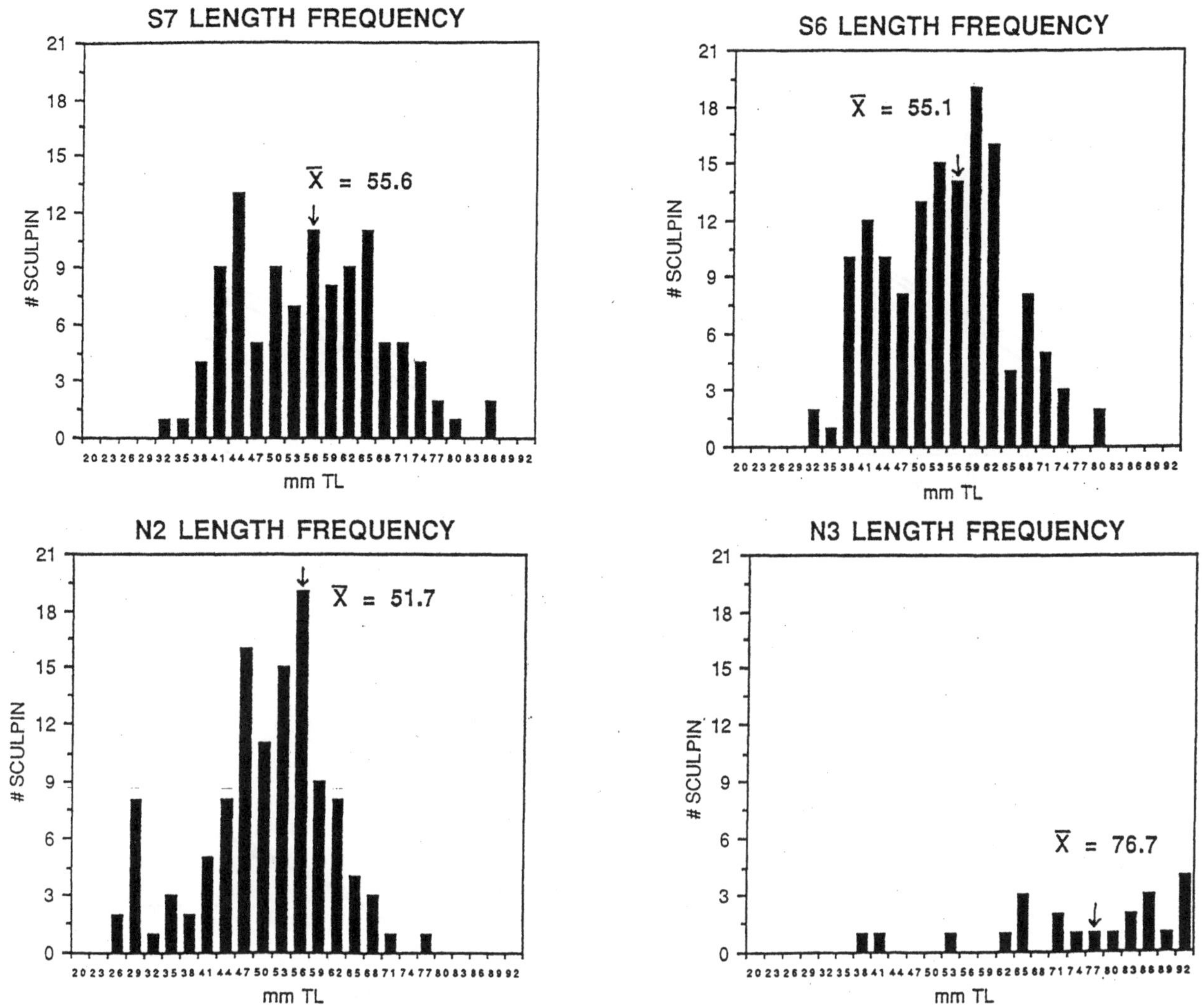

Fig. 3. Length-frequency distribution of sculpin in LT (3a) and NoLT (3b) lakes.

significantly different in other years, although the trend was in the direction of faster growth in NoLT lakes (Table 4).

Fish condition analyzed by log weight/log TL regression resulted in slopes measuring from 2.93 to 3.17, with 3.0 indicating isometric growth (Table 5). Slopes overlapped and showed no significant difference between sculpin condition in LT and NoLT lakes.

Sculpin catch per unit effort (CPUE) ranged from 0.087 to 0.39 fish per trap. The highest CPUE was in NoLT lakes N-2 and S-7, with the remaining NoLT lake, N-3, showing the lowest CPUE overall. Mean CPUE for NoLT lakes (0.29) was very similar to the mean in LT lakes (0.28).

Discussion

Slimy sculpin have adapted to a variety of cold-water habitats, both lacustrine and lotic, as evidenced by their wide-spread distribution in northern North America (Scott & Crossman, 1973). Although some sculpin behaviors undoubtably change with respect to geographic locality, such as nocturnalism versus diurnalism in response to light intensity (Van Vliet, 1964; Emery, 1973;

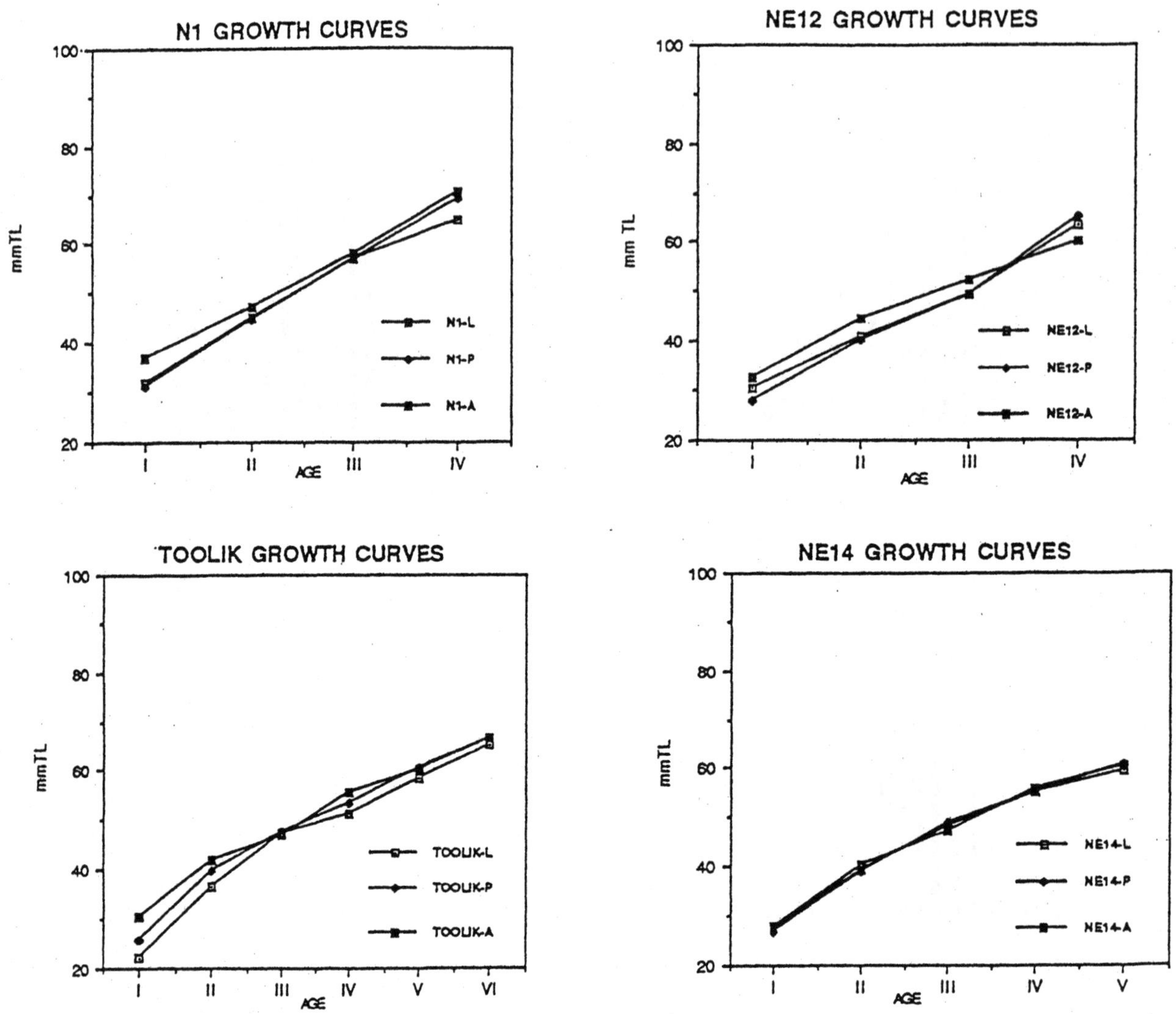

Eriksson, 1978, Müller, 1978a, b; Selgeby, 1988), many life history parameters can be expected to be similar in all habitats. Sculpin life history was not the focus of this study, but it cannot be separated from the ecological implications of the data presented. Slimy sculpin behavior in northern Saskatchewan was studied extensively by Van Vliet (1964). Conducting both field and laboratory experiments, he observed that breeding males establish and aggressively defend nest sites in rocky crevices. During spawning, females were attracted to nest sites and laid eggs on the underside of rocks. Males remained in nests to fan egg masses and protect them from predation. Parental care by the male continued for the entire incubation period (28–29 days at 8-18 °C), and after egg hatching for 2–3 days until the fry began feeding. Van Vliet terms slimy sculpin behavior during the rest of the year as 'solitary.'

Young of the year (yoy) sculpin were trapped in this study beginning on July 4, 1988. Sculpin at earliest capture dates were 8.3 to 10.2 mmTL, had yolk absorbed, a full complement of adult fin rays (sensu Heufelder, 1982), and were estimated to be 5 days old using the criteria of Van Vliet (1964). Assuming a 28–29 day incubation period (Van Vliet, 1964), this sets the earliest spawning date near June 1 for that year and confirms that

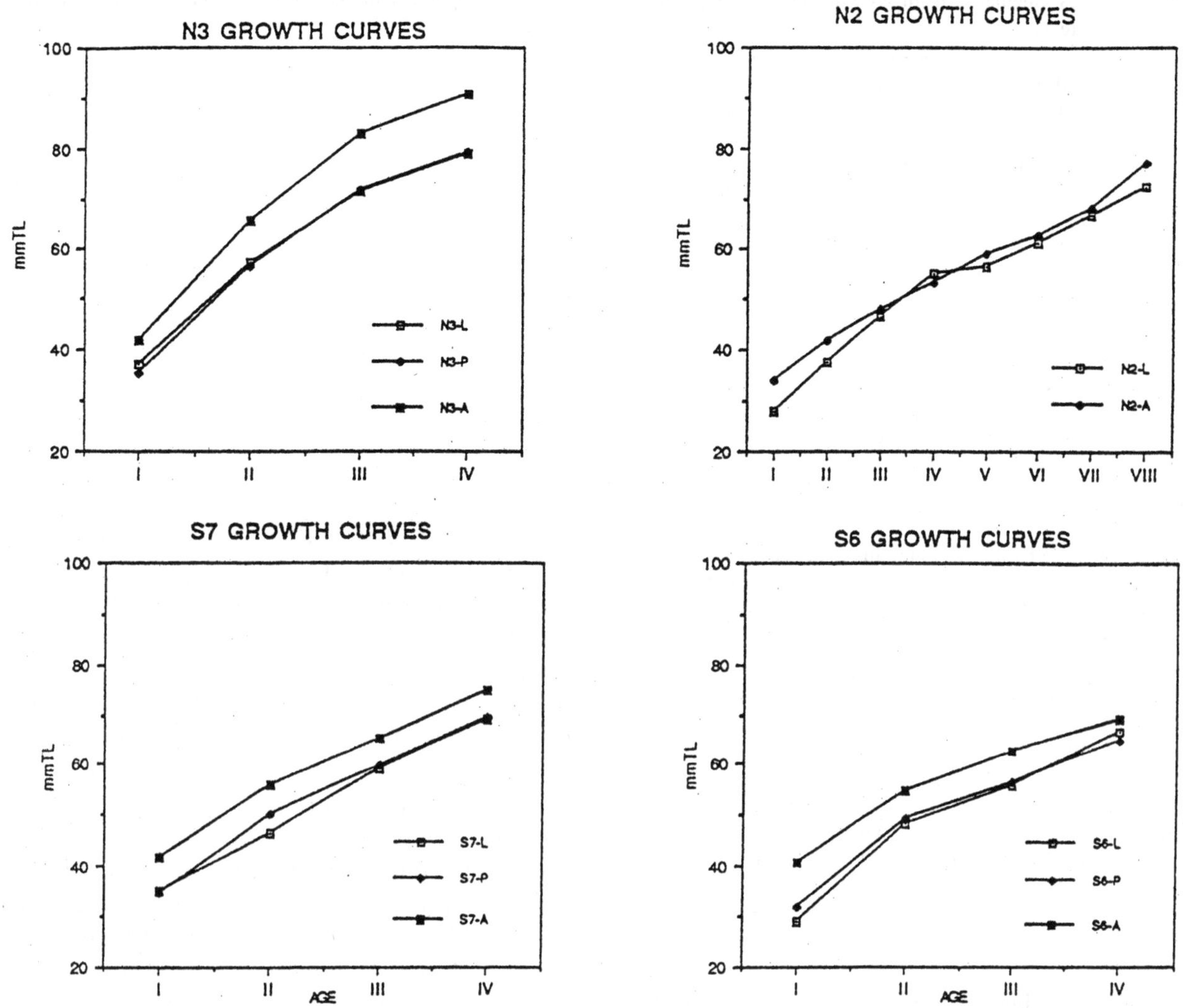

Fig. 4. Sculpin growth curves determined for LT lakes (4a) and NoLT lakes (4b). L = linear regression back-calculation method, P = proportional back-calculation method, and A = method using age-length means.

sculpin in this area are probably spawning at ice-out in the small lakes. The ability to trap yoy fish was important to aging verification (Jearld, 1983) and to add insights to early growth patterns.

Breeding male sculpin are known to leave the nesting area after yoy dispersal (Van Vliet, 1964). This probably results in a shift in habitat for them, and possibly other fish they may displace. Analyzing the adult distribution throughout the sampling period of late June to mid-August showed no seasonal change in habitat distribution which might be attributed to cessation of male breeding behavior. However, the percentage of the population comprised of breeding males is probably small and would likely not to have strongly influenced the observed patterns of sculpin distribution. Also, males may have already left nests when sampling began.

Other seasonal changes are probably initiated near the end of the open-water season. Little is known about the winter behavior of arctic fishes and we can only surmise that both predator and prey undergo major changes from the constant daylight (Eriksson, 1978, Müller, 1978a, b), changing water temperatures, spawning and feeding modes of summer. Age analysis

Table 3. Slopes calculated from growth curves using three methods (see text).

NoLT lakes	Actual TL/age	Linear back-calc	Proportional back-calc
N2	5.74	5.96	–
S6	9.38	11.18	12.06
S7	10.9	11.44	11.21
N3	16.25	14.63	13.90
LT lakes			
Toolik	6.99	7.85	8.11
NE14	8.05	8.41	7.97
NE12	9.16	12.10	10.73
N1	11.17	12.68	11.16

Table 4. Mean (SE) focus to annulus distances (year I) and interannular distances (years II–IV) for sculpin in LT and NoLT lakes. P-values indicate outcome of t-tests on $\log_e$ transformed data for each year.

Growth (year)	NoLT lakes		LT lakes		*P*-value
	(mean)	(SE)	(mean)	(SE)	
Year I	0.44	0.02	0.39	0.02	<0.2
Year II	0.34	0.04	0.30	0.01	>0.5
Year III	0.25	0.03	0.24	0.01	>0.5
Year IV	0.28	0.04	0.19	0.01	<0.05*

of otoliths implies growth during winter is markedly reduced (Van Vliet, 1964), but fish are presumed to be foraging at some level (Müller, 1978a, b; Hershey & McDonald, 1985). The effects of fish predation can differ according to season (Gilinsky, 1984), and during the dark arctic winter, visual feeders like lake trout (Kettle & O'Brien, 1978) may become more reliant on tactile senses, making them less effective piscivores. At the same time, slimy sculpin likely move into deeper water. Mousseau & Collins (1987) reported slimy sculpin in Ontario lakes moved to hypolimnetic waters in August or September. In the arctic, much of the littoral zone freezes solid and sculpin must move off the rocks by winter. Correspondingly, habitat for each sculpin would be reduced.

Table 5. Slopes calculated from log weight/total length data (see text).

NoLT lakes	Log wt/TL regression	SE	*n*	R^2
N2	3.03	0.15	113	0.79
S6	3.00	0	117	0.95
S7	3.17	0.08	70	0.96
N3	2.93	0.11	33	0.95
LT lakes				
Toolik	2.94	0.11	67	0.92
NE14	3.09	0.13	37	0.94
NE12	3.00	0	134	0.96
N1	3.04	0.13	35	0.94

The presence of sculpin across all three zones in both LT and NoLT lakes may be due to intraspecific behavior in sculpin populations. Van Vliet (1964) reported that in aquaria observations, sculpin quickly established a 'pecking order', with the larger, more aggressive fish dominant in terms of food consumption and habitat. Brandt (1986) found smaller, younger slimy sculpin inhabiting shallower depths than larger individuals in Lake Ontario. In arctic lakes, larger sculpin probably use those habitats with more food in order to find sufficient prey to meet their greater metabolic needs. This hypothesis is supported by a significant trend toward larger fish in the sediment and interface zones than in the rock zone of both LT and NoLT lakes (Table 2).

During the summer, adult sculpin in NoLT lakes were most prevalent in the sediment zone. Sculpin distribution in LT lakes shows a significantly smaller percentage of sculpin captured over the sediments, and more sculpin in the interface and rock zones, which afford refugia from predators. This may reflect the necessity for sculpin to balance predator threat against food availability, because chironomid biomass data indicates the sediments offer more food. This result has potentially important bioenergetic implications for sculpins because these arctic sculpin appear to be food limited (McDonald *et al.*, 1982; Hershey & McDonald, 1985).

Fish growth is closely correlated with diet, and

growth in many fish species has been shown to be slow at higher latitudes (Craig & Wells, 1976; Johnson, 1976; Power, 1978). In many cases arctic systems support fewer and smaller prey items than temperate ones. The diet of Toolik-area sculpins is 84% chironomid larvae (Hershey, 1985). A 5-year old slimy sculpin feeding on amphipods in Lake Superior can grow to 2 × the length and 3 × the weight of arctic sculpin, despite colder summer water temperatures (4–5.5 °C) in Lake Superior (data from Selegby, 1988). A diet of mainly chironomid larvae limits arctic sculpin growth, and chironomids are less abundant in rocky areas than on sediments. Goyke & Hershey (1992) found chironomid diversity, density and biomass to be significantly less in the sediments of NoLT lakes than in LT lakes, and attributed this to the impact of sculpin feeding behavior on the sediments in the absence of predators. A reciprocal relationship, less chironomids in the rock zone of LT lakes than in the rock zone of NoLT lakes, appeared to exist, but was not statistically significant. However, rock zone sampling was far less extensive than sediment zone sampling (Goyke & Hershey, 1992), thus there was considerably less power to detect a difference between lake types in the rock zone. A trend toward lower chironomid biomass in the rock zone of LT lakes could indicate sculpins rely more on feeding in rock areas when in the presence of predators. In lakes containing predators, the threat of predation may reduce sculpin individual habitat size, and/or reduce the amount of time sculpin actually spend feeding. Larger overall weights and total lengths from fish captured over the sediments in NoLT lakes suggests sculpin feed more readily and openly in the absence of predators.

The implications of sculpin distribution differences between lake types, in conjunction with resource differences in the rock and sediments zones, suggests that sculpin growth and population structure could be different in lakes with and without predators. TL frequency diagrams showed a tendency for larger-sized sculpin in NoLT lakes. The sexual maturity of slimy sculpin, as in most fish, is largely determined by reaching a minimum size. Aging of Toolik-area sculpin reveals there is variability in the longevity of fish by lake, with fish in some populations living only to age IV and others showing a maximum age of VIII. The maximum attainable size for slimy sculpin in this area appears to be approximately 95 mm TL, regardless of age. This implies very different growth rates among lakes. Age-length growth curves and growth curves by back-calculation gave varying slopes, but did not show a clear distinction between growth rates in LT and NoLT lakes. The condition index of the slope of the log weight/length regression line also shows no pattern.

Examination of interannular distances does indicate that sculpin grew more during year IV in NoLT lakes. This significantly higher year IV growth is consistent with the pattern of greater use of the sediment habitat in these lakes. Higher overall growth rate of all year IV sculpin in NoLT lakes contrasted with smaller mean size of sediment-dwelling fish in the same lakes probably reflects the fact that a higher proportion of the populations are using the sediments when predators are absent. Lower food abundance on the sediments has been attributed to sculpin predation (Goyke & Hershey, 1992), but lack of a consistent concomitant growth response combined with an apparent maximum size regardless of lake type, implies that predator threat determines distribution within a lake, but that predation and food supply are closely related factors affecting growth of sculpin in these arctic lakes.

Acknowledgements

This work was supported by NSF grants No. DPP 8722015 and BSR 870232 to John Hobbie, Marine Biological Laboratory (MBL), Woods Hole, MA, and by subcontract from MBL to AEH and MEM. We thank the entire Toolik Lake group for their field assistance during the project, and offer special thanks to Jason McCrea, Debra Hinterleitner-Anderson, and Andy Goyke for their efforts in the field and laboratory. We are also grateful to Chris Leucke and an anonomous reviewer for helpful comments on the manuscript.

References

Bagenal, T. B. & F. W. Tesch, 1978. Age and growth. Pages 101–136 *in* T. B. Bagenal, editor. Methods for assessment of fish production in fresh waters. International Program Handbook No. 3. Blackwell, Oxford.

Becker, G. C., 1983. Fishes of Wisconsin. University of Wisconsin Press. Madison, Wisconsin.

Brandt, S. B., 1986a. Disappearance of the deepwater sculpin (*Myoxocephalus thompsoni*) from Lake Ontario: the keystone predator hypothesis. J. Great Lakes Res. 12: 18–24.

Brandt, S. B., 1986b. Ontogenetic shifts in habitat, diet, and diel-feeding periodicity of slimy sculpin in Lake Ontario. Trans. am. Fish. Soc. 115: 711–715.

Cailliet, G. M., M. S. Love & A. W. Ebeling, 1986. Fishes. A field and laboratory manual on their structure, identification, and natural history. Wadsworth Publishing, Belmont, California, 144 pp.

Carlander, K. D., 1981. Caution on the use of the regression method of back-calculating lengths from scale measurements. Fisheries 6: 2–4.

Carpenter, S. R., J. F. Kitchell & J. R. Hodgson, 1985. Cascading trophic interactions and lake productivity. Bioscience 35: 634–639.

Cone, R. S., 1989. The need to reconsider the use of condition indices in fishery science. Trans. am. Fish. Soc. 118: 510–514.

Craig, P. C. & J. Wells, 1976. Life history notes for a population of slimy sculpin (*Cottus cognatus*) in an Alaskan arctic stream.J. Fish.Res. Bd Can. 33: 1639–1642.

Crowder, L. B. & W. E. Cooper, 1982. Habitat structural complexity and the interaction between bluegills and their prey. Ecology 63: 1802–1813.

Eddy, S. & J. C. Underhill, 1974. Northern fishes. University of Minnesota Press, Minneapolis, Minnesota, 414 pp.

Emery, A. R., 1973. Preliminary comparisons of day and night habits of freshwater fish in Ontario lakes. J. Fish. Res. Bd Can. 30: 761–774.

Everhart, W. H. & W. D. Youngs, 1981. Principles of fishery science, 2nd edn. Cornell University Press, Ithaca, New York, 349 pp.

Eriksson, L.-O., 1978. Nocturnalism versus diurnalism – dualism within fish individuals. Pp. 69–89 *in* J. E. Thorpe (ed.), Rhythmic activity of fishes. Academic Press, New York.

Gilinski, E., 1984. The role of fish predation and spatial heterogeneity in determining benthic community structure. Ecology 65: 455–468.

Godkin, C. M., W. J. Christie & D. E. McAllister, 1982. Problems of species identification in the Lake Ontario sculpins *Cottus bairdi* and *C. cognatus*. Can. J. Fish. aquat. Sci. 39: 1373–1382.

Goyke, A. P. & Hershey, A. E., 1992. Effects of fish predation on chironomid (*Diptera: Chironomidae*) communities in an arctic ecosystem. Hydrobiologia 240: 203–211.

Hershey, A. E., 1985. Effects of predatory sculpin on the chironomid communities in an arctic lake. Ecology 66: 1131–1138.

Hershey, A. E., 1987. Tubes and foraging behavior in larval Chironomidae: implications for predator avoidance. Oecologia 73: 236–241.

Hershey, A. E., 1990. Snail populations in arctic lakes: competition mediated by predation? Oecologia 82: 26–32.

Hershey A. E. & S. I. Dodson, 1985. Selective predation by a sculpin and a stonefly on two chironomids in laboratory feeding trials. Hydrobiologia 124: 269–273.

Hershey, A. E. & M. E. McDonald, 1985. Diet and digestion rates of slimy sculpin, *Cottus cognatus*, in an arctic lake. Can. J. Fish. aquat. Sci. 42: 483–487.

Heufelder, G., 1982. Family Cottidae, sculpins, pp. 656–676 *in* N. A. Auer (ed.), Identification of larval fishes of the Great Lakes basin with emphasis on the Lake Michigan drainage. Great Lakes Fisheries Comm., Ann Arbor, Michigan, Spec. Publ. 82–3: 744 pp.

Hollander, M. & D. A. Wolfe, 1973. Nonparametric statistical methods. John Wiley & Sons., New York. 503 pp.

Jearld, A., Jr., 1983. Age determination. Pp. 301–324 *in* L. A. Nielson & D. L. Johnson (eds), Fisheries Techniques, American Fisheries Society, Bethesda, Maryland.

Johnson, L., 1976. Ecology of arctic populations of lake trout, *Salvelinus namaycush*, lake whitefish, *Coregonus clupeaformis*, arctic char, *S. alpinus*, and associated species in unexploited lakes of the Canadian Northwest Territories. J. Fish. Res. Bd Can. 33: 2459–2488.

Keenleyside, M. H. A., 1979. Diversity and adaptation in fish behaviour. Vol. 11, Zoophysiology, Springer-Verlag Publishing, New York, 208 pp.

Kerr, S. R., 1971. A simulation model of lake trout growth. J. Fish. Res. Bd Can. 28: 815–819.

Kettle, D. & W.J. O'Brien, 1978. Vulnerability of arctic zooplankton species to predation by small lake trout (*Salvelinus namaycush*). J. Fish. Res. Bd Can. 35: 1495–1500.

Konkle, B. R. & W. G. Sprules, 1986. Planktivory by stunted lake trout in an Ontario lake. Trans. am. Fish. Soc. 115: 515–521.

Luecke, C., 1990. Changes in abundance and distribution of benthic macroinvertebrates after introduction of cutthroat trout into a previously fishless lake. Trans. am. Fish. Soc. 119: 1010–1021.

McDonald, M. E., B. E. Cuker & S. C. Mozley, 1982. Distribution, production, and age structure of slimy sculpin in an arctic lake. Envir. Biol. Fishes 7: 171–176.

McDonald, M. E. & A. E. Hershey, 1989. Size structure of a lake charr population in an arctic lake: influence of angling and implications for fish community structure. Can. J. Fish. aquat. Sci. 46: 2153–2156.

Merrick, G. W., A. E. Hershey & M. E. McDonald, 1991. Lake trout (*Salvelinus namaycush*) control of snail density and size distribution in an arctic lake. Can. J. Fish. aquat. Sci. 48: 498–502.

Merrick, G. W., A. E. Hershey & M. E. McDonald, 1992. Salmonid diet and the size, distribution, and density of benthic invertebrates in an arctic lake. Hydrobiologia 240: 225–233.

Miller, M. C., G. R. Hater, P. Spatt, P. Westlake & D. Yeakel, 1986. Primary production and its control in Toolik Lake, Alaska. Arch. Hydrobiol./ Suppl. 4: 97–131.

Mittelbach, G. G., 1981. Foraging efficiency and body size: a study of optimal diet and habitat use by bluegills. Ecology 62: 1370–1386.

Mittelbach, G. G., 1986. Predator-mediated habitat use: some consequences for species interactions. Pp. 159–169 *in* C. A. Simmenstad & G. M. Cailliet (eds), Contemporary studies on fish feeding. Dr W. Junk, Dordrecht, Netherlands.

Mousseau, T. A. & N. C. Collins, 1987. Polygyny and nest site abundance in the slimy sculpin (*Cottus cognatus*). Can. J. Zool. 65: 2827–2829.

Moyle, P. B., 1977. In defense of sculpins. Fisheries 2: 20–23.

Müller, K., 1978a. Locomotor activity of fish and environmental oscillations. Pp. 1–19 *in* J. E. Thorpe (ed.), Rhythmic activity in fishes. Academic Press, New York.

Müller, K., 1978b. The flexibility of the circadian system of fish at different latitudes. Pp. 91–104 *in* J. E. Thorpe (ed.), Rhythmic activity in fishes. Academic Press, New York.

Petrosky, C. E. & T. F. Waters, 1975. Annual production by the slimy sculpin population in a small Minnesota trout stream. Trans. am. Fish. Soc. 2: 237–244.

Power, G., 1978. Fish population structure in arctic lakes. J. Fish. Res. Bd Can. 35: 53–59.

Power, M. E., 1984. Depth distributions of armored catfish: predator- induced resource avoidance? Ecology 65: 523–528.

Pyke, G. H., H. R. Pulliam & E. L. Charnov, 1977. Optimal foraging: a selective review of theory and tests. Quarterly Review of Biology 52: 137–154.

Ricker, W. E., 1979. Growth rates and models. Pp. 677–743 *in* Fish Physiology, Vol. VIII. Academic Press, New York.

Scott, W. B. & E. J. Crossman, 1973. Freshwater fishes of Canada. Bull. Fish. Res. Bd Can. 184: 1–966.

Selegby, J. H., 1988. Comparative biology of the sculpins of Lake Superior. J. Great Lakes Res. 14: 44–51.

Smock, L. A., 1980. Relationships between body size and biomass of aquatic insects. Freshwat. Biol. 10: 375–383.

Van Vliet, W. H., 1964. An ecological study of *Cottus cognatus* Richardson in northern Saskatchewan. M. Sci. Thesis, Univ. Saskatchewan, Saskatoon, 155 pp.

Ware, D. M., 1973. Risk of epibenthic prey to predation by rainbow trout (*Salmo gairdneri*). J. Fish. Res. Bd Can. 30: 787–797.

Werner, E. E. & D. J. Hall, 1988. Ontogenetic habitat shifts in bluegill: the foraging rate-predation risk trade-off. Ecology 69: 1352–1366.

Werner, E. E., J. F. Gilliam, D. J. Hall & G. G. Mittelbach, 1983a. An experimental test of the effects of predation risk on habitat use in fish. Ecology 64: 1540–1548.

Werner, E. E., G. G. Mittelbach, D. J. Hall & J. F. Gilliam, 1983b. Experimental tests of optimal habitat use in fish: the role of relative habitat profitability. Ecology 64: 1525–1539.

Hydrobiologia **240**: 203–211, 1992.
W.J. O'Brien (ed.), Toolik Lake – Ecology of an Aquatic Ecosystem in Arctic Alaska.

Effects of fish predation on larval chironomid (*Diptera*: *Chironomidae*) communities in an arctic ecosystem

Andrew P. Goyke[1] & Anne E. Hershey
Department of Biology, University of Minnesota-Duluth, Duluth, MN 55812, USA; [1] *Current address: Chesapeake Biological Laboratory, University of Maryland System, P.O. Box 38, Solomons, MD 20688, USA*

Key words: arctic, chironomidae, diversity, predation, slimy sculpin

Abstract

Chironomid communities were analyzed in systems with three types of predator regimes to determine hierarchical effects of predation; ponds without fish present, lakes with slimy sculpin (*Cottus cognatus*) present, and lakes with slimy sculpin present along with burbot (*Lota lota*) and lake trout (*Salvelinus namaycush*). Samples were collected by coring bare sediment habitats in 4 systems of each type near the Toolik Lake field station in northern Alaska. Lakes with burbot and lake trout present in addition to slimy sculpin displayed significantly higher ($P < 0.05$) biomass, density, richness, and diversity. This is likely due to the increased complexity of the predator regime. Ponds without fish had a significantly greater percentage of predacious chironomids present.

Introduction

Predation has been demonstrated to influence habitat use in aquatic systems (Stein & Magnuson, 1976; Sih, 1982; Power & Matthews, 1983; Werner *et al.*, 1983; Power *et al.*, 1985; Mittelbach, 1986), where behavior or distribution will be altered in the face of predation. Minnows have been demonstrated to change their distribution and grazing behavior in the presence of piscivorous bass, which resulted in significant changes in the algal community depending on whether bass were present or absent (Power & Matthews, 1983; Power *et al.*, 1985). Predator avoidance may also influence the foraging efficiency of fish feeding on benthic invertebrates.

Many studies examining the effects of vertebrate predation on freshwater benthic communities have used one predator, commonly sunfish (Centrarchidae) (e.g. Ball & Hayne, 1952; Hayne & Ball, 1956; Hall *et al.*, 1970; Thorp & Bergey, 1981a, b; Gilinsky, 1984) in the system. However, when forced to increase vigilance, move to protective habitats, or limit foraging time and/or intake (Mittelbach, 1986), sunfish, which are also planktivorous (Werner & Hall, 1976), may show an altogether different effect on the benthos than a strictly benthivorous fish. These studies did not address the potential hierarchical effects of predation on the benthic vertebrate predators themselves.

Examining the effects of predation on the benthic community is easier in simple systems with replication as well as systems with differing predator regimes. The Alaskan arctic has areas which meet these criteria. The benthos of many arctic limnetic systems are dominated by larvel chironomids (Oliver, 1968; Welch, 1976), which are important secondary producers in most aquatic systems (Pennack, 1978). Chironomid communities may be influenced by habitat (Crowder & Cooper, 1982; Hershey, 1985a, b), predation by invertebrates (Hilsenhoff, 1967; Hall *et al.*, 1970; Hershey, 1986), and predation by vertebrates (Ball & Hayne, 1952; Hayne & Ball, 1956; Hilsenhoff, 1967; Hall *et al.*, 1970; Crowder & Cooper, 1982; Gilinsky, 1984; Hershey, 1985a).

Chironomid life history is a factor in determining their susceptibility to predation. Predatory chironomids tend to be the largest (A. P. Goyke, personal observation), most mobile (Cuker, 1981) chironomids and do not construct tubes (Wiederholm, 1983). Fish tend to select for the largest prey available (Hall *et al.*, 1970), and can control community structure not only by preying on invertebrates, but by preying on predatory invertebrates. Gilinsky (1984) found predacious chironomids to exhibit strong negative responses to predation.

Paine (1966) suggested diversity is positively related to the number of predators in the system. If lakes follow the principles of island biogeography and those proposed for continental areas (Cody, 1975), larger lakes should have higher diversity due to their larger area for attracting colonists. However, larger lakes also tend to have more predators, thus area and predator diversity are two potentially conflicting factors which may affect chironomid diversity.

To determine the effects of fish predation and predator interactions on chironomid communities, benthic samples were collected from macrophyte-free bare sediment habitats in systems with various predator regimes. Three types of systems were examined; permanent ponds without fish present ('ponds'), lakes with slimy sculpin (*Cottus cognatus* (Richardson)) present ('sculpin lakes'), and lakes with slimy sculpin and sculpin predators present ('trout lakes'). Chironomid biomass, density, richness, diversity, and percentages of predacious chironomids were compared between the three types of systems.

Study area

All lakes and ponds were on rolling tundra in the Alaskan arctic at the Toolik Lake LTER site (68° 37′ north, 149° 36′ west), near the Dalton Highway and trans-Alaskan pipeline (Fig. 1). Toolik Lake is approximately 200 km south of Prudhoe Bay and 25 km north of the Brooks Mountains.

Five species of fish were present in the study area; of these fish slimy sculpin is the primary benthic predator. Chironomids are an important part of slimy sculpin diets, making up 81% of prey items found in the guts of slimy sculpin from Toolik Lake (Hershey, 1985a). Slimy sculpin were generally less than 70 mm in length in Toolik Lake (McDonald *et al.*, 1982); size likely did not offer slimy sculpin a refuge from predation. Burbot (*Lota lota* (L.)) will prey on most available fishes

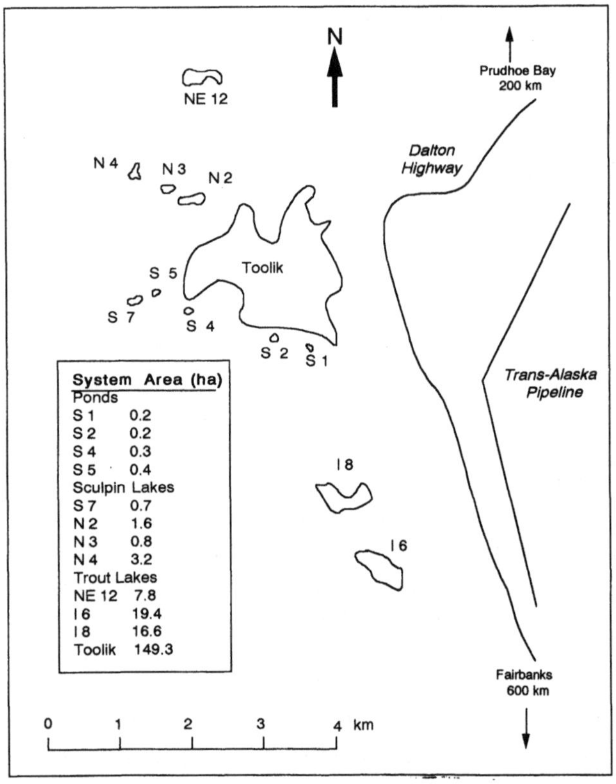

Fig. 1. Location and area of the systems sampled near the Toolik Lake Long Term Ecological Research site in northern Alaska (68° 37′ N, 149° 36′ W).

(Becker, 1983), including slimy sculpin (A. P. Goyke, personal observation). Small lake trout (*Salvelinus namaycush* (Walbaum)) feed on zooplankton (Becker, 1983), while slimy sculpin, burbot (Becker, 1983), and mollusks (Hershey, 1990) are important in the diets of larger lake trout. Round whitefish (*Prosopium cylindraceum* (Pallas)) are bottom feeders whose diets may include snails and chironomid larvae (Becker, 1983), although Merrick (1989) found little evidence of chironomid predation by round whitefish in Toolik Lake. Planktivorous arctic grayling (*Thymallus arcticus* (Pallas)) (Schmidt & O'Brien, 1982) are also present.

The trout lakes (NE 12, I 6, I 8, and Toolik Lake) were the largest systems, with areas ranging between 7.8 to 149 ha. All five fish species were present in the trout lakes (except for NE 12, which did not contain round whitefish). Sculpin lakes (S 7, N 2, N 3, and N 4) had areas ranging from 0.7 to 3.2 ha, while ponds (S 1, S 2, S 4, and S 5) had areas ranging from 0.2 to 0.4 ha. Slimy sculpin and arctic grayling were present in the sculpin lakes, and no fish were present in the ponds. Lake N 2 was divided by a curtain as part of a fertilization experiment; only the control side was sampled for this study. The lakes are oligotrophic to highly oligotrophic (Miller *et al.*, 1986).

The sediment was fine silt throughout the lakes; pond sediment was course and contained much organic material. Ponds were also much shallower than lakes, with maximum depths of less than 1 m, while all lakes were deeper than 6 m.

Materials and methods

Each lake and pond was sampled twice during 1989. The first sampling period was 24 June–7 July, while the second sampling period was 19 July–4 August. Samples from lakes were collected from bare sediment at a depth of approximately 4 m by a diver using SCUBA. Lake samples were collected with core tubes having inside diameters of 93 mm and a cross-sectional area of 75.43 cm^2. The core tubes were inserted 15–20 cm into the sediment, sealed with rubber stoppers, and brought to the surface. Five samples were collected on each date from each lake. One sample was lost on each date from N 2, and one sample was lost from Toolik Lake on the second date.

Samples from ponds were collected from bare sediment in the deepest portion of the pond, approximately 0.5–0.75 m. Pond samples were collected in core tubes with an inside diameter of 48 mm and a cross-sectional area of 18.10 cm^2. The same procedure was used as for lake samples, except four cores were pooled in a bucket for each sample to give a total sampling area of 72.36 cm^2. Five samples were collected from each pond on each date.

Samples were washed through a 200 μm mesh net in the field; early instar larvae were generally not collected. Samples were preserved in 95% ethyl alcohol and later sorted using sugar flotation (Anderson, 1959) after being washed with water through a 70 μm mesh net. After all visible chironomids were removed, the lake samples were washed through a 500 μm sieve and sorted under magnification to remove any remaining chironomids. The copious amounts of organic material in the pond samples made this step prohibitively time consuming. However, several pond samples were completely sorted under magnification, and the sugar flotation was found to be 90–95% effective, agreeing with the values obtained by Gerking (1962). Chironomids were measured for length using a dissecting microscope and Zidas digitizing pad, mounted in Hoyer's solution, and identified to genus (Wiederholm, 1983; Coffman & Ferrington, 1984).

A simple ordination of the 12 systems based on similarity of the fish and chironomid community was performed by comparing each system to Toolik Lake using the following equation (Krebs, 1978):

$$\text{Index of similarity} = \frac{2c}{a + b},$$

where a is the number of species of fish or chironomid genera present in Toolik Lake, b is the number of fish species or chironomid genera in the system being compared, and c is the fish species or chironomid genera present in both systems.

Chironomid biomass, density, genera richness, diversity, and percentage of predacious chironomids were tested between trout lakes, sculpin lakes, and ponds using a nested (systems nested within type of system) analysis of variance (Montgomery, 1984) in the general linear models procedure of SAS (SAS Institute, 1985). Since there was not an interaction between date and type of system ($P = 0.213$ for biomass × date, $P = 0.213$ for density × date, $P = 0.350$ for richness × date, $P = 0.068$ for diversity × date, and $P = 0.710$ for percent predacious chironomids × date), the dates were not analyzed separately. Pond, sculpin lake, and trout lake means were compared using Duncan's multiple range test (Montgomery, 1984). P-values of 0.05 or less were considered significant.

S 7 was not used in testing for biomass or density effects due to the large population of *Corynocera ambigua* (Zetterstedt) present in S 7 but absent from the other systems (Table 2). *C. ambigua* does not fly and hence its dispersal is limited; it may have historically produced populations in other lakes as large as that found in S 7 (Walker & Mathewes, 1988). Values of 4.21 ± 1.10 g m^{-2} (mean ± SE) and $21\,846 \pm 6253$ m^{-2} were obtained from S 7; Cook's distance test for outliers (SAS Institute, 1985) resulted in values of 0.967 for biomass and 0.995 for density.

Biomass was calculated from length by the following length weight regressions (Smock, 1980):

for Tanypodinae Weight = 0.0038 Length$^{2.41}$;
for all other chironomids Weight = 0.0051 Length$^{2.32}$.

A Shannon-Wiener diversity index was calculated using the following equation (Pielou, 1975):

$$H' = -\Sigma p_i \log_{10} p_i,$$

where p_i is the proportion of a genus in a particular sample.

Chironomids were divided into functional groups (predators and collectors) based on feeding specialization (Cummins & Coffman, 1984).

Results

A total of 3349 chironomids were collected in this study. Biomass of each genus collected by system is listed in Table 1. Approximately 5% of the chironomids could not be measured due to broken specimens; and a similar number could not be identified due to poor condition of mounts. *Heterotrissocladius* and *Stictochironomus* were dominant in trout lakes. Sculpin lakes showed individual dominants, with *Corynocera* in S 7, *Psectrocladius* in N 2, *Dicrotendipes* in N 3, and *Stictochironomus* in N 4. *Procladius*, *Psectrocladius*, and *Dicrotendipes* were dominants in the ponds.

Ordination by similarity of fish and chironomid communities (Fig. 2), resulted in three clusters, although the break between trout and sculpin lakes is not as distinct as that between ponds and the other two system types.

Trout lakes had significantly higher biomass than ponds (49% higher) and sculpin lakes (60% higher) (Fig. 3). Mean trout lake biomass was 1.23 ± 0.16 g m^{-2} (mean ± 1 SE), mean pond

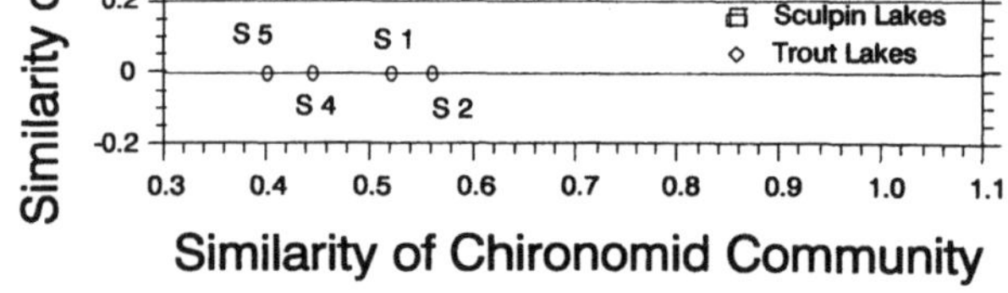

Fig. 2. Ordination of systems used in the study based on similarity of the fish (species present) and chironomid (genera present) communities, with all systems being compared to Toolik Lake. The criteria used is presence or absence of taxa.

Table 1. Biomass ($mg \cdot m^{-2}$) for each genus found in each system, and functional grouping for each genus.

Taxa		*	Ponds				Sculpin Lakes				Trout Lakes			
			S 1	S 2	S 4	S 5	S 7	N 2	N 3	N 4	NE 12	I 6	I 8	Toolik
Subfamily Chironominae														
Tribe Chironomini														
Cladopelma	Kieffer	C				2.6			339.8	85.5				
Cryptochironomus	Kieffer	P			1.1		120.3					0.9		
Dicrotendipes	Kieffer	C	16.4	66.7	138.7	138.5	76.8							
Parachironomus	Lenz												4.4	
Paratendipes	Kieffer	C		78.1										
Polypedilum	Kieffer	C	6.5		3.6	5.5		0.5	4.2	6.2	21.6	8.2	18.2	
Stictochironomus	Kieffer	C		1.3	2.0		23.4	4.8	112.3	381.6	299.1	319.5	293.1	779.9
Tribe Tanytarsini														
Cladotanytarsus	Kieffer	C		22.9	13.1	62.8	1.4		0.7				5.9	
Constempellina	Brundin	C									0.9	6.7		
Corynocera	Zetterstedt	C					3150.6							
Microspectra	Kieffer										2.0	0.2		
Paratanytarsus	Thienemann	C	3.2	2.1	7.2	13.5	23.3	0.4	0.8	0.5	0.2	45.8	7.1	82.9
Stempellinella	Brundin	C					2.8	5.0	8.7	7.2	2.7		12.3	48.6
Tanytarsus	v.d. Wulp	C	29.8	18.0	123.6	6.0	174.1		27.2			16.6	79.3	48.6
Subfamily Diamesinae														
Protanypus	Kieffer	C								12.8	29.3	84.8		2.8
Subfamily Orthocladinae														
Corynoneura	Winnertz	C			5.1									
Cricotopus	v.d. Wulp	C		4.1	21.9	33.9							14.9	
Heterotrissocladius	Spark	C			0.7	5.8		73.0	9.0	47.3	226.5	309.1	288.8	642.0
Paracladius	Hirvenoja	C	6.7	8.0				12.7	2.4	4.6			74.2	9.6
Parakieferiella	Thienemann	C	0.9	6.8			5.7		1.3	1.8	0.2	24.3	19.3	21.4
Psectrocladius	Kieffer	C	11.1	151.7	191.6	154.2		114.6	7.4	8.3	91.1	30.7	3.9	14.9
Zalutshia	Lipina	C					8.4				10.1	14.5	5.1	17.3
Subfamily Prodiamesinae														
Monodiamesa	Kieffer	C						78.9			123.6	31.8	311.9	123.2
Subfamily Tanypodinae														
Ablabesmyia	Johannsen	P	6.8	15.0	100.8	3.9								
Larsia	Fittkau	P	1.8		35.4	43.4						16.4		
Procladius	Skuse	P	188.9	245.8	159.5	245.1	623.9	1.9	12.0	149.0	155.0	49.6	6.3	105.8

* Functional Group: C = collector, P = predator (from Cummins & Coffman, 1984).

biomass was 0.63 ± 0.07 g m^{-2}, and mean sculpin lake biomass was 0.49 ± 0.07 g m^{-2}.

Trout lakes had significantly higher density than ponds (30% higher) and sculpin lakes (51% higher) (Fig. 4). Mean trout lake density was 5364 ± 364 m^{-2}, mean pond density was 3774 ± 513 m^{-2}, and mean sculpin lake density was 2623 ± 374 m^{-2}.

Genera richness was significantly different between all three types of systems (Fig. 5). Trout lakes had a mean of 7.2 ± 0.4 genera present, ponds had 5.8 ± 0.4 genera present, and sculpin lakes had 4.5 ± 0.3 genera present.

Shannon-Wiener diversity indices were significantly different between all three types of systems (Fig. 6). Trout lakes had a mean H′ of 0.66 ± 0.03,

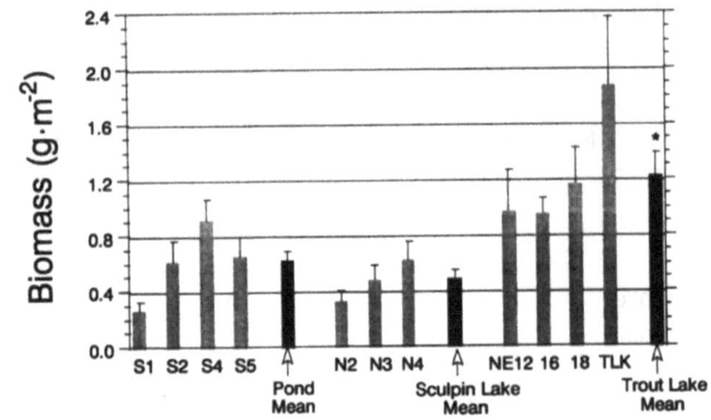

Fig. 3. Mean chironomid biomass (g m^{-2}) for each system and for ponds, sculpin lakes, and trout lakes. Error bars represent ± 1 SE; * indicates means which are significantly ($P < 0.05$) different using Duncan's multiple range test.

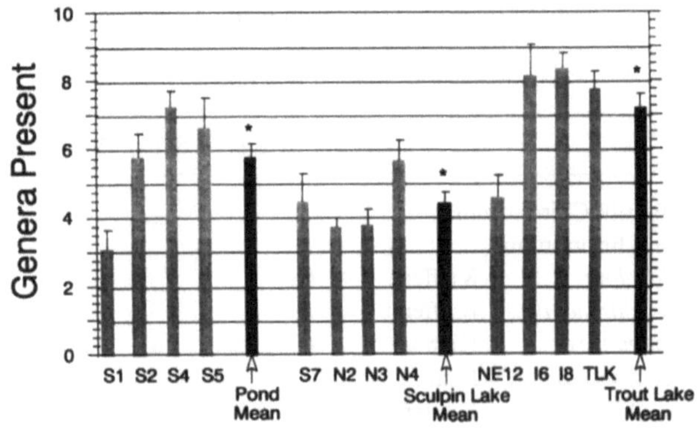

Fig. 5. Mean number of chironomid genera present for each system and for ponds, sculpin lakes, and trout lakes. Error bars represent ± 1 SE; * indicates means which are significantly ($P < 0.05$) different using Duncan's multiple range test.

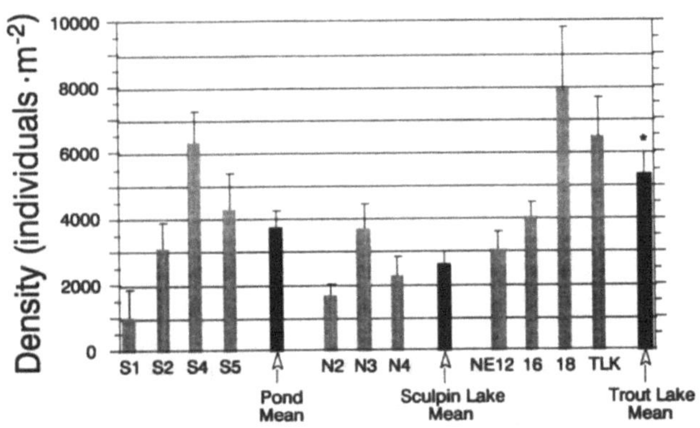

Fig. 4. Mean chironomid density (n m^{-2}) for each system and for ponds, sculpin lakes, and trout lakes. Error bars represent ± 1 SE; * indicates means which are significantly ($P < 0.05$) different using Duncan's multiple range test.

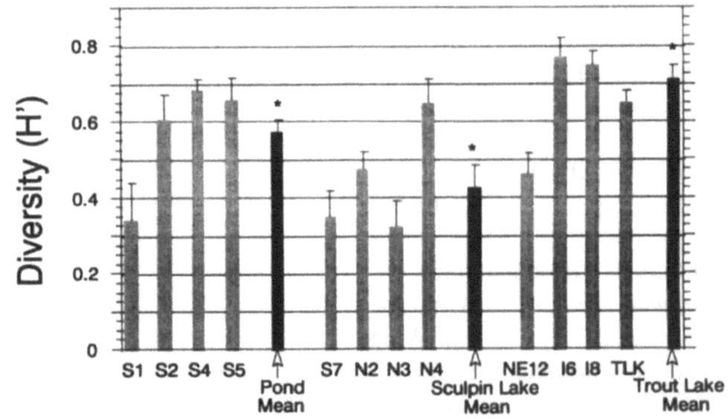

Fig. 6. Mean chironomid diversity (Shannon Wiener H′) for each system and for ponds, sculpin lakes, and trout lakes. Error bars represent ± 1 SR; * indicates means which are significantly ($P < 0.05$) different using Duncan's multiple range test.

ponds had a mean H′ of 0.58 ± 0.03, and sculpin lakes had a mean H′ of 0.45 ± 0.04.

A linear regression of the diversity index (H') by $\log_{10}$ of system area (Fig. 7) resulted in the equation

$$H' = 0.536 + 0.0737 \log_{10} \text{area}; R^2 = 0.17.$$

The coefficient of determination of 0.17 indicates there is a poor linear relationship between diversity and system area.

Predators and collectors were the only functional groups obtained (Table 1). Ponds had a significantly higher percentage of predacious chironomids (by biomass) present than sculpin lakes (72% higher) and trout lakes (75% higher) (Fig. 8). Predacious chironomids accounted for $44.47 \pm 3.52\%$ of the biomass in ponds, $12.29 \pm 2.16\%$ of the biomass in sculpin lakes, and $10.92 \pm 2.11\%$ of the biomass in trout lakes.

Discussion

Observational evidence suggests that slimy sculpin will modify their behavior in the presence of fish predators (Cuker, 1981). Sculpin lakes, when compared to trout lakes, had a greater proportion of slimy sculpin on the more profitable

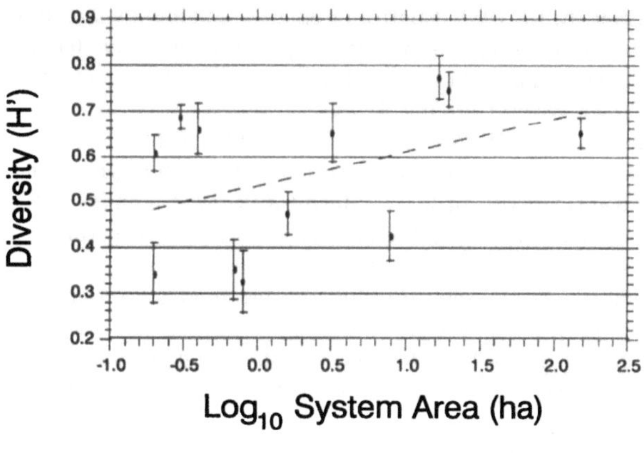

Fig. 7. Plot of chironomid diversity versus $\log_{10}$ system area (ha). Error bars represent ± 1 SE. A linear regression produced and R^2-value of 0.17.

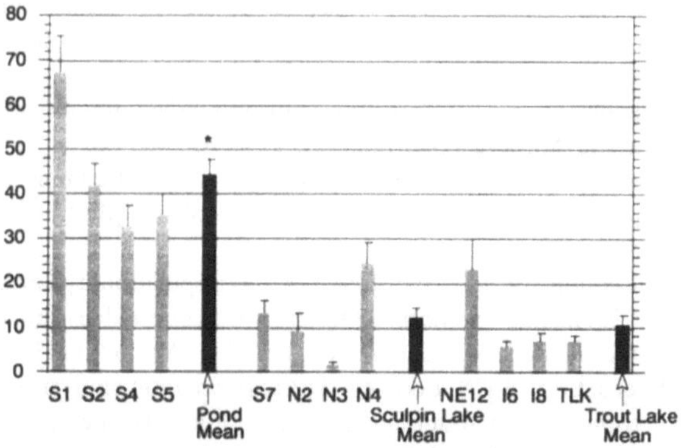

Fig. 8. Mean percentage of predacious chironomids (by biomass) for each system and for ponds, sculpin lakes, and trout lakes. Error bars represent ± 1 SE; * indicates means which are significantly ($P < 0.05$) different using Duncan's multiple range test.

bare sediment as opposed to rocks (K. L. Hanson, personal communication), indicating predators are influencing the distribution of their prey. Biomass and density of slimy sculpin was greatest at the rock mud interface in Toolik Lake (McDonald *et al.*, 1982), suggesting slimy sculpin are balancing the greater food availability on bare sediments with the refuge afforded by the rocks. The differences in chironomid communities between sculpin lakes and trout lakes appears linked to the changes in slimy sculpin foraging induced by lake trout and burbot, although the mechanism (predation, predator avoidance, both, or neither) has not been determined.

Dramatic differences were found in the algal community of stream pools with bass present or absent (Power & Matthews, 1983; Power *et al.*, 1985). The distribution and behavior of minnows, which graze the algae, were altered in the presence of bass; they would not forage near a tethered bass and seemed to vacate pools containing bass. Similarly, the chironomid community is demonstrating differences depending on whether burbot and lake trout are present or absent. Predation pressure on the benthos by sculpin appears reduced in the when piscivores are present.

Previous studies of the effects of fish predation on chironomid communities did not examine the effects of piscivory on the chironomid predators. Additionally, many of these studies used sunfish as the benthivore; sunfish will likely grow too large to ingest (Seaburg & Moyle, 1964) and are able to switch to more profitable (non-benthic) diets should conditions warrant (Werner & Hall, 1976). Slimy sculpin are obligate benthivores and did not have size refuge from predation in this study.

Comparisons between lakes and ponds need to be considered in light of the differences in their physical composition (such as depth and sediment type). The effects of these physical variables on the chironomid communities in this study are unknown. Tundra ponds, by their nature, represent a harsh environment and may exclude some species from becoming permanent residents or decrease abundance of those species which are present; ponds may also favour other species. Regardless, ponds represented the best available fish-less comparison to the lakes containing fish.

The similarity in biomass and density for sculpin lakes and ponds indicates that predation, whether vertebrate or invertebrate, can influence these factors. This may explain why Thorpe & Bergey (1981b) were unable to find a significant vertebrate predation effect on chironomid communities in fish exclusion experiments. Gilinsky (1984) found predatory invertebrates to increase in fish exclusion cages, although Thorpe & Bergey (1981b) did not find an increase in invertebrate predators in their exclusion cages. The similarity of response of the chironomid communities in sculpin lakes and ponds indicates both vertebrate and invertebrate predators seem to have significant influence on the composition and abundance of the chironomid community.

Diversity does not appear to be related to the area of a system; several ponds had diversity similar to that found in Toolik Lake. Since only one similar habitat type was examined in each systems, diversity is not related to the number of different habitats available, and appears related to the complexity of the predator regime. The higher mean diversity in trout lakes supports Paine's (1966) hypothesis that greater predator diversity will enhance overall diversity.

Acknowledgements

We would like to thank Jason McCrea, Kristi Hanson, and Milton Vollmer for their help with this project. Hollie Collins, Kang James, Ben Cuker, and an anonymous reviewer provided comments which improved this paper. This work was supported by NSF grant No. DPP 8320544 to John Hobbie and Bruce Peterson, Marine Biological Laboratory (MBL), Woods Hole, Massachusetts, by subcontract from MBL to AEH.

References

Anderson, R. O., 1959. A modified flotation technique for sorting bottom fauna samples. Limnol. Oceanogr. 4: 223–225.

Ball, R. C. & D. W. Hayne, 1952. Effects of the removal of the fish population on the fish-food organisms of a lake. Ecology 33: 41–48.

Becker, G. C., 1983. Fishes of Wisconsin. University of Wisconsin Press. Madison, Wisconsin, USA.

Cody, M. L., 1975. Towards a theory of continental species diversity. In M. L. Cody & J. M. Diamond (ed.), Ecology and evolution of communities. Harvard University Press, Cambridge, Massachusetts, USA: 214–257.

Coffman, W. P. & L. C. Ferrington, 1984. Chironomidae. In R. W. Merritt & K. W. Cummins (ed.), An introduction to the aquatic insects of North America. Kendall/Hunt, Dubuque, Iowa, USA/ 551–645.

Crowder, L. B. & W. E. Cooper, 1982. Habitat structural complexity and the interaction between bluegills and their prey. Ecology 63: 1802–1813.

Cuker, B. E., 1981. Control of epilithic community structure in an arctic lake by vertebrate predation and invertebrate grazing. PhD Thesis. North Carolina State University, Raleigh, North Carolina, USA:

Cummins, K. W. & W. P. Coffman, 1984. Table 25A, Summary of ecological and distributional data for Chironomidae (Diptera). In R. W. Merritt & K. W. Cummins (eds), An introduction to the aquatic insects of North America. Kendall/Hunt, Dubuque, Iowa, USA: 644–652.

Gerking, S. D., 1962. Production and food utilization in a population of bluegill sunfish. Ecol. Monogr. 32: 31–78.

Gilinsky, E., 1984. The role of fish predation and spatial heterogeneity in determining benthic community structure. Ecology 65: 455–468.

Hall, D. J., W. E. Cooper & E. E. Werner, 1970. An experimental approach to the production dynamics of freshwater animal communities. Limnol. Oceanogr. 15: 839–928.

Hayne, D. W. & R. C. Ball, 1956. Benthic productivity as influenced by fish predation. Limnol. Oceanogr. 1: 162–175.

Hershey, A. E., 1985a. Effects of predatory sculpin on the chironomid communities in an arctic lake. Ecology 66: 1131–1138.

Hershey, A. E., 1985b. Littoral chironomid communities in an arctic Alaskan lake. Holarct. Ecol. 42: 483–487.

Hershey, A. E., 1990. Snail populations in arctic lakes: competition mediated by predation? Oecologia (Berl.) 82: 26–32.

Hilsenhoff, W. L., 1967. Ecology and population dynamics of *Chironomus plumosis* (Diptera: Chironomidae) in Lake Winnebago, Wisconsin. Ann. ent. Soc. Am. 60: 1183–1194.

Krebs, C. J., 1978. Ecology: The experimental analysis of distribution and abundance, second edition. Harper and Row, New York, NY, USA.

McDonald, M. E., B. E. Cuker & S. C. Mozley, 1982. Distribution, production, and age structure of slimy sculpin in an arctic lake. Envir. Biol. Fishes 7: 171–176.

Merrick, G. W., 1989. Lake trout (*Salvelinus namaycush*) and benthic community ecology in an arctic ecosystem. M.S. thesis. University of Minnesota-Duluth, Duluth, Minnesota, USA.

Miller, M. C., G. R. Hater, P. Spatt, P. Westlake & D. Yeakel, 1986. Primary production and its control in Toolik Lake, Alaska. Arch. Hydrobiol./Suppl. 4: 97–131.

Mittelbach, G. G., 1986. Predator-mediated habitat use: some consequences for species interactions. In C. A. Simmenstad & G. M. Cailliet (ed.), Contemporary studies on fish feeding. Dr. W. Junk, Dordrecht, Netherlands: 159–169.

Montgomery, D. C., 1984. Design and analysis of experiments. John Wiley & Sons, New York, New York, USA.

Oliver, D. R., 1968. Adaptations of arctic Chironomidae. Ann. Zool. Fenn. 5: 111–118.

Paine, R. T., 1966. Food web complexity and species diversity. Am. Nat. 100: 65–75.

Pielou, E. C., 1975. Ecological diversity. John Wiley & Sons. New York, New York, USA.

Power, M. E. & W. J. Matthews, 1983. Algae-grazing minnows (*Campstoma anomallum*), piscivorous bass (*Micropterus* spp.), and the distribution of attached algae in a small prairie-margin stream. Oecologia 60: 328–332.

Power, M. E., W. J. Matthews & A. J. Stewert, 1985. Grazing minnows, piscivorous bass, and stream algae: Dynamics of a strong interaction. Ecology 66: 1448–1456.

SAS Institute, 1985. SAS users guide: statistics, version 5 edition. SAS Institute, Cary, North Carolina, USA.

Schmidt, D. & W. J. O'Brien, 1982. Planktivorous feeding ecology of arctic grayling (Thymallus arcticus). Can. J. Fish. aquat. Sci. 39: 475–482.

Seaburg, K. G. & J. B. Moyle, 1964. Feeding habits, digestive rates, and growth of Minnesota warmwater fishes. Trans. am. Fish. Soc. 93: 269–285.

Sih, A., 1982. Foraging strategies and the avoidance of predation by an aquatic insect, *Notenecta hoffmanni*. Ecology 63: 786–796.

Smock, L. A., 1980. Relationship between body size and biomass of aquatic insects. Freshwat. Biol. 10: 375–383.

Stein, R. A. & J. J. Magnuson, 1976. Behavioral response of crayfish to a fish predator. Ecology 57: 751–761.

Thorp, J. H. & E. A. Bergey, 1981a. Field experiments on responses of a freshwater, benthic macroinvertebrate community to vertebrate predators. Ecology 62: 365–375.

Thorp, J. H. & E. A. Bergey, 1981b. Field experiments on the interactions between vertebrate predators and larval midges (Diptera: Chironomidae) in the littoral zone of a reservoir. Oecologia (Berl.) 50: 285–290.

Walker, I. R. & R. W. Mathewes, 1988. Late-quaternary fossil Chironomidae (Diptera) from Hippa Lake, Queen Charlotte Islands, British Columbia, with special reference to *Corynocera* Zett. Can. Ent. 120: 739–451.

Welch, H. E., 1976. Ecology of Chironomidae (Diptera) in a polar lake. J. Fish. Res. Bd Can. 33: 227–247.

Werner, E. E. & D. J. Hall, 1976. Niche shifts in sunfishes: experimental evidence and significance. Science (Washington, D.C.) 191: 404–406.

Werner, E. E., J. F. Gilliam, D. J. Hall & G. G. Mittelbach, 1983. An experimental test of the effects of predation risk on habitat use in fish. Ecology 64: 1540–1548.

Wiederholm, T. (ed.), 1983. Chironomidae of the holarctic region. Ent. Scand. Suppl. 19: 1–457.

Hydrobiologia **240**: 213–218, 1992.
W.J. O'Brien (ed.), Toolik Lake – Ecology of an Aquatic Ecosystem in Arctic Alaska.

Cost of predation avoidance in young-of-year lake trout (*Salvelinus namaycush*): growth differential in sub-optimal habitats

Michael E. McDonald,[1] Anne E. Hershey[2] & W. John O'Brien[3]
[1] *Dept. of Chemical Engineering;* [2] *Dept. of Biology, University of Minnesota-Duluth, Duluth, MN 55812, USA;* [3] *Dept. of Systematics and Ecology, University of Kansas, Lawrence, KA 66045, USA*

Key words: lake trout, predation risk, zooplankton, diet, bioenergetics, habitat, lake

Abstract

Selection of habitat to avoid predation may affect the diet of young-of-year (YOY) lake trout (*Salvelinus namaycush*). YOY lake trout may use inshore habitat to avoid predation; this habitat may be sub-optimal for growth. To test this, YOY lake trout were penned in nearshore and offshore pelagic areas of two arctic lakes. Toolik Lake had a lake trout population, the other lake, S6, did not. YOY lake trout in Toolik Lake lost weight, but those offshore lost less weight. The YOY lake trout in Lake S6 gained weight and those offshore gained more weight.

The primary diet item of the YOY lake trout in both lakes during this experiment was the zooplankter *Diaptomis probilofensis*; it was also one of the most abundant species. However, its density inshore in Lake S6 was similar to inshore and offshore densities in Toolik Lake. The increased availability of alternative zooplankton prey in Lake S6 may account for the growth differential of YOY lake trout in Lake S6 relative to Toolik Lake.

Bioenergetic modeling of YOY lake trout suggests that growth similar to that in the offshore of Lake S6 would be necessary for successful recruitment. If the reduced zooplankton availability in Toolik Lake leads to the reduced growth of YOY in the inshore and offshore pelagic areas, then these fish will be more susceptable to winter predation/starvation. For YOY lake trout to survive in Toolik Lake they most likely shift to feeding on benthic prey before the end of their first summer.

Introduction

Piscivores can directly influence the early life history of fishes through size-selective predation. The risk of predation may affect selection of habitat by small fishes (Mittlelbach & Chesson, 1987), thus indirectly affecting the small fishes' diet. If growth potential and mortality differ among habitats, then fishes forced into sub-optimal habitats may be trading off short term survival for a longer period of vulnerability to predation due to slower growth. Thus, the mechanisms influencing survival and recruitment of fishes are likely dependent upon the size of the individuals within the population.

Unexploited lake trout (*Salvelinus namaycush*) communities in oligotrophic arctic lakes have been likened to climax forests (Johnson, 1972, 1976). Johnson (1972) observed a bimodal distribution for unexploited lake trout populations in the Canadian arctic and suggested that this distribution resulted from the larger lake trout forcing smaller lake trout into sub-optimal habitat to avoid predation. These smaller lake trout were found in the nearshore littoral areas (Johnson,

1972, 1976). Presumably, their growth was retarded in the nearshore habitat relative to that in the offshore pelagic areas of the lake. Risk of predation by lake trout also appears to significantly affect the distribution (Hanson *et al.*, 1992; McDonald & Hershey, 1992) and growth rate (McDonald & Hershey, 1992) of slimy sculpin (*Cottus cognatus*) in arctic lakes.

Young-of-year (YOY) lake trout are almost exclusively invertebrate feeders, typically consuming zooplankton and aquatic insects (Martin & Olver, 1980). As lake trout get larger they undergo ontogenetic diet changes to larger invertebrates and fishes (Martin & Olver, 1980), but it is likely that the availability of zooplankton to YOY lake trout may be pivotal to their initial growth and recruitment success.

In Alaskan arctic lakes we have also observed the pattern of small lake trout being present primarily in the nearshore areas (McDonald & Hershey, personal observation), although sampling of offshore areas is logistically more difficult (O'Brien *et al.*, 1979; Merrick *et al.*, 1992). But, these lakes are relatively depauperate in both species and abundance of zooplankton (O'Brien *et al.*, 1979), and lake trout are not very effective zooplankton predators (Kettle & O'Brien, 1978). If food in the nearshore is reduced or has a higher acquisition cost, then YOY lake trout, by remaining in the nearshore to avoid predation, may pay a substantial penalty in terms of lost growth.

To assess the magnitude of this lost growth in the nearshore areas, we examined the late season growth of YOY lake trout in nearshore and offshore areas in a lake containing lake trout and in a lake without lake trout. Growth changes were modeled to examine the implications of habitat use for the survival of individual YOY lake trout.

Methods

Two lakes in the Alaskan Arctic Longterm Ecological Research Site were used for this study. Toolik Lake, a 140 ha lake (See McDonald *et al.* 1982 for more complete description), contains lake trout, grayling (*Thymallus arcticus*), round whitefish (*Prosopium cylindraceum*), slimy sculpin, and burbot (*Lota lota*) (McDonald & Hershey, 1989, 1992). Lake S6, a 1 ha lake, at the time contained only grayling and slimy sculpin, although lake trout have been recently introduced (see Merrick *et al.*, 1991). In both of these lakes, an inshore and an offshore YOY lake trout holding pen (1 m in diameter and 1.3 m in depth, volume about 4 m^3) was anchored. The top of the pen floated at the lake surface and the pens were held open by a semi-rigid frame. The pens were constructed of open weave 4.8 mm delta mesh nylon netting to retain the YOY lake trout, but to allow free movement of zooplankton.

YOY lake trout (about 0.4 g) were collected by seining in a nearby lake trout lake, Lake I8 (an inlet lake to Toolik Lake). Lake trout were held for a week and fed zooplankton, prior to the start of the experiment. Lake trout were individually marked with a unique dorsal or anal fin clip, weighed, and placed in the mesh holding pens on 13 August 1986. Due to limited availability of YOY lake trout, only three fish were put in each of the inshore pens and four fish were put in each of the offshore pens. After 15 days the pens were pulled out of the water and the fish recovered and preserved in 95% ethanol. In the laboratory, lake trout within each treatment were individually identified, and weighed. Weight change during the course of the experiment was determined. Effects of habitat, lake, and habitat × lake interactions on YOY lake trout growth were tested using a two-way ANOVA.

At the start of the experiment, surface water temperatures in both lakes were taken and three replicate zooplankton tows were made next to each holding pen. A tow consisted of 0.3 m diameter zooplankton net (159 μ mesh) pulled from a depth of 2 m to the surface. Inshore and offshore zooplankton samples from both lakes were preserved in 95% ethanol. In the laboratory zooplankton were identified and enumerated. The densities of the zooplankton species collected were determined. Effects of habitat, lake, and habitat × lake interaction on zooplankton density was determined for each zooplankton species by two-way ANOVA. All count data were transformed by adding 0.5 and taking the square root (Snedechor & Cochran, 1967).

Stomach contents from each YOY lake trout in each pen in each lake were also examined in the laboratory. Identifiable diet items were enumerated. YOY lake trout diet differences relative to habitat, lake, and habitat × lake interactions were examined by two-way ANOVA. For all tests, when $p < 0.05$ differences were considered significant.

A bioenergetics model for Toolik Lake lake trout has been developed (McDonald, unpubl. data) based on the generic model of Hewitt & Johnson (1987). The generic model has been used for a number of species, including lake trout (see Hewitt & Johnson, 1987), and has been subjected to sensitivity analysis (Bartell *et al.*, 1986). Seasonal temperature data for Toolik Lake (M. Miller, Dept. of Biology, University of Cincinati, OH, unpubl. data) was used in the model, and the hatch out date for YOY lake trout was estimated to be 15 June (approximate ice-out). Using back-calculated weight at age data for Toolik Lake lake trout (McDonald, unpubl. data), the yearly growth pattern of YOY lake trout, which have successfully recruited to older age groups, was determined from the bioenergetics model. Using the observed experimental weight changes (see RESULTS), the model was used to estimate the yearly growth for the experimental YOY lake trout inshore and offshore in Toolik Lake and in Lake S6 to assess the likelihood of their surviving to the next spring. All YOY lake trout were assumed to be eating *Diaptomis* sp. (see RESULTS) with a caloric content of 550 cal g^{-1} wet weight (Cummins & Wuycheck, 1971).

Results

Surface temperatures inshore and offshore in both Toolik Lake and Lake S6 were 13 °C at the beginning of the experiment. Mean total zooplankton densities were >4.7 fold higher in Lake S6 than in Toolik Lake (Fig. 1). The dominant species were *Bosmina longirostris*, *Diaptomis probilofensis*, and *Cyclops scutifer*. *Daphnia*

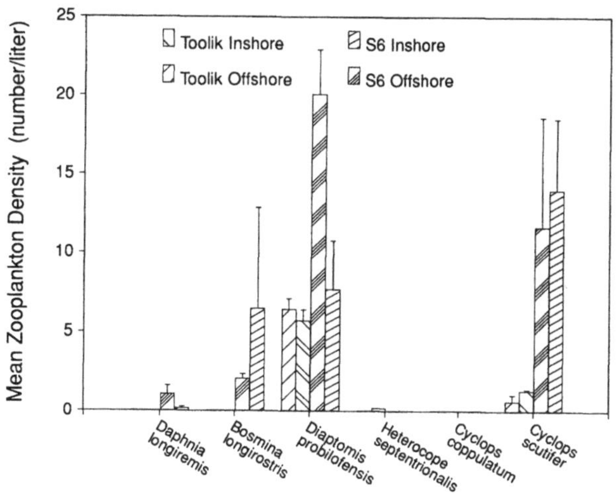

Fig. 1. Zooplankton densities (mean ±1 SE) inshore and offshore in Toolik Lake and in Lake S6.

longiremis, *B. longirostris*, *D. probilofensis*, and *C. scutifer* were all significantly more abundant in Lake S6 than in Toolik Lake. *D. probilofensis* was also significantly more abundant in the offshore habitat; the interaction of lake and habitat also had a significant affect on *D. probilofensis* density. *Heterocope septentrionalis* was significantly more abundant in Toolik Lake than in Lake S6. The lake × habitat interaction for *H. septentrionalis* was also significant with higher densities being found inshore in Toolik Lake.

There were significant differences in YOY lake trout growth between lakes; in Toolik Lake they lost weight in both the inshore and offshore pens (Fig. 2). This was significantly different from the weight gained by lake trout in Lake S6. The weight change of the inshore lake trout in both lakes was not significantly different from the weight change of the offshore lake trout. But, the offshore environment did provide the highest weight gains in Lake S6 and the lowest weight loss in Toolik Lake. The effects of habitat × lake interactions on growth was also found to be nonsignificant.

D. probilofensis was the main diet item of the YOY lake trout in both lakes (Fig. 3). YOY lake trout in Lake S6 had a higher number of prey items in their stomachs; *D. longiremis* was the second most abundant diet item. There were significant differences between lakes in the number

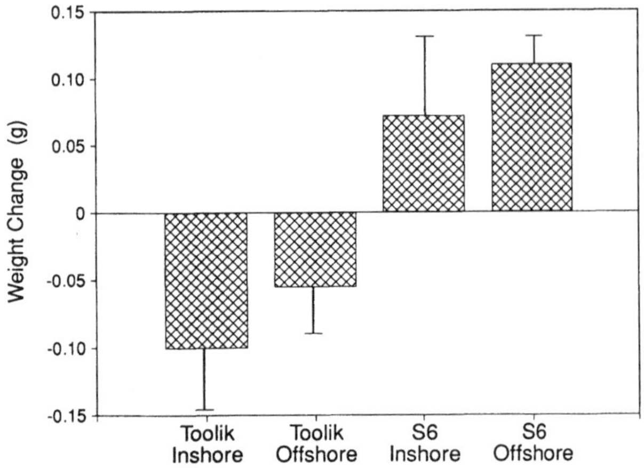

Fig. 2. Mean weight change (mean ± 1 SE) in YOY lake trout during the experimental period inshore and offshore in Toolik Lake and Lake S6.

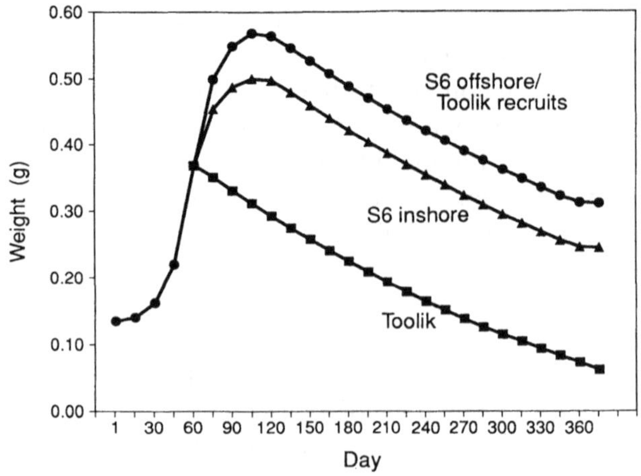

Fig. 4. Bioenergetics growth estimates for penned YOY lake trout in Toolik Lake and Lake S6. Growth of YOY lake trout, which do recruit in Toolik Lake, was also estimated and tracks the S6 offshore growth.

of *D. probilofensis* and *D. longiremis* included in the lake trout's diet. For the inshore lake trout in Lake S6, *B. longirostris* was also an abundant diet item. Two of the YOY lake trout from offshore Lake S6 also had 2 and 3 chironomids, respectively, in their stomachs in addition to zooplankton.

The growth of Lake S6 offshore lake trout coincided with the bioenergetic estimate for YOY lake trout from Toolik Lake which survived to older ages (Fig. 4). In the inshore area of Lake S6, lake trout grew about 12% slower, than in the offshore area of Lake S6. Toolik lake trout, because they were losing weight in the pens, were assumed to have finished their growth for the season and their weight loss was assumed to be at the winter rate (Fig. 4).

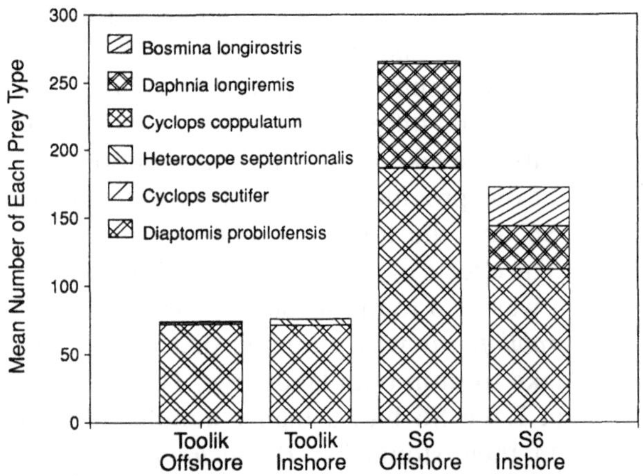

Fig. 3. Mean diet composition for YOY lake trout in the inshore and offshore habitats in Toolik Lake and Lake S6.

Discussion

D. probilofensis was the most common zooplankter collected in either lake (Fig. 1). The other two abundant zooplankton, *B. longirostris* and *C. scutifer* are approximately 70% of the size of *D. pribilofensis* (Kettle & O'Brien, 1978; O'Brien *et al.*, 1979). Small lake trout have been shown to feed on *D. pribilofensis* in experimental systems; because of their pigmentation, they may be >3 fold more visible to lake trout than a similarly sized cladoceran (Kettle & O'Brien, 1978). However, *D. pribilofensis* can be highly evasive of capture (39.3% chance of evasion), relative to most cladocerans, which are unable to actively evade fish predation (Kettle & O'Brien, 1978).

O'Brien *et al.* (1979) have previously suggested that because of the sparse densities of zooplankton in the lakes in the Toolik area, zooplanktivorous fishes would have to search intensively for food. At the beginning of the experiment, *D. pribilofensis* densities inshore in Toolik Lake were 5.7 l^{-1} and offshore were 6.4 l^{-1}; in Lake S6, *D. pribilofensis* densities inshore were 7.7 l^{-1} and offshore were 20.1 l^{-1}. The difference in the abundance of *D. pribilofensis* offshore in S6 might account for the observed increased growth of the YOY lake trout in that area (Fig. 2), but *D. longiremis* and chironomids may also have been important diet items. Chironomids, while small in number in YOY lake trout diets, are much larger than zooplankton, and thus may have been important in the growth of the fish in the offshore of Lake S6. Densities of *D. probilofensis* in Toolik Lake were not dissimilar from those inshore in S6, yet the YOY lake trout in Toolik all lost weight, while those inshore in S6 gained weight. Since the holding pens were open to zooplankton, a localized depletion within a pen relative to the area outside the pens was not likely. The total density of zooplankton in Toolik Lake was only about 22% of that in inshore Lake S6; inshore Lake S6's total zooplankton density was about 79% of that in offshore Lake S6 (Fig. 1). Thus, the difference in growth of YOY lake trout in the different lakes and habitats may have been due to differences in the availability of zooplankton (Figs 1, 3).

Use of inshore habitat by YOY lake trout appears sub-optimal (Fig. 2), but the growth differential (about 0.002–0.003 g d^{-1}) between inshore and offshore habitats was not significant (but, the power of the test was low due to the small sample size). In Toolik Lake, even in the offshore area, the YOY lake trout lost weight, suggesting that at this time of year the offshore pelagic zone also may not provide sufficient food for positive growth. Predation risk for YOY lake trout may increase late in the season as they likely have to search further, and in different habitats, for suitable food in Toolik Lake.

Lake trout, which have successfully recruited in Toolik Lake, have a growth rate estimated to be similar to that achieved by the offshore YOY's in Lake S6 (Fig. 4). Costs of reduced growth may be substantial in the overwinter survival of YOY lake trout. If the reduced food availability late in the summer in the nearshore and offshore pelagic zone of Toolik Lake leads to reduced growth rates, then those YOY lake trout feeding on zooplankton may be more susceptable to size-dependent sources of mortality (see Crowder *et al.*, 1987). Predation may become especially important to small size lake trout in winter with ice cover (to a depth of 2 m on Toolik Lake) substantially reducing the availability of nearshore refugia. An ontogenetic shift (Werner & Gilliam, 1984) most likely occurs in recruiting YOY lake trout before mid-August. Presumably, they switch from feeding on zooplankton to feeding on benthic invertebrates. Benthic invertebrates in the nearshore rocky areas are greatly reduced relative to the offshore soft sediments (Hanson *et al.*, 1992). Slimy sculpin use the nearshore rock and soft sediment benthos extensively, especially in the presence of large lake trout (see Hanson *et al.*, 1992, McDonald & Hershey, 1992) and may negatively interact with YOY lake trout, restricting access to these areas. Adult lake trout are tightly coupled to the benthos (O'Brien *et al.*, 1979; Merrick *et al.*, 1992); YOY lake trout may also be coupled to the benthos by the end of their first summer of life. The survival of YOY Toolik Lake lake trout may depend upon finding a predation refuge with a suitable benthic food supply.

Conclusions

The nearshore habitat in arctic lakes may be sub-optimal for YOY lake trout growth. Differences between lakes and habitats appear to depend upon the density of a suitable zooplankton prey. Bioenergetics modeling of YOY lake trout suggests that the limited availability of zooplankton in the nearshore and offshore pelagic zone of Toolik Lake would likely be insufficient to allow successful recruitment to older age classes. For successful recruitment, YOY lake trout in Toolik would likely have to exploit a different habitat; presumably, this is the interface between the nearshore rocky area and the deeper soft sediments. Here, YOY lake trout would have access to larger prey, soft sediment chironomids, and the rocks would provide a refuge from predation.

Acknowledgements

This work was supported by NSF grants No. DPP 8722015 and BSR 870232 to John Hobbie, Marine Biological Laboratory (MBL), Woods Hole, MA, by sub-contracts from MBL to MEM, to AEH, and to WJO'B. We thank B. Loveless and one anonymous reviewer for their comments on the manuscript.

References

Bartell, S. M., J. E. Breck, R. H. Gardner & A. L. Brenkert, 1986. Individual parameter perturbation and error analysis of fish bioenergetics models. Can. J. Fish. aquat. Sci. 43: 160–168.

Crowder, L. B., M. E. McDonald & J. A. Rice, 1987. Understanding recruitment of Lake Michigan fishes: the importance of size-based interactions between fish and zooplankton. Can. J. Fish aquat. Sci. 44 (Suppl. 2): 141–147.

Cummins, K. W. & J. C. Wuycheck, 1971. Caloric equivalents for investigations in ecological energetics. Mitt. int. Ver. Limnol. 18: 158 pp.

Hanson, K. L., A. E. Hershey & M. E. McDonald, 1992. A comparison of slimy sculpin (*Cottus cognatus*) populations in arctic lakes with and without piscivorous predators. Hydrobiologia 240: 189–201.

Hewitt, S. W. & B. L. Johnson, 1987. A generalized bioenergetics model of fish growth for microcomputers. University of Wisconsin Sea Grant Institute, Madison, WIS-SG-87-245. 47 pp.

Johnson, L., 1972. Keller Lake: characteristics of a culturally unstressed salmonid community. J. Fish. Res. Bd Can. 29: 731–740.

Johnson, L., 1976. Ecology of arctic populations of lake trout, *Salvelinus namaycush*, lake whitefish, *Coregonus clupeaformis*, arctic char, *S. alpinus*, and associated species in unexploited lakes of the Canadian Northwest Territories. J. Fish. Res. Bd Can. 33: 2459–2488.

Kettle, D. & W. J. O'Brien, 1978. Vulnerability of arctic zooplankton species to predation by small lake trout (*Salvelinus namaycush*). J. Fish. Res. Bd Can. 35: 1495–1500.

Martin, N. V. & C. H. Olver, 1980. The lake charr, *Salvelinus namaycush*. In E. K. Balon (ed.), Charrs. Dr W. Junk Publishers, The Hague, Netherlands: 205–277.

McDonald, M. E., B. E. Cuker & S. C. Mozley, 1982. Distribution, production, and age structure of slimy sculpin in an arctic lake. Envir. Biol. Fishes 7: 171–176.

McDonald, M. E. & A. E. Hershey, 1989. Size structure of a lake trout (*Salvelinus namaycush*) population in an arctic lake: influence of angling and implications for fish community structure. Can. J. Fish. aquat. Sci. 46: 2153–2156.

McDonald, M. E. & A. E. Hershey, 1992. Shifts in abundance and growth of slimy sculpin in response to changes in the predator population in an arctic Alaskan lake. Hydrobiologia 240: 219–223.

Merrick, G. W., A. E. Hershey & M. E. McDonald, 1991. Lake trout (*Salvelinus namaycush*) control of snail density and size distribution in an arctic lake. Can. J. Fish. aquat. Sci. 48: 498–502.

Merrick, G. W., A. E. Hershey & M. E. McDonald, 1992. Salmonid diet and the size, distribution, and density of benthic invertebrates. Hydrobiologia 240: 225–233.

Mittelbach, G. G. & P. L. Chesson, 1987. Predation risk: indirect effects on fish populations. In W. C. Kerfoot & A. Sih (eds), Predation. University Press of New England, Hanover, NH: 315–332.

O'Brien, W. J., C. Buchanon & J. Haney, 1979. Arctic zooplankton community structure: exceptions to some general rules. Arctic 32: 237–247.

Snedecor, G. W. & W. G. Cochran, 1967. Statistical methods, 6th edn. Iowa State Press, Ames, IA. 593 pp.

Werner, E. E. & J. F. Gilliam, 1984. The ontogenetic niche and species interactions in size-structured populations. Annu. Rev. Ecol. Syst. 15: 393–425.

Hydrobiologia **240**: 219–223, 1992.
W.J. O'Brien (ed.), Toolik Lake – Ecology of an Aquatic Ecosystem in Arctic Alaska.

Shifts in abundance and growth of slimy sculpin in response to changes in the predator population in an arctic Alaskan lake

Michael E. McDonald & Anne E. Hershey
Dept. of Chemical Engineering, University of Minnesota-Duluth, Duluth, MN 55812, USA; Department of Biology, University of Minnesota-Duluth, Duluth, MN 55812, USA

Key words: slimy sculpin, lake trout, predation, benthos, habitat

Abstract

Lake trout (*Salvelinus namaycush*) in Toolik Lake are tightly coupled to the benthos, since they have no pelagic forage fishes. Slimy sculpins (*Cottus cognatus*) are a prey of lake trout and the soft sediment chironomids are an important prey for the sculpin. Our previous work showed that the median size of lake trout in Toolik Lake had decreased significantly between 1977 and 1986, and smaller lake trout are likely to be less effective as sculpin predators. Using our historic data on the slimy sculpin population from 1978, we took advantage of the recent change in the predator community to examine for subsequent changes in the sculpin community. Between 1978 and 1987, the percentage of slimy sculpin caught in the soft sediments has increased (25% to 39.5%). In 1987 there was a significant difference in the mean weight of sculpin caught on different substrates. The mean individual weight of sculpins increased from the nearshore rock area to the rock/soft-sediment interface to the soft sediments. There was no difference in mean individual weight with habitat in 1978. The mean total length at age for slimy sculpins during this time has also increased significantly. We suggest that the risk of predation while foraging in the soft sediments has declined. The increased use of the soft sediment area appears to have contributed to their increased growth, likely due to greater food abundance.

Introduction

Regulation of temperate lake ecosystem processes by predatory fishes is dependent upon piscivores harvesting zooplanktivorous forage fishes (Shapiro, 1980; Shapiro & Wright, 1984; Carpenter & Kitchell, 1984; Crowder *et al.*, 1987). In arctic lakes, which lack pelagic forage fishes, the effects of a top predator may be very different. In these systems there appears to be a strong benthic-pelagic coupling; adult lake trout (*Salvelinus namaycush*) feed extensively on benthic prey (O'Brien *et al.*, 1979; Merrick *et al.*, 1992). While changes in the predator population in temperate lake systems may result in changes in trophic interactions which cascade (sensu Carpenter *et al.*, 1985) through the ecosystem, the consequences of changes in the predator population of arctic lakes may be very different.

Toolik Lake is a 140 ha lake on the north slope of the Brooks Range in arctic Alaska (see McDonald *et al.*, 1982). The lake consists of a rocky nearshore area, rocky reefs, and a deep, soft sediment bottom (McDonald *et al.*, 1982). The fish community of Toolik Lake consists of lake trout, burbot (*Lota lota*), grayling (*Thymalus arcticus*), round whitefish (*Prosopium cylindraceum*), and slimy sculpin (*Cottus cognatus*) (O'Brien *et al.*, 1979; McDonald & Hershey, 1989).

Slimy sculpin in Toolik Lake are long-lived and slow-growing (McDonald *et al.*, 1982). They rely on a diet of larval chironomids (Hershey & McDonald, 1985) and can control bare sediment chironomid density (Hershey, 1985). Both density and production of slimy sculpin in Toolik Lake in 1978 were highest at the interface between the nearshore rocky area and the deeper soft sediments; presumably, this was due to access to food resources on the soft sediments and to predator protection offered by the rocks (McDonald *et al.*, 1982).

Lake trout in Toolik Lake do eat slimy sculpins, but they occur in only about 12% of the lake trout examined (Merrick *et al.*, this volume). Lake trout in Lake Ontario have been suggested as keystone predators on competing sculpin populations (Brandt, 1986). Lake trout in Toolik Lake were surveyed in 1977 (O'Brien *et al.*, 1979), and by 1986 the median size of lake trout had significantly decreased (578 g to 313 g), presumably due to increased sports angling (McDonald & Hershey, 1989). These smaller-sized individuals are less likely to feed on slimy sculpins (Merrick *et al.*, 1992).

With the reduction in lake trout size, we hypothesized that slimy sculpin should increase their use of the deeper soft sediment areas and should show increased growth. Because of historic data on slimy sculpin in Toolik Lake (McDonald *et al.*, 1982), we had the opportunity to examine changes in the slimy sculpin population with the reduction in predation pressure by lake trout. We repeated the 1978 sampling effort (McDonald *et al.*, 1982) during the summer of 1987 to determine if changes in the slimy sculpin distribution and growth had occurred.

Methods

We used the sampling technique and original sampling sites of McDonald *et al.* (1982) for collecting slimy sculpins. This technique consisted of three jar traps tethered together with three replicate sets on each of four substrate types. These substrate types were: near-shore rocky (0.5 m), mid-rock (1.5 m), rock/soft-sediment interface (3.5–5.0 m), and soft-sediments (7.0 m). These sites were trapped continuously from 26 July to 12 August 1987 and traps were sampled every third day.

Slimy sculpin collected from each substrate were preserved in 95% ethanol for transport to the laboratory and short term storage (<4 h). In the laboratory slimy sculpin were weighed, total length measured, and sagital otoliths removed and stored in glycerol for later aging. Microscopic aging of otoliths was done *in toto*; mean total length at each age of slimy sculpin caught in 1987 was back-calculated from otolith length (see McDonald *et al.*, 1982).

The expected catch of slimy sculpin for each substrate type in 1987 was based on the percent catch on each substrate in 1978 and on the total number in the 1987 catch. To test the hypothesis that there was no change in slimy sculpin distribution between 1978 and 1987, the squared differences between the observed and expected catch of slimy sculpin on each substrate type in 1987 was summed to yield a single chi-square value for all habitats (Snedecor & Cochran, 1967). Changes in slimy sculpin density between just the rock/soft-sediment interface and the soft sediments from 1978 to 1987 was similarly tested by chi-square (Snedecor & Cochran, 1967). Mean biomass of slimy sculpin found on each substrate type in 1978 (from McDonald *et al.*, 1982) and in 1987 was examined separately using ANOVA. Mean total length at age for slimy sculpin caught in 1987 was compared to that for 1978 (from McDonald *et al.*, 1982) by Wilcoxon signed rank test. Only age groups represented in both 1978 and 1987 were included in the analysis (age groups I–VII, see RESULTS). Differences for all tests were considered significant if $p < 0.05$.

Results

In both 1978 and 1987 >75% of all slimy sculpin caught were on, or near, the soft sediments (rock/

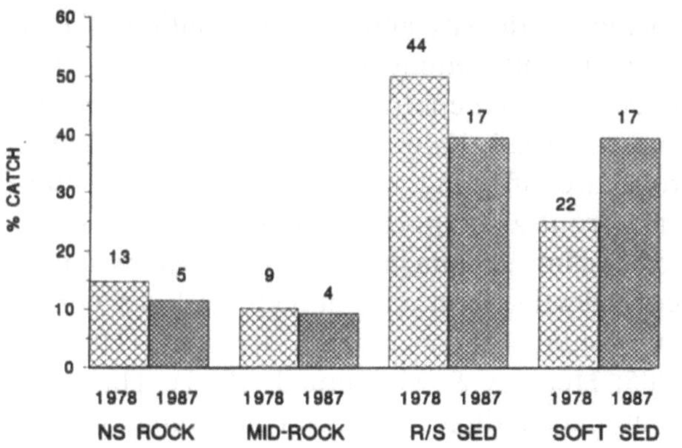

Fig. 1. Per cent of total catch on each substrate type (0.5 m nearshore rocks – NS ROCK, 1.5 m rocky area – MID-ROCK, 3 m to 5 m rock/soft sediment interface – R/S SED, and 7 m soft sediment – SOFT SED) in 1978 (from McDonald *et al.*, 1982) and in 1987. Total number of fish caught in each year on each substrate is given above histograms. Catch at the rock/soft sediment interface and on the soft sediments in 1987 was significantly different than that predicted from the 1978 catch.

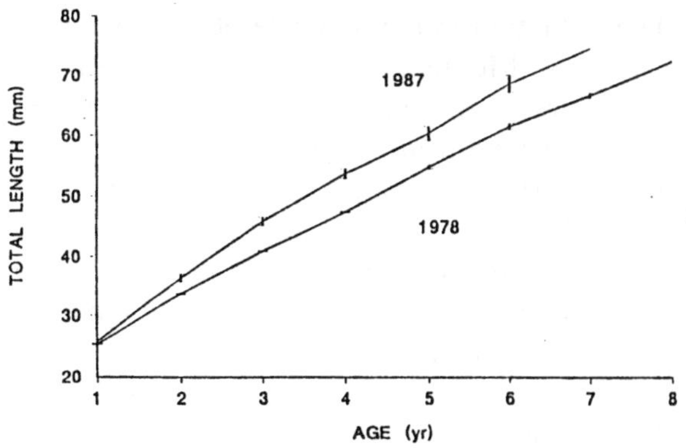

Fig. 3. Mean back-calculated total length (± 1 SE) at each age class for slimy sculpin in 1978 (from McDonald *et al.*, 1982) and 1987. In pair-wise comparison across all age groups, slimy sculpin caught in 1987 were significantly larger.

soft-sediment interface and soft sediments, Fig. 1). There was no significant difference in the catch of slimy sculpin in the different substrate types in 1987 relative to 1978 ($X^2 = 4.99$, $df = 3$). However, there was a significant change in slimy sculpin density between the soft sediments and

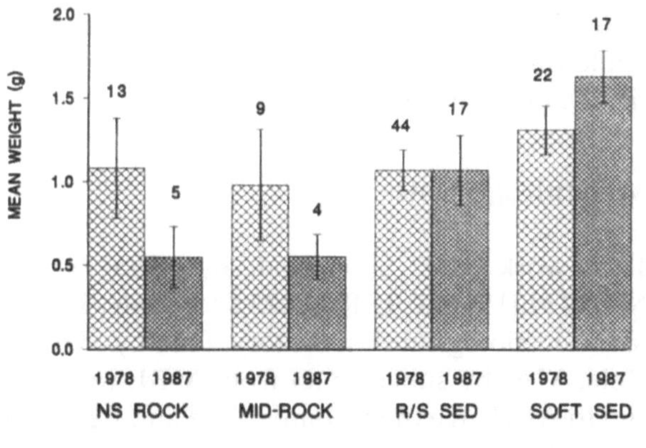

Fig. 2. Mean weight (± 1 SE) of slimy sculpin caught on each substrate (0.5 m nearshore rocks – NS ROCK, 1.5 m rocky area – MID-ROCK, 3 m to 5 m rock/soft sediment interface – R/S SED, and 7 m soft sediment – SOFT SED) in 1978 (from McDonald *et al.*, 1982) and 1987. Total number of fish caught on each substrate in each year is given above histograms. Mean slimy sculpin weight was significantly different with habitat in 1987.

the rock/soft-sediment interface between 1978 and 1987 ($X^2 = 4.65$, $df = 1$).

There were no significant differences in the mean weight of slimy sculpin on the different substrate types in 1978 (Fig. 2). However, in 1987 there was a significant difference in mean slimy sculpin biomass due to habitat ($F = 4.55$). In 1987 the slimy sculpin increased in size with depth from the nearshore rocks to the deeper, soft sediments (Fig. 2).

Slimy sculpins caught in 1987 were significantly larger at age than those caught in 1978 (Fig. 3). Increases in mean total length at age for 1987 slimy sculpin occurred in age groups I–VII and ranged from 1.8% (I) to 13.3% (IV). No sculpin in age group VIII was collected in 1987.

Discussion

The size of the slimy sculpin population in Toolik Lake was estimated and its life history characteristics examined in 1978 (McDonald *et al.*, 1982). Since that time, the slimy sculpin's interaction with its benthic food source has been examined in the nearshore rocks (Cuker, 1983; Hanson *et al.*, 1992) and in the soft sediments, where they control chironomid density (Hershey, 1985; Goyke & Hershey, 1992).

The reduction in median size of the lake trout since 1978 (McDonald & Hershey, 1989) has probably reduced predation on slimy sculpin, since larger lake trout are more likely to exhibit piscivory (Merrick *et al.*, 1992). The significant shift in the slimy sculpin population from the rock/soft-sediment interface out onto the soft sediments between 1978 to 1987 (Fig. 1) may be the result of reduced predation risk while foraging.

McDonald *et al.* (1982) hypothesized that sculpin were concentrated at the rock/soft-sediment interface in order to forage out on the soft sediments, while taking advantage of the predation refuge provided by the rocks. The observed changes in the slimy sculpin distribution and growth, coupled with the observed shift in the lake trout population to a smaller median size (McDonald & Hershey, 1989), provides strong support for the earlier hypothesis.

Toolik Lake slimy sculpin live longer and grow slower than those in north temperate, boreal, and more southerly arctic areas (McDonald *et al.*, 1982). Van Vliet (1964) suggested that water temperature and food supply may limit the growth of more northerly slimy sculpin populations. Hershey (1985) found that 84% of the diet of slimy sculpins in the Toolik Lake area was chironomids and that sculpin could reduce chironomid populations in the bare sediments. Goyke & Hershey (1992) found chironomid density, biomass, and diversity to be higher in lakes containing lake trout and slimy sculpin, than in lakes containing only slimy sculpin. With increased access to the soft sediment chironomid community, the growth rate of slimy sculpins in 1987 was significantly higher than 1978 (Fig. 3). This suggests that while temperature may limit the overall physiological capacity to grow, the Toolik Lake slimy sculpin in 1978 were likely food limited as a result of restricted foraging on the soft sediments due to risk of predation.

Unlike temperate systems, Toolik Lake lacks a pelagic forage fish, and the top predator, lake trout, feeds primarily on snails and, to a much lesser extent, slimy sculpin (Merrick *et al.*, 1992). Lake trout can influence snail populations, controlling the density and size distribution of *Lymnaea* on soft sediments (Merrick *et al.*, 1991). Since *Lymnaea* controls recruitment of the other common snail, *Valvata* (Hershey, 1990), the lake trout are affecting snail community structure. Lake trout also appear to influence the distribution of slimy sculpin in Toolik Lake. Slimy sculpin feed selectively on chironomids and can control bare sediment chironomid densities (Hershey, 1985; Hershey & McDonald, 1985). Thus, lake trout may indirectly affect chironomid populations. It appears that in these systems with strong benthic-pelagic coupling, changes in the predator population have consequences for the structure of the entire benthic community.

Conclusion

The soft sediment habitat offers an increased food supply to slimy sculpin, but access has likely been limited by the risk of lake trout predation. A significant reduction in the median size of lake trout in Toolik Lake by sport fishing harvest likely reduced predation pressure on the sculpin. In response to this reduction, slimy sculpin appear to have increased their use of the soft sediments and their growth has increased. Thus, in arctic lakes with strong benthic-pelagic coupling, even moderate changes in a top predator population may ripple through the benthic community, and may significantly alter its structure.

Acknowledgements

This work was supported by NSF grants No. 8722015 and BSR 870232 to John Hobbie, Marine Biological Laboratory (MBL), Woods Hole, MA, by subcontract from MBL to MEM and to AEH. We thank James Rice and Benjamin Cuker for review comments on the manuscript.

References

Brandt, S. B., 1986. Disappearance of the deepwater sculpin (*Myoxocephalus thompsoni*) from Lake Ontario: the keystone predator hypothesis. J. Great Lakes Res. 12: 18–24.

Carpenter, S. R. & J. F. Kitchell, 1984. Plankton community structure and limnetic primary production. Am. Nat. 124: 159–172.

Carpenter, S. R., J. F. Kitchell & J. R. Hodgson, 1985. Cascading trophic interactions and lake productivity. Bioscience 35: 634–639.

Crowder, L. B., M. E. McDonald & J. A. Rice, 1987. Understanding recruitment of Lake Michigan fishes: the importance of size-based interactions, ontogenetic bottlenecks and zooplankton composition. Can. J. Fish. aquat. Sci. 44: 141–147.

Cuker, B., 1983. Competition and coexistence among the grazing snail *Lymnea*, Chironimidae, and microcrustacea in an arctic epilithic lacustrine community. Ecology 64: 10–15.

Goyke, A. P. & A. E. Hershey, 1992. Effects of fish predation on larval chironomid (*Diptera: Chironomidae*) communities in an arctic ecosystem. Hydrobiologia 240: 203–211.

Hanson, K. L., A. E. Hershey & M. E. McDonald, 1992. A comparison of slimy sculpin (*Cottus cognatus*) populations in arctic lakes with and without piscivorous predators. Hydrobiologia 240: 189–201.

Hershey, A. E., 1985. Effects of predatory sculpin on the chironomid communities in an arctic lake. Ecology 66: 1131–1138.

Hershey, A. E., 1990. Snail populations in arctic lakes: interactions of predation and competition. Oecologia 82: 26–32.

Hershey, A. E. & M. E. McDonald, 1985. Diet and digestion rates of the slimy sculpin, *Cottus cognatus*, in an arctic lake. Can. J. Fish. aquat. Sci. 42: 483–487.

McDonald, M. E. & A. E. Hershey, 1989. Size structure of a lake trout (*Salvelinus namaycush*) population in an arctic lake: influence of angling and implications for fish community structure. Can. J. Fish. aquat. Sci. 46: 2153–2156.

McDonald, M. E., B. E. Cuker & S. C. Mozley, 1982. Distribution, production, and age structure of slimy sculpin in an arctic lake. Envir. Biol. Fish. 7: 171–176.

Merrick, G. W., A. E. Hershey & M. E. McDonald, 1991. Lake trout (*Salvelinus namaycush*) control of snail density and size distribution in an arctic lake. Can. J. Fish. aquat. Sci. 48: 498–502.

Merrick, G. W., A. E. Hershey & M. E. McDonald, 1992. Salmonid diet and the size, distribution, and density of benthic invertebrates in an arctic lake. Hydrobiologia 240: 225–233

O'Brien, W. J., C. Buchanan & J. Haney, 1979. Arctic zooplankton community structure: exceptions to some general rules. Arctic 32: 237–247.

Shapiro, J., 1980. The importance of trophic level interactions to the abundance and species composition of algae in lakes. In L. Mur & J. Barica (eds), Hypereutrophic Ecosystems. Dr W. Junk Publishers, The Hague: 105–115.

Shapiro, J. & D. I. Wright, 1984. Lake restoration by biomanipulation. Freshwat. Biol. 14: 371–383.

Snedecor, G. W. & W. G. Cochran, 1967. Statistical methods, 6th ed. Iowa State Press, Ames, IA. 593 pp.

Van Vliet, W. H., 1964. An ecological study of *Cottus cognatus* Richardson in northern Saskatchewan. M.S. thesis, University of Saskatchewan, Saskatoon, Saskatchewan. 155 pp.

Hydrobiologia **240**: 225–233, 1992.
W.J. O'Brien (ed.), Toolik Lake – Ecology of an Aquatic Ecosystem in Arctic Alaska.

Salmonid diet and the size, distribution, and density of benthic invertebrates in an arctic lake

Glenn W. Merrick,[1] Anne E. Hershey[1,*] & Michael E. McDonald[2]
[1] *Department of Biology;* [2] *Department of Chemical Engineering; University of Minnesota/Duluth, Duluth, MN 55812, USA (* author for correspondence)*

Key words: arctic, benthic, lake trout, predation, snails

Abstract

Size selective predation on molluscs was apparent for lake trout (*Salvelinus namaycush*) and round whitefish (*Prosopium cylindraceum*), but not for arctic grayling (*Thymallus arcticus*), in the Toolik Lake region of arctic Alaska during the summer of 1986. Lake trout consumed significantly larger molluscs of all taxa than did round whitefish, and selected larger molluscs than were available on either rocky or soft-sediment habitats. Round whitefish were not size-selective on the snail *Lymnaea*, but were size-selective on the snail *Valvata* and on clams from the soft sediments. Round whitefish consumed fewer and smaller *Lymnaea* compared to lake trout. Because lake trout ate more *Lymnaea* and also tended to select larger, reproductive-sized individuals, this fish could potentially have a more detrimental impact on the *Lymnaea* population. Finally, differences in *Lymnaea* densities and size distributions between lakes with and without lake trout suggest that these fish may be responsible for the pattern of distribution, size, and density observed for *Lymnaea* in Toolik Lake and other area lakes.

Introduction

The effects of predation on the community structure of freshwater benthic epifaunal invertebrates is not well established. Existing evidence suggests that predation by fish may alter the relative abundance and species composition of some benthic communities (Crowder & Cooper, 1982; Gilinsky, 1984; Hershey, 1985), but not others (Thorp & Bergey, 1981a, b). Fish predation can influence density or standing crops of benthic macroinvertebrates (Crowder & Cooper, 1982; Hershey, 1985). Macan (1966) found that while trout predation led to decreased densities of a few benthic invertebrates, most species were unaffected.

It has been hypothesized that fish may influence benthic community composition by positively selecting larger invertebrates (Hall *et al.*, 1970; Crowder & Cooper, 1982). Werner & Hall (1977) demonstrated that size-selective predation is correlated with fitness in sunfish. In a growth efficiency model for piscivorous lake trout, Kerr (1971) suggested that size composition of the prey resource is a primary determinant of fish growth. In response to Kerr (1971), Konkle & Sprules (1986) asserted that lake trout must have access to increasingly larger food items, even when rare, to achieve large body size, and, if true, one may expect fish to select larger prey items. Selective predation by fish on benthic invertebrates also may be influenced by factors such as activity, exposure, and density of prey (Ware, 1973; Crowder & Cooper, 1982; Werner *et al.*, 1983; Gilinsky, 1984; Hershey, 1985). Shell morphology is reported to influence the vulnerability of snails to fish that crush shells (Vermeij & Covich, 1978; Stein *et al.*, 1984). Trout, however, ingest their prey whole, and large size is an important determinant in prey selection (Kerr, 1971; Ware, 1982, 1973; Konkle & Sprules, 1986).

Lake trout (lake charr, *Salvelinus namaycush*) are most commonly classified as piscivores but have a more diverse diet (e.g.; Martin, 1966; Johnson, 1972; Smith, 1972; Eddy & Underhill, 1974; O'Brien *et al.*, 1979; Konkle & Sprules, 1986). Prey availability appears to determine diet, and lake trout exploit almost any food that is found in abundance (Scott & Crossman, 1973). In arctic lakes, lake trout are typically viewed as a top predator and are reported to feed predominantly on fish (Johnson, 1976), including lake whitefish (*Coregonus clupeaformis*), lake cisco (*Coregonus artedii*), and slimy sculpin (*Cottus cognatus*).

Populations of lake trout in the Toolik Lake region of arctic Alaska show strong demersal foraging habits. Lakes in the Toolik area differ from many other lakes with lake trout populations since they lack a pelagic forage fish (Bendock & Burr, 1981). As juveniles, lake trout in Toolik Lake consume zooplankton and chironomids (Kettle & O'Brien, 1978; O'Brien *et al.*, 1979), but benthic prey, especially snails, dominate the adult lake trout diet (O'Brien *et al.*, 1979).

In the Toolik Lake region, populations of the pulmonate snail *Lymnaea elodes* are considerably less dense in lakes with lake trout in comparison to *Lymnaea* densities in lakes without these fish (Hershey, 1990). Round whitefish (*Prosopium cylindraceum*) and arctic grayling (*Thymallus arcticus*) commonly co-occur with lake trout in arctic lakes. Round whitefish forage almost exclusively on benthic invertebrates (Scott & Crossman, 1973) and prey on *Lymnaea*, as well as other common benthic invertebrates in Toolik Lake (O'Brien *et al.*, 1979). Grayling, although primarily planktivorous, also eat small benthic invertebrates in Toolik Lake (O'Brien *et al.*, 1979).

To evaluate the potential for lake trout to structure the benthic community of a lake which lacks a pelagic forage fish, we investigated the diet selectivity of lake trout in Toolik Lake. We compared lake trout diets to those of arctic grayling and round whitefish, which typically are not piscivorous. We examined size selectivity of these species by comparing the mean sizes of the dominant invertebrates in their diet with those available in Toolik Lake. The mean size of these invertebrates was also determined in five other area lakes; three contained no lake trout and two had lake trout present. In the two other lakes with lake trout, we also examined aspects of lake trout diet for which we had comparable data.

Study site

Toolik Lake is a small complex glacial kettle lake (148 ha) lying in an irregular basin in the northern foothills of the Brooks Range at 720 m elevation approximately 216 km south of Prudhoe Bay in arctic Alaska (Miller *et al.*, 1986). Toolik Lake has a maximum depth of 25 m, and a mean depth of approximately 7 m (McDonald *et al.*, 1982). It is ice-free from late June until late September or early October. Rocky shoals, which divide the lake into five basins and cover approximately 25% of the lake bottom, are scoured by ice each year. The remainder of the lake bottom is composed of soft sediments and is sparsely covered by diminutive macrophytes in some areas (Hershey, 1985).

The fish species present in Toolik Lake include lake trout (*Salvelinus namaycush*), round whitefish (*Prosopium cylindraceum*), arctic grayling (*Thymallus arcticus*), slimy sculpin (*Cottus cognatus*), and burbot (*Lota lota*) (O'Brien *et al.*, 1979; Bendock & Burr, 1981; McDonald *et al.*, 1982; Hershey & McDonald, 1985; Hershey, 1985). Other lakes in the Toolik area support some combination of these five species, but smaller lakes usually have fewer than five (Bendock & Burr, 1981). Our study sites also include nearby lakes S6, S7, and N2, which have grayling and sculpin, and Lake I8 and Itigaknit Lake, which have lake trout. Lake I8 supports all five species of fish; Itigaknit certainly has sculpin and grayling in addition to lake trout, but may also contain burbot and round whitefish.

Methods

A plastic sampling frame ($2\ m^2$) was used to assess relative densities and size distributions of *Lymnaea* in three lakes with and three lakes without lake trout. Sampling was conducted over unvegetated soft sediments in less than 1 m of water. Lakes S6 and S7 were sampled on 1 July 1986, lakes N1 and N2 on 9 July 1986, and lakes I8 and Toolik on 1 August 1986. All *Lymnaea* were collected, counted, measured with vernier calipers to the nearest 0.1 mm, and released in the area of capture. Mean density and size distributions for *Lymnaea* in lakes with trout and without trout were compared using a t-test for independent means. For this and all other statistical comparisons, means are reported ± 1 standard error, but variances were examined for homogeneity and transformed if necessary [using $\log_e (x + 1)$]. All statistical comparisons were regarded as significant if $p \leq 0.05$.

Salmonids for diet analysis were captured with a 30 m experimental gillnet and by angling. From June 13 until June 26, the gillnet was operated near the main inlet to Toolik Lake over soft sediments. After June 26, the gillnet was used on a variety of rocky shoals on the perimeter of Toolik's larger basins. A gillnet was set in Itigaknit Lake on 20–22 July 1986 (3 net days), and in Lake I8 on 13–15 August, 1986. Net depth ranged from approximately 2 to 8 m. Lake trout and grayling were also caught by angling in the same locations where gillnets were used.

Following capture, fish were weighed, measured, and sexed. Stomachs were then removed and preserved in ethanol. All prey items were enumerated. All molluscs were measured using a digitizing pad in conjunction with a dissecting microscope and camera lucida. *Lymnaea* were measured to the nearest 0.1 mm, and all other molluscs were measured to the nearest 0.01 mm.

Diet breadth was assessed by presence or absence of each prey type and expressed in terms of percent occurrence for the fish population of interest. Lake trout and round whitefish diets from rocky shoal and river inlet habitats were compared to determine if fish diets varied by habitat. Large (>400 mm) and small lake trout (≤ 400 mm) from rocky shoals were compared to determine if fish size within a single habitat influenced diet breadth. Lake trout were assigned to small and large size classes based on data suggesting that fish <400 mm ate zooplankton more commonly than larger fish (O'Brien *et al.*, 1979). The size distribution of lake trout also showed a natural break at about 400 mm TL. These data were not evaluated statistically.

To determine whether relative abundances of prey items in fish diets varied among fish species, the mean number of each prey type was calculated for lake trout, round whitefish, and grayling. Comparison of means to determine minimum significant differences was done using Tukey's studentized range test (SAS User's Guide: Statistics, 1985). Because grayling consumed very few molluscs, they were not included in statistical comparisons for molluscan prey. Similarly, round whitefish and grayling were compared with respect to mean number of *Grensia* (caddisfly) consumed; lake trout and round whitefish were examined with respect to mean number of chironomids consumed; and comparison was made between lake trout and grayling with respect to mean number of zooplankton consumed.

To examine size selective predation on molluscs by lake trout and round whitefish in rocky (dominated by the snail *Lymnaea*) and soft-sediment (dominated by the snail *Valvata* and fingernail clams) habitats, mean mollusc size eaten was compared to the mean size available by t-tests for independent means. Round whitefish and lake trout were also compared with each other with regard to size of molluscs eaten using t-tests for independent means. To determine if larger fish selected larger molluscs, shell length was regressed on total fish length. A positive, significant slope would indicate that larger fish were selecting larger molluscs.

To determine if the lake trout sampled from Toolik Lake, Lake I8, and Itigaknit Lake possessed significantly different size distributions, mean total lengths were compared with Tukey's studentized range for minimum significant difference (SAS User's Guide: Statistics, 1985). To determine if lake trout in the three lakes were preying on molluscs of similar size, the mean sizes of mollusc selected by lake trout in each lake were compared in a two way ANOVA (Sokal & Rohlf, 1981). Because there was a significantly positive relationship between trout total length and the size of *Lymnaea* eaten in Toolik Lake (see Results), lake trout from Toolik Lake, Itigaknit Lake, and Lake I8 were divided into three matching size classes (320–399 mm, 400–479 mm, 480–559 mm). This allowed us to determine if fish of the same size class in different lakes selected the same size *Lymnaea*. A size class was added in addition to the size classes used in the Toolik Lake comparison because most of the Itigaknit lake trout were considerably larger than those in Toolik; it would not be appropriate to compare diet selection between fishes in different lakes if the fish were not of similar sizes. Mean *Lymnaea* size selected by lake trout in each lake for each size class was compared with Tukey's studentized range test for minimum significant differences (SAS User's Guide: Statistics, 1985).

Results

Nearshore *Lymnaea* density was $0.4 \pm 0.4/m^2$ (mean ± SE) in lakes containing lake trout compared to $8.9 \pm 1.3/m^2$ in lakes without lake trout; means were significantly different ($p < 0.001$, Fig. 1). The mean size of adult snails in troutless lakes (22.4 ± 0.5 mm) was significantly ($p < 0.02$) larger than the mean size from lakes with trout (18.0 ± 1.2 mm; Fig. 1).

Total length of lake trout caught in Toolik Lake was 423 ± 10 mm, with a maximum length of 607 mm. These lake trout fed heavily upon benthic organisms, especially molluscs (Table 1). The most commonly encountered organisms in trout stomachs were the snails *Lymnaea* (70% of lake trout sampled) and *Valvata* (58%) (Table 1). Other invertebrates in lake trout diets included chironomid larvae and pupae (50%), caddisflies (*Grensia*) (36%), several small fingernail clam species (32%), and zooplankton (30%) (Table 1). Fish were found in the stomachs of 27% of the lake trout sampled (Table 1).

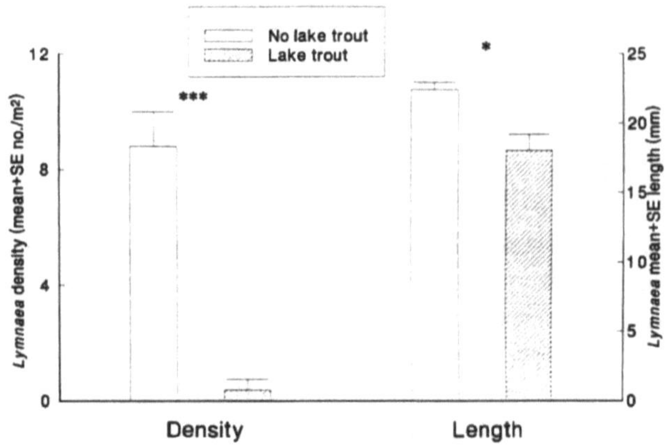

Fig. 1. Density and size distribution of *Lymnaea* at three sites in three lakes without lake trout and three lakes with lake trout. Asterisks indicate significant differences between lakes with and without lake trout (**, $p < 0.02$, $p < 0.001$).

Toolik Lake round whitefish total length was 428 ± 4 mm. These fish were strongly demersal, with molluscs being the dominant food group found in their stomachs (83%; Table 1).

Grayling in Toolik Lake were the least demersal fish considered. Grayling total length was 327 ± 29 mm. The dominant forage items found

Table 1. Percent occurrence of prey types in the stomachs of lake trout, round whitefish, and graying during the summer of 1986.

	Lake trout ($n = 74$)	Round whitefish ($n = 71$)	Grayling ($n = 10$)
Mollusca	79.7%	83.1%	70.0%
Lymnaea	70.3%	36.6%	10.0%
Valvata	58.1%	43.7%	60.0%
Sphaeriidae	32.4%	31.0%	0.0%
All fish	27.0%	0.0%	0.0%
slimy sculpin alone	12.2%	0.0%	0.0%
Chironomidae	50.0%	45.1%	30.0%
Trichoptera	36.5%	67.6%	50.0%
Zooplankton	29.7%	0.0%	70.0%
Terrestrial Insects	5.8%	1.4%	20.0%

in Toolik Lake grayling stomachs were zooplankton (70%), *Valvata* (60%), *Grensia* (50%), and chironomids (30%) (Table 1).

Some differences in diet breadth data for Toolik Lake were observed when fish of similar size (>400 mm), but different locality were compared. Lake trout captured over soft sediments near the lake's main inlet had more clams, fish, and chironomids, and fewer zooplankton in their stomachs than trout captured over rocky shoals (Table 2). Percent occurrence of *Lymnaea*, *Valvata*, and *Grensia* in lake trout stomach contents were similar between capture locations (Table 2). This comparison was not made for trout ≤400 mm as none were captured at the inlet gillnet sampling location. Whitefish diets were also slightly different by habitat (Table 2). Fish captured near Toolik Lake's main inlet contained more clams, chironomids, and trichopteran larvae and had less *Lymnaea*, than fish captured over shoals (Table 2). Percentage of whitefish consuming *Valvata* was similar in both areas.

Differences in diet breadth were apparent between large trout and small lake trout from the rocky shoal (Table 3). Larger fish more commonly preyed on molluscs, less commonly on *Grensia* and zooplankton than small fish. Large trout did not feed on sculpin at this sampling site (Table 3).

Lake trout in Toolik Lake ate significantly more *Lymnaea* than round whitefish and grayling, but round whitefish ate significantly more *Grensia*

Table 2. Percent occurrence of prey types in the stomachs of large lake trout (>400 mm) and round whitefish (>379 mm) from two gillnetting sites in Toolik Lake.

	Shoal		Inlet	
	LT ($n = 13$)	RW ($n = 25$)	LT ($n = 23$)	RW ($n = 38$)
Mollusca	84.6%	84.0%	91.3%	82.6%
Lymnaea	76.9%	60.0%	78.3%	23.9%
Valvata	69.2%	44.0%	73.9%	43.5%
Sphaeriidae	30.8%	12.0%	47.8%	41.3%
All fish	0.0%	0.0%	52.2%	0.0%
Chironomidae	15.4%	8.0%	87.0%	65.2%
Trichoptera	30.8%	52.0%	26.1%	76.1%
Zooplankton	38.5%	0.0%	0.0%	0.0%

Table 3. Percent occurrence of prey types in the stomachs of large (>400 mm) and small (≤400 mm) lake trout gillnetted in Toolik Lake on a shoal.

	Large Lake Trout ($n = 13$)	Small Lake Trout ($n = 17$)
Mollusca	90.9%	47.4%
Lymnaea	83.6%	36.8%
Valvata	67.3%	36.8%
Sphaeriidae	41.8%	5.3%
All fish	30.9%	15.8%
Slimy Sculpin alone	10.9%	15.8%
Aquatic Insects	69.1%	73.7%
Zooplankton	9.1%	89.5%

than lake trout and grayling (Table 4). *Valvata* consumption by lake trout was high but variable and did not differ significantly from consumption by whitefish (Table 4).

Lake trout consumed significantly larger *Lymnaea*, *Valvata*, and clams than did round whitefish in Toolik Lake (Fig. 2). Compared with molluscs available on the sediments (*Valvata* and clams) and rocks (*Lymnaea*), lake trout ate significantly larger *Lymnaea* (13.0 ± 0.2 mm vs. 5.6 ± 0.8 mm), *Valvata* (4.5 ± 0.04 mm vs. 2.9 ± 0.1 mm), and clams (5.2 ± 0.1 mm vs. 2.6 ± 0.1 mm) (Fig. 2).

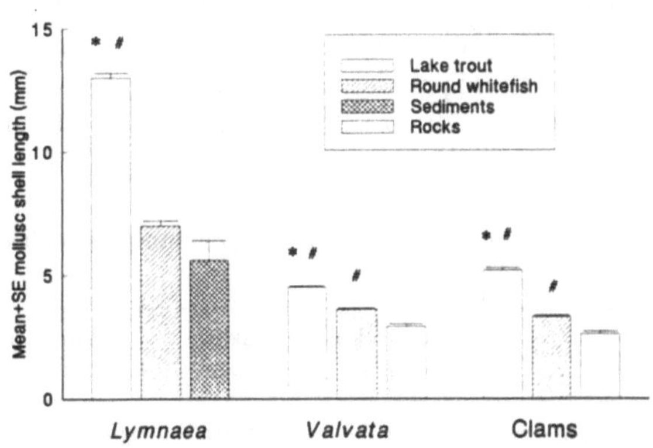

Fig. 2. Comparison of size of molluscs consumed by lake trout and round whitefish and sampled from substrates in Toolik Lake. Asterisks (*) indicate significant differences in sizes of molluscs consumed between lake trout and round whitefish. Pound symbols (#) indicate significant differences between the mean size mollusc of a given taxon eaten by the designated fish species and the mean size of that mollusc on the substrate.

Table 4. Mean ($\pm$ SE) number of prey items in the stomachs of lake trout ($n = 74$), round whitefish ($n = 71$), and grayling ($n = 10$) from Toolik Lake. One asterisk indicates significant difference between indicated mean and lowest compared mean. Two asterisks indicate significant difference between indicated mean and both compared means ($p < 0.05$, Tukey's test).

Prey taxon	Lake trout	Round whitefish	Grayling
Lymnaea	**20.8 ± 3.9	6.4 ± 1.8	0.1 ± 0.1
Valvata	86.3 ± 69.3	40.5 ± 13.0	2.6 ± 1.0
Sphaeriidae	10.8 ± 4.5	13.1 ± 3.9	–
All fish	0.3 ± 0.07	–	–
Sculpin	0.2 ± 0.06	–	–
Trichoptera	3.2 ± 1.6	**27.8 ± 5.9	9.8 ± 4.5
Chironomidae	17.5 ± 4.5	12.3 ± 5.1	2.8 ± 2.2
Zooplankton	135.2 ± 49.7	–	298 ± 175.5

Round whitefish did not eat significantly larger *Lymnaea* (7.0 ± 0.2 mm vs. 5.6 ± 0.8 mm) than were sampled on Toolik Lake rocks, but did eat significantly larger *Valvata* (3.6 ± 0.05 mm vs. 2.9 ± 0.1 mm) and clams (3.3 ± 0.05 mm vs. 2.6 ± 0.1 mm) than were sampled on the sediments (Fig. 2).

Based on regression analyses of prey and fish length, larger lake trout selected significantly larger *Lymnaea* ($Y = 0.014X + 6.49$, $n = 395$, $p < 0.001$) and clams ($Y = 0.004X + 3.08$, $n = 0.036$, $p = 0.036$), but not larger *Valvata* ($Y = 0.001X + 4.224$, $n = 403$, $p = 0.335$) with increasing total length. However, round whitefish total length was not related to prey size selection for any of these prey items (*Lymnaea*: $Y = -0.002X + 7.95$, $n = 179$, $p = 0.777$; *Valvata*: $Y = 0.001X + 3.32$, $n = 436$, $p = 0.60$; $Y = -0.004X + 4.97$, $n = 252$, $p = 0.051$).

Table 5. Percent occurrence of prey types in the diets of lake trout from three area lakes.

	Toolik Lake ($n = 74$)	Lake I8 ($n = 11$)	Itigaknit Lake ($n = 14$)
Mollusca	79.7%	72.7%	92.9%
Lymnaea	71.6%	45.4%	78.6%
Valvata	59.5%	72.7%	50.0%
Sphaeriidae	32.4%	36.4%	42.9%
All fish	27.0%	27.3%	0.0%
Slimy Sculpin	12.2%	27.3%	0.0%
Chironomidae	50.0%	36.4%	35.7%
Limnephilidae	36.5%	27.3%	14.3%
Zooplankton	29.7%	27.3%	0.0%

The total length of lake trout sampled from Itigaknit Lake (541 ± 12 mm) was significantly larger than from Toolik Lake (423 ± 10 mm) and Lake I8 (427 ± 35 mm). Of the lake trout captured in these lakes, 73% ate at least one species of mollusc (Table 5). Lake trout captured in Itigaknit Lake differed from trout of Toolik and I8 in that they did not have fish or zooplankton in their stomachs (Table 5). All lake trout captured in Itigaknit Lake were in the large size class. These fish ate significantly larger *Lymnaea* (21.04 ± 0.6 mm) than similar-sized lake trout from Toolik Lake (13.9 ± 0.3 mm) and Lake I8 (11.5 ± 0.5 mm). Mean *Lymnaea* shell lengths selected by trout from I8 and Toolik lakes did not differ significantly for any fish size class. *Valvata* selected by lake trout in Toolik Lake (4.5 ± 0.04 mm), I8 (4.1 ± 0.1 mm), and Itigaknit Lake (4.0 ± 0.1 mm) were of similar size. Clams selected by lake trout in Itigaknit Lake were significantly smaller (3.6 ± 0.1 mm) than clams eaten by trout in Toolik Lake (5.2 ± 0.1 mm) and Lake I8 (4.9 ± 0.2 mm).

Discussion

The summer diets of lake trout, round whitefish, and grayling in Toolik Lake suggest different predatory roles for these species: lake trout had more diverse diets than round whitefish, and arctic grayling had a much weaker demersal orientation than either of the other fishes. Differences in diet breadth between lake trout and round whitefish were associated with *Lymnaea* and Trichoptera (predominantly the larvae of *Grensia* sp.). This was largely due to increased use of caddisfly larvae and reduced use of *Lymnaea* by round whitefish in the inlet during June. The differences observed in percent of lake trout and round whitefish consuming chironomids between the two gillnet locations may be explained by the large chironomid emergence which was observed near the inlet in late June during the sampling period. There are major differences in relative abundances of benthos on rocks (Cuker, 1983) and soft sediments (Hershey, 1985; Hanson *et al.*, this issue). Higher densities of sphaeriid clams are found in the dense macrophyte beds near the inlet than on the rocky shoal (Hershey, pers. obs.). Thus, the relatively high percent occurrence of clams in the diets of both lake trout and whitefish in the inlet habitat when compared to the rocky shoal site may be due to availability.

In considering size-selective predation on *Lymnaea* by round whitefish and lake trout, it is important to note that multiple year classes of *Lymnaea* are present throughout the year in Toolik Lake and that reproductive size for *Lymnaea* is approximately 15.0 mm (Hershey, 1990). Lake trout consume significantly more and larger *Lymnaea* than round whitefish. Because only lake trout consume *Lymnaea* of reproductive size, the potential for lake trout to have a significant impact on the *Lymnaea* population in Toolik Lake is greater than for round whitefish. This size selectivity by lake trout on *Lymnaea* was observed for all sizes of lake trout, with larger fish selecting increasingly larger *Lymnaea*. We suggest that the reduction in the number of reproductively-sized *Lymnaea* due to lake trout predation is sufficient to cause the observed differences in *Lymnaea* populations between lakes with and without lake trout (see Merrick *et al.*, 1991).

Although lake trout and round whitefish selected significantly larger *Valvata* and clams than occurred in benthic samples from Toolik Lake, regression analysis showed that the size of *Valvata* did not significantly increase with increasing fish size. Lake trout did appear to select larger clams with increasing fish size. However, the slope of the regression line for trout predation on clams ($m = 0.004$) was shallow, indicating only a very slight tendency for larger lake trout to select larger clams.

The mean size of *Valvata* preyed on by lake trout was smaller than the reproductive size for *Valvata* (5.0 mm; Hershey, 1990) but closer to the reproductive size than the mean size preyed on by round whitefish. Thus, whitefish have little potential to control *Valvata* population density, but trout may exert some influence. However, Hershey (1990) has shown that *Valvata* are competitively inferior to *Lymnaea* and were only abundant on soft sediments when *Lymnaea* density was very low (as in Toolik Lake). Thus competition rather than predation is likely a more important mechanism limiting *Valvata* population density. Reproductive size for the sphaeriid clams in Toolik Lake has not been determined.

Lake trout's selection of larger *Valvata* or sphaeriid clams with increasing fish size was very small, but they did choose a significantly larger mean size for each prey type than round whitefish. Therefore, we hypothesize that selectivity for large *Valvata* and clams is a function of the lake trout's inability to detect or capture small molluscs. Round whitefish with their small subterminal mouth and stronger demersal orientation (Scott & Crossman, 1973) may be more effective predators on the smaller prey. Conversely, the failure of round whitefish to eat large *Lymnaea* is probably due to gape limitation.

Lake trout captured in Itigaknit Lake were significantly larger than trout captured in Toolik Lake and Lake I8; this may be due to reduced angling because of its more remote location. Toolik Lake and Lake I8 are both located in close proximity to the Dalton Highway and receive moderate fishing pressure during the summer. Other mechanisms which could contribute to fish size, such as higher lake productivity, have not been evaluated, although comparative data which do exist suggest that at least Toolik and I8 are similarly unproductive (M. C. Miller, Univ. of Cincinnati, pers. comm.). Diet breadth of the lake trout in these three lakes appears to be very similar, with the only differences found in the lack of sculpin and zooplankton in the diets of lake trout from Itigaknit. The absence of zooplankton was likely due to the lack of small lake trout in the sample; the absence of sculpin may reflect the small sample size (n = 14). Lake trout in the 480–559 mm size class in Itigaknit Lake ate significantly larger *Lymnaea* than trout of the same size class in Toolik Lake and Lake I8. This may be related to the amount of available refuge for *Lymnaea* associated with a particular lake. Itigaknit Lake, in addition to reduced angling pressure, also has a complex morphometry and extensive beds of well-developed submerged macrophytes (Hershey, pers. obs.) which may provide more refuge for *Lymnaea* and increase their changes of attaining a larger size. Lake trout size differences between lakes may also be due to larger prey items available in Itigaknit Lake. Toolik Lake and Lake I8, with less well-developed macrophyte beds relative to total lake area, may not provide *Lymnaea* with as much opportunity to grow as large as in Itigaknit Lake.

Conclusion

Size-selective predation on molluscs by lake trout and round whitefish was apparent during the summer of 1986. Lake trout consumed more and larger *Lymnaea* than did round whitefish; their tendency to select individuals of reproductive size suggests greater potential for lake trout to affect *Lymnaea* populations. Predation on *Valvata* and probably sphaeriid clams by both fish species appeared to be a function of availability on the sediments without regard to fish size. Finally, the pattern of distribution of *Lymnaea* across Toolik Lake rocky and soft-sediment habitats, coupled with the size and density distribution of *Lymnaea* in lakes with and without lake trout, is consistent with the hypothesis that lake trout determine *Lymnaea* distribution and abundance in lakes of the Toolik area.

Acknowledgements

The field assistance and advice of L. Deegan, C. Harth, J. Helfrich, M. Miller, B. Moller, and J. O'Brien was invaluable. We would also like to thank H. Collins, C. Leucke, J. Schuldt, and J. Thorpe for editorial comments and advice, and R. Regal for statistical advice. This work was supported by NSF grant no. DPP 8320544 to J. Hobbie and B. Peterson, Marine Biological Laboratory (MBL), Woods Hole, Massachusetts, by subcontract from MBL to AEH, and through an equipment grant to AEH from the University of Minnesota Graduate School.

References

Bendock, T. & J. Burr, 1981. Index to North Slope streams and lakes survey. Alaska Fish and Game Report.

Crowder, L. B. & E. C. Cooper, 1982. Habitat structural complexity and the interaction between bluegills and their prey. Ecology 63: 1802–1813.

Cuker, B. E., 1983. Competition and coexistence among the grazing snail *Lymnaea*, Chironomidae, microcrustacea in an arctic epilithic lacustrine community. Ecology 64: 10–15.

Eddy, S. & J. C. Underhill, 1974. Northern Fishes. Minneapolis, Mn. Univ. of Mn. Press. 414 pp.

Gilinsky, E., 1984. The role of fish predation and spatial heterogeneity in determining benthic community structure. Ecology 65: 455–468.

Hall, D. J., W. E. Cooper & E. E. Werner, 1970. An experimental approach to the production dynamics and structure of freshwater animal communities. Limnol. Oceanogr. 15: 839–848.

Hanson, K. L., A. E. Hershey & M. E. McDonald, 1992. A comparison of slimy sculpin (*Cottus cognatus*) populations in arctic lakes with and without piscivorous predators. Hydrobiologia 240: 189–201.

Hershey, A. E., 1985. Experimental effects of predatory sculpin on the chironomid communities of two arctic littoral zone habitats. Ecology 66: 1131–1138.

Hershey, A. E., 1990. Snail populations in arctic lakes: Competition mediated by predation? Oecologia 82: 26–32.

Hershey, A. E. & M. E. McDonald, 1985. Diet and digestion rates of the slimy sculpin, *Cottus cognatus*, in an Alaskan arctic lake. Can. J. Fish. Aquat. Sci. 42: 483–487.

Johnson, L., 1972. Keller Lake: characteristics of a culturally unstressed salmonid community. J. Fish. Res. Bd Can. 29: 731–740.

Johnson, L., 1976. Ecology of arctic populations of lake trout, *Salvelinus namaycush*, lake whitefish, *Coregonus clupe-*

aformis, arctic char, *S. alpinus*, and associated species in unexploited lakes of the Canadian Northwest Territories. J. Fish. Res. Bd Can. 33: 2459–2488.

Kerr, S. R., 1971. A simulation model of lake trout growth. J. Fish. Res. Bd Can. 28: 815–819.

Kettle, D. & W. J. O'Brien, 1978. Vulnerability of Arctic zooplankton species to predation by small Lake Trout (*Salvelinus namaycush*). J. Fish. Res. Bd Can. 35: 1495–1500.

Konkle, B. R. & W. G. Sprules, 1986. Planktivory by stunted lake trout in an Ontario Lake. Trans. Am. Fish. Soc. 115: 515–521.

Macan, T. T., 1966. The influence of predation on a moorland fishpond. Arch. Hydrobiol. 61: 432–452.

Martin, N. V., 1966. The significance of food habits in the biology, exploitation, and management of Algonquin Park, Ontario, lake trout. Trans. Am. Fish. Soc. 81: 415–422.

McDonald, M. E., B. E. Cuker & S. C. Mozley, 1982. Distribution, production, and age structure of slimy sculpin in an arctic lake. Envir. Biol. Fishes 7: 171–176.

Merrick, G. W., A. E. Hershey & M. E. McDonald, 1991. Lake trout (*Salvelinus namaycush*) control of snail density and size distribution in an arctic lake. Can. J. Fish. aquat. Sci. 48: 498–502.

Miller, M. C., G. R. Hater, P. Spatt, P. Westlake & D. Yeakel, 1986. Primary production and its control in Toolik Lake, Alaska. Arch. Hydrobiol./Suppl. 4: 97–131.

O'Brien, W. J., C. Buchanan & J. F. Haney, 1979. Arctic zooplankton community structure: Exceptions to some general rules. Arctic 32: 237–247.

SAS User's Guide: Statistics, Version 5 Edition, 1985. SAS Institute Inc., Cary, N.C.

Scott, W. B. & E. J. Crossman, 1973. Freshwater Fishes of Canada. Bulletin 184. J. Fish. Res. Bd Can.

Smith, S. H., 1972. Factors in the ecological succession in oligotrophic fish communities of the Laurentian Great Lakes. J. Fish Res. Bd Can. 29: 717–730.

Sokal, R. R. & F. J. Rohlf, 1981. Biometry, W. H. Freeman and Company, San Francisco.

Stein, R. A., C. G. Goodman & E. A. Marshall, 1984. Using time and energetic measures of cost in estimating prey value for fish predators. Ecology 65: 702–715.

Thorp, J. H. & E. A. Bergey, 1981a. Field experiments on responses of a freshwater, benthic macroinvertebrate community to vertebrate predators. Ecology 62: 365–375.

Thorp, J. H. & E. A. Bergey, 1981b. Field experiments on interactions between vertebrate predators and larval midges (Diptera: Chironomidae) in the littoral zone of a reservoir. Oecologia 50: 285–290.

Vermeij, G. L. & A. P. Covich, 1978. Coevolution of freshwater gastropods and their predators. Am. Nat. 112: 833–843.

Ware, D. M., 1972. Risk of epibenthic prey to predation by rainbow trout (*Salmo gairdneri*). J. Fish. Res. Bd Can. 30: 787–797.

Ware, D. M., 1973. Predation by rainbow (*Salmo gairdneri*): The influence of hunger, prey density, and prey size. J. Fish Res. Bd Can. 29: 1193–1201.

Werner, E. E. & D. J. Hall, 1977. Competition and habitat shift in two sunfishes (Centrarchidae). Ecology 58: 869–876.

Werner, E. E., J. F. Gilliam, D. J. Hall & G. G. Mittelbach, 1983. An experimental test of the effects of predation risk on habitat use in fish. Ecology 64: 1540–1548.

Hydrobiologia **240**: 235–245, 1992.
W.J. O'Brien (ed.), Toolik Lake – Ecology of an Aquatic Ecosystem in Arctic Alaska.

Simulation model of the planktivorous feeding of arctic grayling: laboratory and field verification

W. John O'Brien & Barbara I. Evans*
Department of Systematics and Ecology, University of Kansas, Lawrence, Kansas 66045, USA;
**Current address: Department of Psychology, Stanford University, Stanford, CA 94305, USA*

Introduction

Interest in the feeding of planktivorous fish began in earnest with the awareness that their feeding could structure the body size and species composition of zooplankton communities (Hrbacek, 1962; Brooks & Dodson, 1965). There have been numerous studies subsequent to this work further demonstrating this capability (O'Brien, 1987; Paloheimo & Fulthorpe, 1987; Skoptsov & Kruppenikova, 1983; Spencer & King, 1984). Virtually all of these studies confirm that planktivorous fish focus their feeding on the larger prey present but the mechanism(s) involved in this size selective feeding is less clear (Eggers, 1977; Wright & O'Brien, 1984).

Werner & Hall (1974) and Confer & Blades (1975) developed a model of planktivorous fish size selective feeding based largely on the size of the search space for different sizes and species of zooplankton. Fish were hypothesized to be able to locate more zooplankton for which there was a large search space. This has been referred to as the reactive field model (RFM), and must indeed be part of the mechanism involved in size selective feeding (Eggers, 1982 and Wright & O'Brien, 1984). Werner & Hall (1974) further suggested that planktivorous fish foraged optimally by choosing that prey that maximized energy intake while minimizing handling. They did not specify a tactical mechanism. However, O'Brien *et al.* (1976) suggested that choice of that prey that appeared largest was an important mechanism in selective feeding of planktivorous fish. Several papers have pointed out, however, that the specific predictions of the apparent size model are not easily distinguished from the RFM predictions (Wetterer & Bishop, 1985). Combining both approaches and incorporating the effect of prey motion, Wright & O'Brien (1984) got good agreement between predicted diet of white crappie and observed diet of field collected white crappie. Thus the mechanisms involved in diet composition in planktivorous fish seem well understood. However, there have been less theoretical and modeling efforts involving the feeding rates of planktivorous fish.

In fact until recently it would have been difficult to construct accurate models of planktivorous fish feeding rate because crucial elements of the predation cycle of these fish had not yet been recognized. Evans & O'Brien (1988) showed that the foraging strategy of the planktivorous fish they studied, arctic grayling, did not fit into the previously assumed cruise search strategy. O'Brien & Evans (1991) showed 4 other fish species did not use a cruise search strategy either. Rather these fish utilized a saltatory search strategy (O'Brien *et al.*, 1989). Saltatory search differs from cruise search in that with saltatory search the fish moves in a stop and go pattern, where search for zooplankton prey occurs only during the frequent pauses and movement is used to move the fish into previously unsearched water (O'Brien *et al.*, 1989) (Fig. 1). Many other species and taxa which search for small, dispersed prey also use a saltatory search strategy (O'Brien *et al.*, 1990).

In this paper we quantify the components of the saltatory search cycle of arctic grayling and determine the dimensions of its search space. We

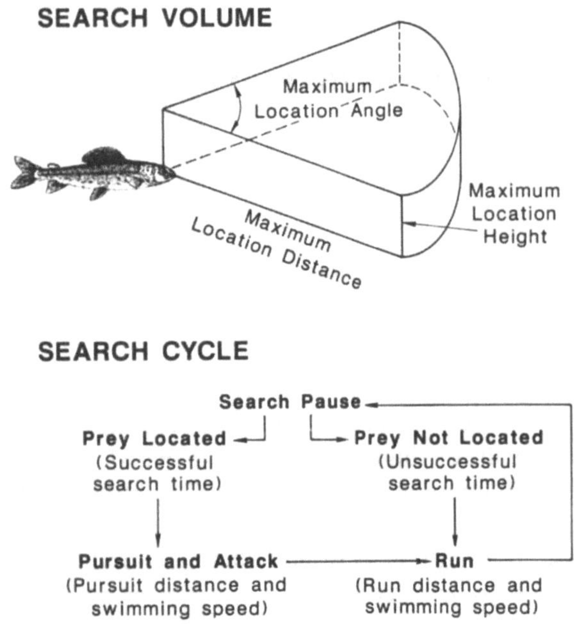

Fig. 1. Search cycle and search volume for arctic grayling feeding on zooplankton.

use these to develop a detailed, mechanistic simulation model of its plankton feeding on the large zooplankter *Heterocope septentrionalis* (2.5–3.5 mm), a common food of lake dwelling grayling (O'Brien *et al.*, 1979). We then compare the predictions of the model to feeding rates obtained in a laboratory arena and fish cages emplaced in Toolik Lake, a large lake in arctic Alaska. The laboratory and field research was conducted at the Toolik Lake Research Station situated on the shores of Toolik Lake, Alaska (68° N, 143° W) while the model of grayling planktivorous feeding rates was developed at the University of Kansas. Most of the details of the methods used to quantify the search cycle and dimensions of the search space have been given in Evans (1986) and Evans & O'Brien (1988); however, we reiterate the highlights and detail any differences here.

Methods to determine components of the search cycle and dimensions of the search volume

The basic technique in developing the model was to determine the time involved in each step of the search cycle and measure the dimensions of the search space. These objectives were achieved by videotaping the feeding behavior of an individual grayling as it fed on the large zooplankter *Heterocope septentrionalis* (3–4 mm). We also estimated the search space for small prey. The fish were not fed for 24 h prior to observation and a total of four fish (30–40 cm TL) were studied over 2 summers. The videotaping of fish feeding in the horizontal plane was done in a 122 cm square tank filled to a depth of 30 cm using a JVC 100 camera mounted 1.5 m above the tank. All taping was done at light intensities above 300–400 lx. Measurements were made at temperatures of 5, 10 and 15 °C. Aspects of the behavior of the fish while feeding were narrated onto the audio portion of the tape by an observer viewing their feeding from the side. The feeding behavior was analyzed using a Panasonic 6300 VCR with a Panasonic NV-A505 remote controller.

The specific components of the search cycle quantified are shown in Figure 1 and include measurements of search time, run and pursuit distances, angles and speeds. A minimum of 20 observations were gathered for each component at each condition studied. Successful and unsuccessful search times were measured by counting tape frames during which the fish were stationary during a feeding bout. The length of run distance and the dimensions of the search volume determined include location distance, location angle, and location height (Fig. 1). These dimensions are measured by observing a large number of pursuits as the fish fed on *Heterocope*. The methods to determine location angles and distances in the horizontal plane are as given in Evans & O'Brien (1987). However, location height was determined in a narrow arena filled to a depth of 50 cm with the sides made of 2 inch dense styrofoam with one side (80 cm) made with plexiglas. The feeding of the fish in the vertical plane was videotaped from the side and only those pursuits parallel to the plexiglas window were analyzed to reduce parallax problems. Maximum location distance and height, and maximum location angle were determined by taking that distance or angle in each dimension at which the 90% longest distance or greatest angle fell.

Results of determination of components of the search cycle and dimensions of the search volume

The search cycle for arctic grayling feeding on zooplankton is shown in Fig. 1. The distances, speeds, and times involved have all been measured. Successful search pause duration with grayling feeding on large and small prey at 10 °C has been published in Evans & O'Brien (1988) and averages 0.07 s when feeding on large prey. Typically unsuccessful search is slightly longer than successful search (O'Brien *et al.*, 1989) so we assessed 0.1 s of search time in the model regardless of success or lack of it (Table 1). If search is successful, grayling pursue the located prey; but if search is unsuccessful grayling swim forward, termed a run, into unsearched water. The swimming speed while grayling were undertaking these two activities is given in Evans & O'Brien (1988). The speed of these two behaviors is remarkably similar across various temperatures. The values used in the model are given in Table 1. Run lengths when grayling are feeding on large and small prey are shown in Fig. 2. While there is a trend towards slightly longer run lengths at warmer temperatures within prey size these are not statistically different. In the model we used a run length of 12 cm and swimming speeds of 8 cms^{-1} or 22 cms^{-1} (Table 1).

Table 1. Dimensions, times and efficiencies used in the model simulation. Swimming speeds used are those measured for both small and large prey (Evans & O'Brien, 1988). The other components are for large prey.

Components of the search cyle	
Search	0.1 s
Run length	12 cm
Swimming speed	8 cm/s or 22 cm/s
Attack efficiency	50%
Dimensions of the search volume	
Maximum location angle	160°
Maximum location height	12 cm
Maximum location distance	32 cm

Search efficiency

Distance from fish cm	% Area of search volume	Probability of location
0–4	0.016	1.0
4–8	0.047	1.0
8–12	0.078	0.65
12–16	0.109	0.40
16–20	0.141	0.25
20–24	0.172	0.15
24–28	0.203	0.08
28–32	0.234	0.05
	$\Sigma = 1.00$	

Attack towards a prey always follows successful search and pursuit but in the case of *H. septentrionalis*, attack is not always successful. The observed likelihood of successful attack is shown in Fig. 3. The value used in the model was 50%, an extrapolation of the graph to 15 °C (Table 1). Schmidt & O'Brien (1982) found a slightly higher value of 60% at 15 °C.

The horizontal and vertical pursuits fields which are used to map out the search volume for fish feeding on large and small prey are shown in Figs 4 and 5. The overall horizontal shape of the search volumes does not differ much between large and small prey. There are a few more pursuits for large prey past 90° than small prey but the shape is quite similar. A value of 160° was used for the maximum location angle for large prey. The vertical search shapes were not as similarly shaped as the large and small horizontal search shapes because the small prey search space seems more wedge shaped than does the large prey vertical search space. A value of 12 cm was used for the maximum location height for large prey and the search volume was considered wedge-shaped as shown in Fig. 1. A value of 32 cm was used for the maximum location distance (Table 1).

Particularly noticeable for horizontal pursuits towards large prey is that pursuits are not evenly distributed throughout the search space but in fact clustered in the 4 to 8 cm region around the fish and the density of pursuits per unit area decline further away from the fish. This is shown in Fig. 6 where the likelihood of a pursuit per square cm of location space drops off exponentially. We used this decline in pursuits with increasing distance from the fish in the model to define the

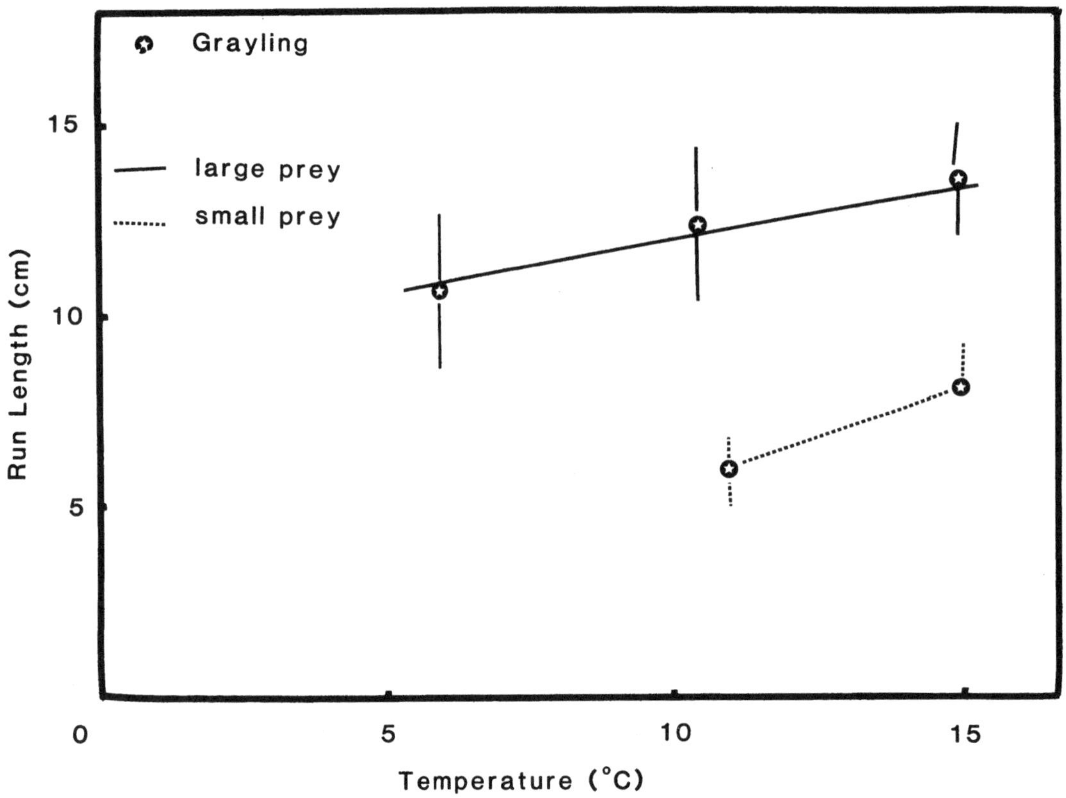

Fig. 2. Run lengths with arctic grayling feeding on large and small zooplankton prey at different temperatures. The vertical bars are the standard error. The lines connecting the means were drawn by hand and are just to help show the relationship.

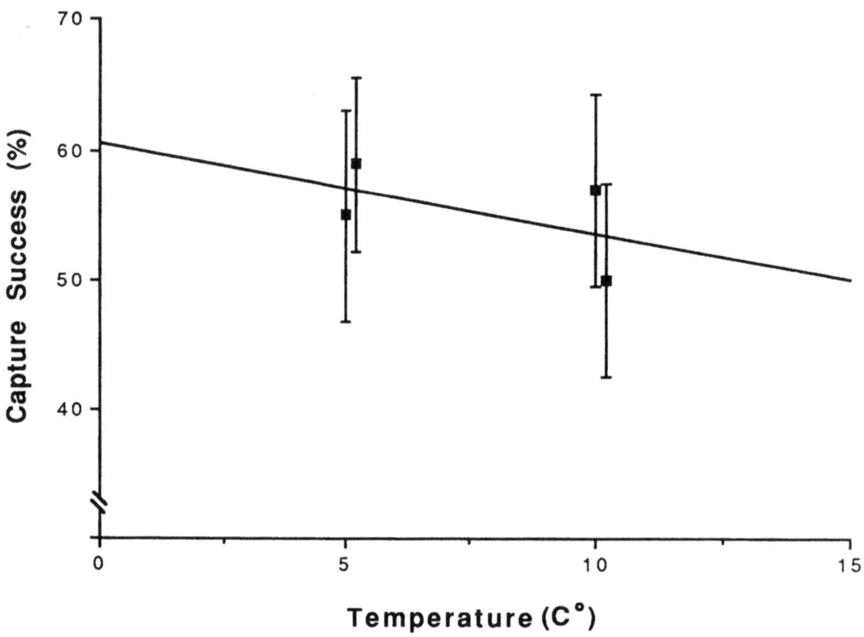

Fig. 3. Frequency of successful attack of *H. septentrionalis* at two temperatures by two different fish. The vertical bars are the standard error. The solid line was fitted by eye.

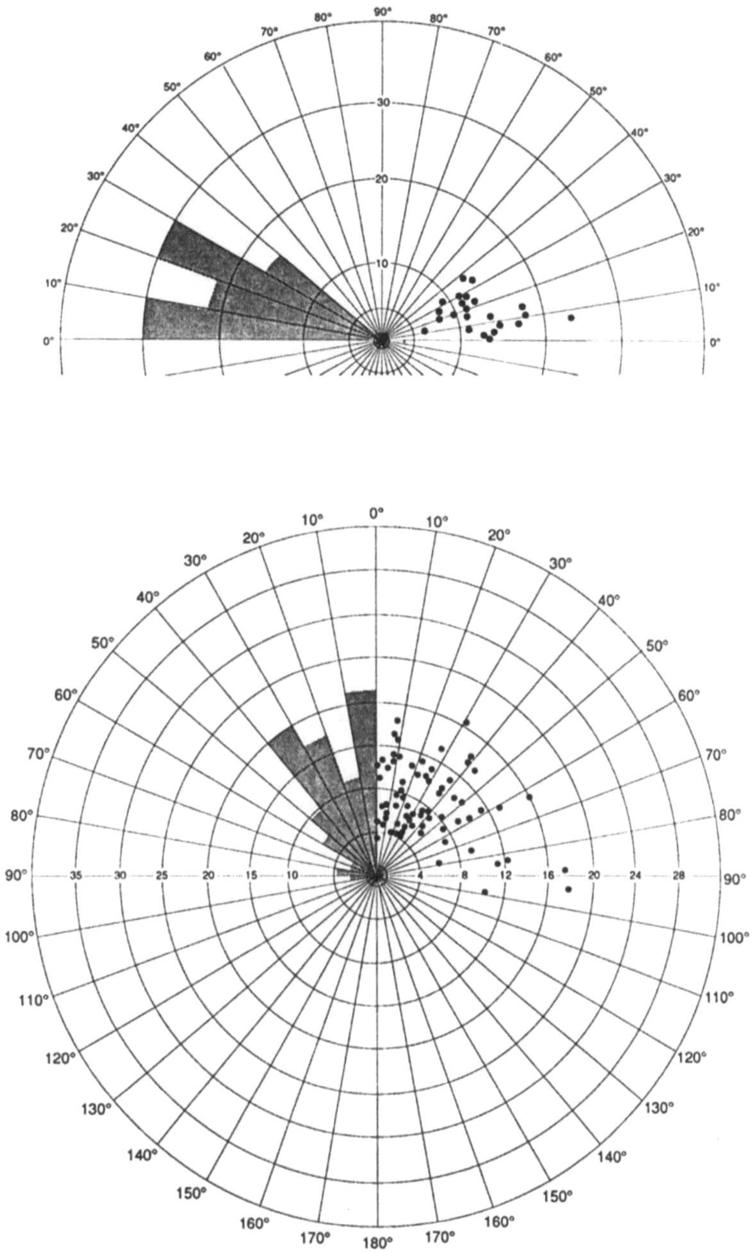

Fig. 4. Distribution of pursuits in the vertical and horizontal plane by arctic grayling feeding on small zooplankton prey. Each dot represents an observed attack on a large or small prey. The nose of the fish should be visualized at 0 cm and the long axis of the fish should be visualized as running along the 0°–180° line. In the lower graph of the horizontal plane all the left points were conflated to the right.

probability of location of a prey situated a specific distance from the fish at the moment of search (see below).

Methods of feeding rate experiments

Two separate means of measuring the feeding rate of plankton-feeding arctic grayling were undertaken. In the first, grayling fed on *H. septen-*

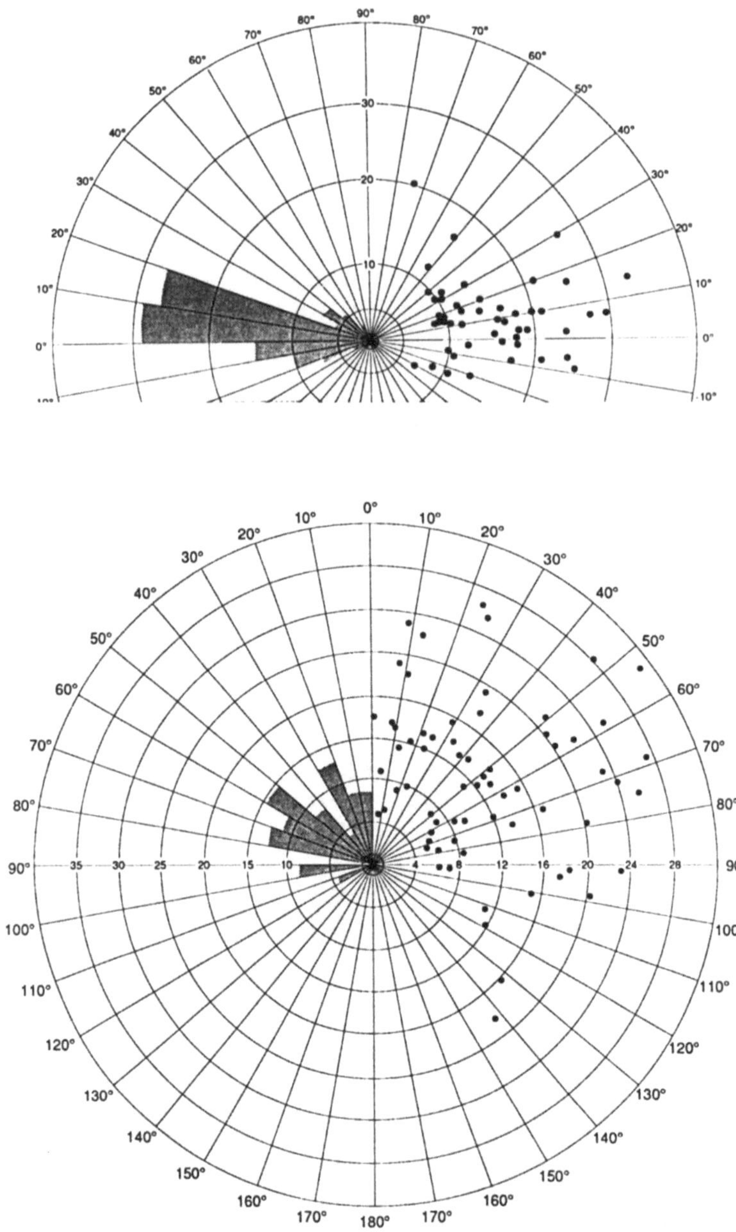

Fig. 5. Distribution of pursuits in the vertical and horizontal plane by arctic grayling feeding on large zooplankton prey. Each dot represents an observed attack on a large or small prey. The nose of the fish should be visualized at 0 cm and the long axis of the fish should be visualized as running along the 0°–180° line. In the lower graph of the horizontal plane all the left points were conflated to the right.

trionalis at an initial density of 0.1 per liter at varied light intensities in the same tank that the videotape observations of the search cycle were made. Grayling were confined to a corner of the arena and 50 *H. septentrionalis* poured into the water. After a brief period of time to allow the copepods to become distributed throughout the tank the fish were released and allowed to feed. At first feeding, time was kept and an observer counted the number of attacks towards a prey. The attack towards a prey is especially distinct when grayling feed on *H. septentrionalis*. Feeding

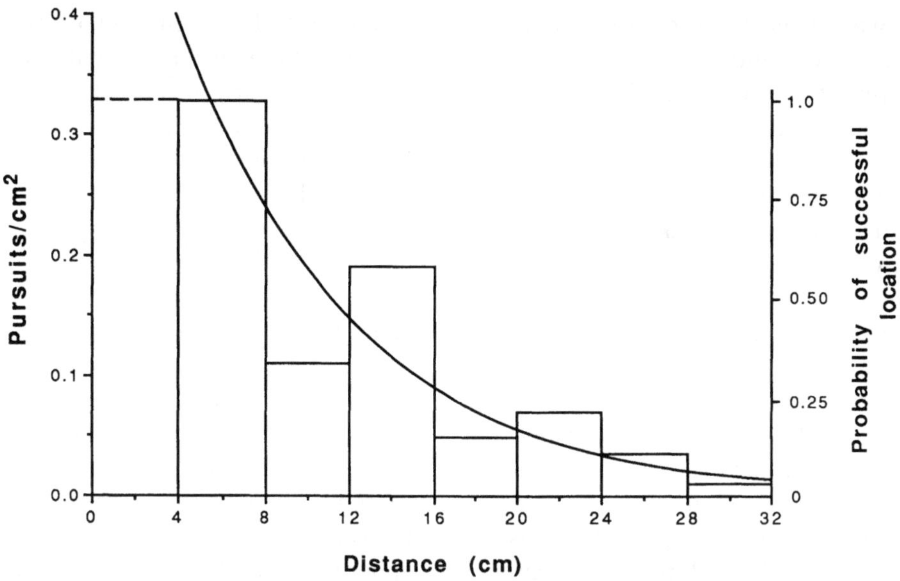

Fig. 6. The distribution of pursuits per cm away from the fish. The histogram shows the actual distribution and the solid line is a smoothed approximation fitted by eye. The probabilities used in the model and shown in Table 1 were obtained from the figure.

rate was observed and these rates occurring from 4 to 8 minutes before prey depletion became a major factor were used in subsequent calculations.

A second means of assessing grayling feeding rate was done in cages placed just under the surface of Toolik Lake well away from shore. The cages were constructed of a 1 m^3 cube using 1 cm square nylon mesh fitted around a frame made with small PVC pipe. The fish were placed in the cage for 24 hours inside a small container made of plankton netting to prevent them from feeding. To begin an experiment the fish were released from the plankton net container into the cage and a snorkeller or an observer using a periscope determined when the fish began to feed and the time noted by a second observer in a nearby boat.

While the fish were feeding a 1 m vertical plankton tow was taken beside the cage using a 335 μm mesh net. After a period of 10–17 minutes the cage was lifted out of the water, the fish captured and its stomach contents flushed out into an enamel pan. The resulting zooplankton were counted as was the plankton from the plankton tow using a Wild M-5 dissecting microscope. The density of zooplankton was used as input to the model and the number of zooplankton in the stomach, generally just *H. septentrionalis*, was compared to the predicted feeding rate from the model.

Results of feeding rate experiments

Feeding rates of grayling foraging in the laboratory arena are given in Table 2 and averaged 6.8 prey per minute. These are estimated ingestion

Table 2. Simulation of arena feeding rates. Prey density was $0.1\,l^{-1}$. The observed was attacks prey and was divided by 2 to correct for 50% *Heterocope* evasion.

Observed attacks/2 min^{-1}	Predicted prey min^{-1}
8.5	5.7
9.0	7.0
8.7	6.7
7.9	7.3
7.5	6.6
9.7	6.0
4.2	7.6
1.32	8.0
4.2	4.7
$\bar{X} = 6.8$	$\bar{X} = 6.6$

rates because what was observed were attacks towards prey and the number of attacks was divided by 2 to reflect the observed 50% capture efficiency of arctic grayling feeding on *Heterocope* at 15 °C.

Feeding rates of grayling foraging on *Heterocope* in cages suspended in the lake are shown in Table 3. In only 1 of the 8 experiments did grayling feed on anything but *Heterocope* or very occasionally *Daphnia middendorffiana*, another large zooplankton species. This focus of feeding on large zooplankton occurred even though small zooplankton were collected in the net tow beside the cage and were often far more abundant than large zooplankton. The feeding rate on *Heterocope* is shown graphically in Fig. 7. There is a slight but significant positive slope of the regression line indicating increased feeding rate with increasing prey density.

Methods to develop the model

The model uses several stochastic algorithms to determine if prey are located and eaten. The first determines the likelihood of prey being in the search volume and if present a second stochastic algorithm determines at what distance the prey is situated. The first algorithm simply computes the probability of one or more prey being situated in the search volume:

$$\text{probability of 1 or more prey} = 1 - e^{-(vd)}, \quad (1)$$

where *vd* is the visual density. Visual density is equal to the prey density per liter times the size of the search volume in liters. The second algorithm factors in the geometry involved in the likelihood of where a prey might be situated in the search volume. There is much more space in the search volume further away from the fish because the search space is a sector of a truncated hemisphere (Table 1). For example from Table 1 a prey would have a 1.6% probability of being in the area 0–4 cm from the fish but a 23.4% probability of being in the area 28–32 cm away. Once the distance is determined then the probability of location at that distance is applied (Fig. 6, Table 1). The probability of location declines as the prey is situated further from the fish (Koopman, 1980).

The probability of location is determined from

Table 3. Conditions and results from cage feedings rate experiments. 'S' refers to total small prey density. 'H' referes to *Heterocope*.

Date	Size of fish (cm)	Prey type	Density #/l	Duration of experiment min	Feeding rate #prey/min
7/16/85	20	H	0.2	16	4.07
		S	9.6		–
7/17/85	20	H	0.3	15	1.53
		S	9.6		–
7/31/85	20	H	0.36	15	2.06
		S	4.3		–
8/4/85	14	H	0.08	17	1.47
		S	4.3		1.36
	12	H	0.08	17	0.4
		S	4.4		12.8
7/22/86	10	H	0.25	10	1.8
		D	11.2		–
8/5/86	11	H	0.03	10	1.3
		D	4.68		–
	10	H	0.03	10	0.9
		S	4.68		–
8/13/86	9	H	0.135	15	0.93
		D	12.2		–
8/13/86	15	H	0.135	15	1.4
		S	12.6		–

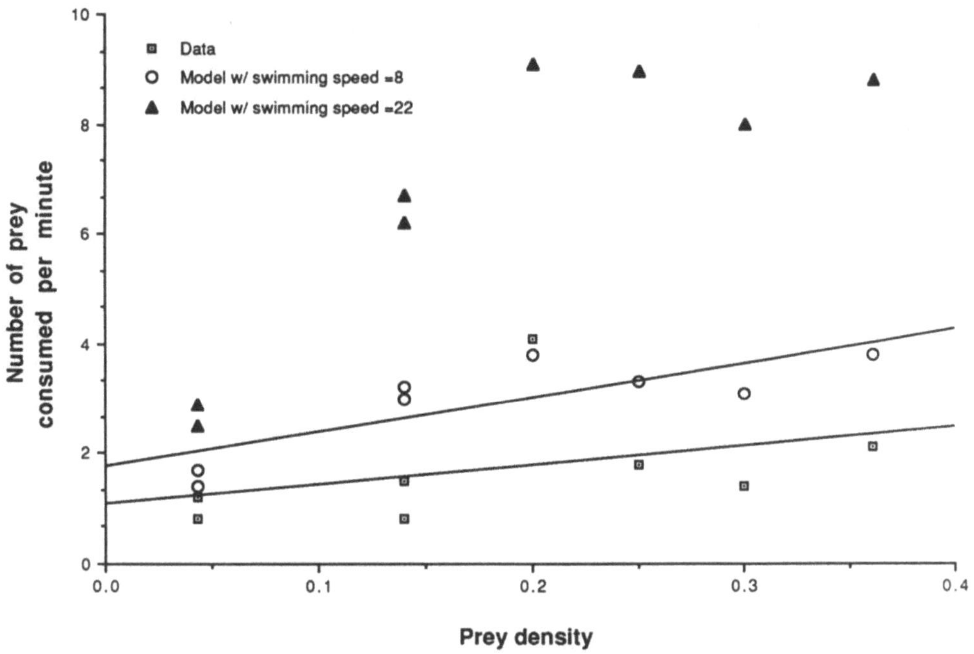

Fig. 7. Observed and predicted feeding rates of arctic grayling confined to a 1 m^3 mesh cage, feeding on lake zooplankton at ambient densities. The upper line is the regression through the model predictions at a swimming speed of 8 cm/s. The lower line is the regression through the observed feeding rates.

laboratory observations by assuming prey situated from 4–8 cm were located with 100% efficiency. Thus declining pursuits per cm^2 of search space is a measure of decreasing search efficiency. The solid curving line in Fig. 6 represents this decreasing location probability. For example a prey situated 0–4 cm away would always be located but one situated at 28–32 would have only a 5% chance of being located (Fig. 6, Table 1). If the prey is not located another run is initiated. If the prey is located in the model, pursuit to the distance where the prey was situated is simulated and the prey attacked. When simulating feeding on *H. septentrionalis* a 50% evasion was implemented during the attack and ingestion phase.

The model was run to a specified length of time (between 3 and 10–17 minutes) to match the duration of the feeding bout being simulated. The model keeps track of the time spent in all the components of the search cycle and the number of prey consumed. The model continually adjusts the density of prey in the cage or arena as a function of the feeding of the fish. The overall feeding rate is determined by dividing the number of prey ingested by the total time.

Results of the model

The predictions of the model simulating the feeding in the laboratory arena are given in Table 2. The model very accurately predicts the mean of 9 feeding experiments and a '*t*' test can not distinguish the means at the 0.05 level. The various simulations all using the same input of 0.1 *Heterocope* per liter show a good bit of variability but so does the observed feeding rate.

The predictions of the model simulating the feeding in the cages suspended in the lake are shown in Fig. 7. Using as input the values given in Table 1, and a swimming speed of 22 cm s^{-1} as observed in the arena when grayling were feeding on *Heterocope* (Evans & O'Brien, 1988) the simulation predicts a feeding rate 5 times greater than observed. However, we observed that the fish in the cage did not swim as fast as fish in the arena did when feeding on *Heterocope*. When the simulation was run using the swimming speed observed when grayling feed on small zooplankton, 8 cm/s, the fit is much closer but still a bit high.

Discussion

The saltatory search cycle of arctic grayling is fairly well known (Evans & O'Brien, 1988). There is a distinct shift when grayling are feeding on small zooplankton versus large zooplankton. The search pause declines, run length and run and pursuit speeds all increase dramatically when large versus small prey are present. There is also an increase in the search volume primarily through increased maximum search distance and maximum search height. The maximum search angle does not change very much. The estimates of search volume seem quite accurate. Notice especially the good agreement in Figs 4 and 5 between the maximum location distance estimated on the vertical and horizontal aspect for both large and small prey. The vertical experiments were done in different years with different fish than the horizontal measurements yet the maximum distance is almost identical.

A major departure of the model from the commonly used disk equation (Holling, 1959) is that it encorporates the saltatory search strategy. The model simulates search only during pauses and simulates no search during the movement phase. Furthermore the model provides no time for handling because grayling do not pause after an attack. However, Evans & O'Brien (1986) have argued that white crappie, which do pause after an attack, do not use the pause to handle the consumed prey but use it to search for another prey. Some zooplanktivorous fish may take time to handle prey but grayling do not. There is some evidence that golden shiners feed on 5–10 zooplankton prey and then pause to handle them (O'Brien & Evans, 1991). Occasionally, arctic grayling would pause during a feeding bout and open the mouth while flaring the operculars. We did not measure either the frequency or duration of this behavior; but if it is occurring during the feeding rate experiments this may be part of the reason the model predicts faster feeding rates than observed since this time is not incorporated into the model.

The fit of the model to the feeding rate in the laboratory is good. The fish were clearly showing the same general type of behavior during these feeding rate trials as when the components of the search cycle were measured and the fit by the model is almost perfect. We only used the first 3 minutes of these experiments so as to minimize prey depletion although the model actually takes prey depletion into account. However, in the model, prey depletion is uniformly applied throughout the cage or arena whereas in the cages and arena there was doubtless local depletion. The fit of the model to the feeding trials in the cages is poor especially if the fish are modeled using the large prey swimming speeds. However, *Heterocope* densities were generally low in the lake and the fish did not seem to swim as fast as they did in the laboratory when feeding on just *Heterocope*. When the slower swimming speed observed when grayling fed on small zooplankton is used in the simulation the fit is much better. Unfortunately we did not obtain a precise estimate of fish swimming speed during the cage feeding experiments.

References

Brooks, J. L. & S. I. Dodson, 1965. Predation, body size and the composition of the plankton. Science 150: 28–35.

Confer, J. L. & P. I. Blades, 1975. Omnivorous zooplankton and planktivorous fish. Limnol. Oceanogr. 20: 571–579.

Eggers, D. M., 1982. Planktivore preference by prey size. Ecology 63: 381–390.

Eggers, D. M., 1977. The nature of prey selection by planktivorous fish. Ecology 58: 46–59.

Evans, B. I., 1986. Strategies and tactics of search behavior in salmonid and centrarchid planktivorous fish. Ph.D. University of Kansas, 83 pp.

Evans, B. I. & W. J. O'Brien, 1986. An analysis of the feeding rate of white crappie. Dev. Envir. Biol. Fishes 7: 299–306.

Evans, B. I. & W. J. O'Brien, 1988. A re-analysis of the search cycle of a planktivorous salmonid. Can. J. Fish. aquat. Sci. 45: 187–192.

Holling, C. S., 1959. Some characteristics of simple types of predation and parasitism. Can. Ent. 91: 385–398.

Hrbacek, J., 1962. Species composition and the amount of the zooplankton in relation to fish stock. Rozpr. Cesk. Adad. Ved. 72: 1–116.

Koopman, B., 1980. Search and screening. Pergamon Press.

O'Brien, W. J., 1987. Planktivory by freshwater fish: thrust and parry in the pelagia. pp: 3–16. In Kerfoot, W. C. & A. Sih (eds). Predation. Direct and Indirect Impacts on Aquatic Communities. University Press of New England.

O'Brien, W. J., N. A. Slade & G. L. Vinyard, 1976. Apparent size as the determinant of prey selection by bluegill sunfish (*Lepomis macrochirus*). Ecology 57: 1304–1310.

O'Brien, W. J., B. I. Evans & H. I. Browman, 1989. Flexible search tactics and efficient foraging in planktivorous fish. Oecologia 80: 100–110.

O'Brien, W. J., H. I. Browman & B. I. Evans, 1990. Search strategies of foraging animals. American Scientist. 78: 152–160.

O'Brien, W. J. & B. I. Evans, 1991. Saltatory search behavior in five species of planktivorous fish. Verh. int. Ver. Limnol. 24: 2371–2376.

Paloheimo, J. E. & R. R. Fulthorpe, 1987. Factors influencing plankton community structure and production in freshwater lakes. Can. J. Fish. aquat. Sci. 44: 650–657.

Schmidt, D. & W. J. O'Brien, 1982. Planktivorous feeding ecology of arctic grayling (*Thymallus arcticus*). Can. J. Fish. aquat. Sci. 39: 475–482.

Skoptsov, V. G. & T. V. Kruppennikova, 1983. Role of planktophagus fish in formation of structure of plankton community of a lake. Sov. J. Ecol. 13: 322–326.

Spencer, C. N. & D. I. King, 1984. Role of fish in regulation of plant and animal communities in eutrophic ponds. Can. J. Fish. aquat. Sci. 41: 1851–1855.

Werner, E. E. & D. J. Hall, 1974. Optimal foraging and the size selection of prey by the bluegill sunfish (*Lepomis macrochirus*). Ecology 55: 1042–4052.

Wetterer, J. K. & C. J. Bishop, 1985. Planktivore prey selection: the reactive field volume model vs. the apparent size model. Ecology 66: 457–464.

Wright, D. I. & W. J. O'Brien, 1984. The development and field test of a tactical model of the planktivorous feeding of white crappie (*Pomoxis annularis*). Ecol. Monogr. 54: 65–98.

Hydrobiologia **240**: 247–258, 1992.
W.J. O'Brien (ed.), Toolik Lake – Ecology of an Aquatic Ecosystem in Arctic Alaska.

The effects of river fertilization on mayfly (*Baetis* sp.) drift patterns and population density in an arctic river

Debra Hinterleitner-Anderson,[1,2] Anne E. Hershey[1,*] & Jeffrey A. Schuldt[3]
[1] *Biology Department, University of Minnesota-Duluth, Life Sciences Building, Duluth, MN 55812, USA (* author for correspondence);* [2] *Present address: RR3 Box 3190A, Hayward, WI 54843, USA);* [3] *Department of Forest Resources, University of Minnesota-St. Paul, St Paul, Minn., USA*

Abstract

The life history, drift behavior, and benthic density of *Baetis* (Ephemeroptera) were examined in the arctic Kuparuk River in conjunction with a whole river fertilization experiment to determine if river fertilization affected *Baetis* drift and life history parameters. Drift was significantly higher in the control section of the river than in the fertilized, which suggested that the control section was a less suitable habitat than the fertilized section. There was no strong linear or exponential relationship between drift and benthic density, suggesting neither of these models are adequate to describe density independence versus density dependence. However, drift patterns in the control and fertilized sections suggest that drift is more of a function of absolute food supply than space or density of individuals. Drift was also sampled along a river transect at 6-hour intervals over a diel cycle. No diel periodicity in arctic summer drift was observed. The number of adults, number of eggs/female, and egg volume were calculated for adult *Baetis* collected in drift samples. There was no clear linear relationship between *Baetis* female dry mass and the number of eggs produced per female. However, a strong linear relationship was observed between individual egg volume and *Baetis* female dry mass, suggesting that larger females tended to produce larger eggs rather than more eggs.

Introduction

The mechanisms initiating and influencing stream invertebrate drift have received considerable attention from stream ecologists (e.g., review papers by Waters, 1972; Muller, 1982) resulting in many different hypotheses and considerable debate. Waters (1965) defined drift as being composed of three major components: behavioral drift, catastrophic drift, and constant drift. Behavioral drift is the result of some behavioral characteristic of the animal, such as a response to changes in light intensity, which causes diel periodicity in mayfly drift (Tanaka, 1960; Waters, 1962). Catastrophic drift is that which occurs as a result of floods or other physical disturbances, whereas constant drift is composed of occasional individuals of all species that for various reasons lose their hold on the bottom and drift (Waters, 1965). This study is concerned with behavioral drift of *Baetis* in the Kuparuk River in arctic Alaska. Sampling was not conducted during any physical disturbances so as not to obtain catastrophic drift effects; the effect of constant drift was assumed to be consistent throughout the experimental conditions in the river.

The influence of population density on behavioral drift has recently been studied in an attempt to determine whether population regulation in streams is a function of population density. Although density-dependent drift has been shown for some stream invertebrates (Walton *et al.*,

1977; Wiley, 1981), it has not been shown for *Baetis* and other Ephemeroptera which dominate stream drift (Bohle, 1978; Corkum, 1978; Wiley & Kohler, 1984; Kohler, 1985).

Several hypotheses concerning stream drift suggest density-dependent population regulation. Muller (1954, 1982) proposed the colonization cycle hypothesis which suggests that as larvae grow, competition for space results in increased drift and colonization of suitable downstream habitats. Muller (1954, 1978) also proposed that adult aquatic insects compensate for the gradual downstream movements of their larvae by flying upstream to oviposit, thus upstream reaches would not be depleted of drifting species. Waters (1966) proposed the excess production hypothesis where, as an insect population density approaches carrying capacity, increasing competition results in increased drift as a means of removal and population control. Both the hypotheses of Muller (1954, 1982) and Waters (1966) suggest that drift is a density dependent agent of population regulation, with the means of regulation being competition between individuals for space, food or both (Hildebrand, 1974).

Food distribution and abundance appear to affect *Baetis* drift rates, at constant population density. Abundant food supply resulted in high local concentrations of *Baetis rhodani* and lack of drift, but depletion of food resulted in a more uniform distribution followed by increased drift (Bohle, 1978). Increased activity in searching for food at low food levels may result in increased drift (Hildebrand, 1974), and individuals may abandon habitats when their estimate of habitat quality falls to some threshold level (Kohler, 1985).

Diel periodicity of drift has been widely documented in many parts of the world (e.g. Tanaka, 1960; Waters, 1962). In arctic Alaska, Miller & Stout (1986) showed diel periodicity for stoneflies and chironomids in September once there were definite periods of darkness, but no periodicity during the summer period of continuous light. Since the mechanisms producing diel variation in drift are poorly understood (review by Wiley & Kohler, 1984), it is important that this aspect of *Baetis* behavior in the arctic summer be documented. This is not only important in understanding *Baetis* drift and feeding behavior, but also important in understanding how photoperiod affects life-history (review by Sweeney, 1984).

Study area

The study area is the Kuparuk River, located on the North Slope of the Alaskan Brooks Range (68° 38′ N, 149° 24′ W). This fourth order stream flows northward into the Arctic Ocean, draining rolling tundra. The Kuparuk, with its substratum of boulders and cobbles, is frozen solid until late May or early June, when discharge is typically high due to snowmelt. In the year of this study, discharge was also high in early August due to substantial rainfall. The Kuparuk is characterized by a series of pools and riffles. The specific area of study was comprised of eight riffles separated by pools. The study area is intersected by the Trans-Alaskan Pipeline and the Dalton Highway which allowed easy access.

From 1 July to 15 August in 1983–1989, a section of the Kuparuk River was fertilized with phosphoric acid as part of a multi-investigator ecosystem level experiment (see Peterson *et al.*, 1985). Approximately 10 $\mu g\,l^{-1}$ of phosphorus was added to the river at the same site (0 meters) during the first two summers of the fertilization. The phosphorus addition site was then moved to a location approximately 600 meters downstream in order to evaluate carry-over effects on the ecosystem. Epilithic algal biomass has significantly increased in the fertilized vs. the control section of the river since the start of the fertilization in 1983. Details and results of the enrichment are presented in Peterson *et al.* (1985), Hershey *et al.* (1988) and Hershey & Hiltner (1988).

Baetis, univoltine in the Kuparuk, is a major component in the drift and grazer community and the overwhelmingly dominant mayfly (Hershey, unpublished). It is also a major source of food for arctic grayling (*Thamallus articus*) (Hershey, unpublished), the only fish species in the river (B. J. Peterson, pers. comm.).

Methods

Response of Baetis *drift patterns to river fertilization*

On June 30, July 13, July 24 and Aug. 12, 1988, drift samples were taken at four riffles in the fertilized section (1000 m, 1400 m, 2000 m, 2500 m from 1983–1984 fertilization site) and at four riffles in the control section (−100 m, −400 m, −500 m, −1000 m) of the Kuparuk River. At each site two fifteen minute drift samples were taken above and two below the riffle with standard drift nets (0.14 m^2, 350 μm mesh). Current readings were taken at the middle of the net opening with a Marsh McBirney current meter. Current readings were corrected for percent of net area submerged in order to calculate the number per m^3 (no./m^3) of *Baetis* nymphs in the drift. Drift samples were concentrated through a 150 μm mesh net and preserved in 95% ethanol.

Baetis benthic density was sampled on June 30, July 8, July 24, and Aug. 8, 1988 at approximately the same river sites by scrubbing four rocks into a plastic basin for an estimated upper rock area of 363 ± 18 cm^2. The contents were sieved through the 150 μm mesh net and preserved in 95% ethanol (Hershey & Hiltner, 1988). Using count data and the area sampled, benthic density in no./m^2 was estimated.

To obtain count data, drift and benthic samples were sorted in a glass pan over a light table. Samples estimated to contain greater than 50 *Baetis* were split with a Folsom sample splitter until a sample size of about 50 *Baetis* was obtained. *Baetis* were then sorted from the resulting sample fraction. The samples or fractions were then examined using a dissecting microscope to make sure all *Baetis* were removed. Lengths of *Baetis* were measured using a camera lucida on a dissecting microscope to project an image onto a digitizing pad.

A two-way analysis of variance was used to compare *Baetis* drift density between control and fertilized sections and between sampling dates. All drift estimates were $\log_e$ transformed to correct for heterogeneity of variances. Net drift across riffles was calculated as the mean drift of the two nets below each riffle minus the mean drift of the two nets above each riffle. Net drift in the control versus fertilized sections on each sampling date was compared using a two-way ANOVA on $\log_e$ transformed data to test the hypothesis that *Baetis* had a greater propensity to leave riffles in the control section due to lower abundance of food. A two-way ANOVA was also used to determine the effect of date and treatment on benthic density. Linear and exponential models were used to describe the relationship between drift and benthic density to determine if drift was density independent versus density dependent (Hildebrand, 1974):

Linear: Drift density = $\beta_o + \beta_1$ (Benthic density);

Exponential: $\log_e$ (Drift density) = $\beta_o + \beta_1$ (Benthic density).

Baetis *diel drift*

From 0955 hrs. on July 23 to 1350 hrs. on July 24, 1988 drift samples were taken at six-hour intervals to determine diel periodicity of *Baetis* drift. Drift was sampled at two fertilized riffles (1000 m, 1400 m) and at two control riffles (−100 m, −400 m). As in the seasonal survey, two fifteen minute drift samples were taken above and two below each riffle with standard drift nets. Five river transects were completed over the two dates. The time of each sample and weather conditions present at each site were recorded.

Current readings, sample processing, and analyses were completed using the same procedures described in the preceding section on the response of *Baetis* drift patterns to river fertilization. A two-way ANOVA was used to make comparisons between drift in control and fertilized sections and time of day. The dependent variable used was the mean no./m^3 *Baetis* drifting during each sampling period.

Life history parameters for Baetis

Adult *Baetis* were collected in the drift samples described above. Total number/m^3 of adult *Baetis* sampled was used as an estimate of numbers ovipositing. Oviposition can be distinguished from emergence because mayflies emerge as subimagos and then go through a final molt before egg laying occurs (e.g., Borror *et al.*, 1981). Adult *Baetis* lengths were measured using a camera lucida, dissecting microscope, and digitizing pad. Female *Baetis* dry mass was estimated from body length measurements using the predictive equation:

$$ln\mathrm{W} = -5.714 + 3.2\ln\mathrm{L},$$

where W = dry mass, L = length (Smock, 1980).

Gravid females were gently split open and the number of eggs per female was counted. A random subsample of five eggs was taken; egg lengths were measured using the camera lucida and digitizing pad. The widths of sixteen eggs were also measured to determine length/width ratios for eggs. The average of the sixteen length/width ratios was used to back calculate widths for all eggs measured. Egg volume was calculated using the standard formula for the volume of an ellipsoid ($4ab^2/3$, where 2a = length and 2b = width).

Relationships of *Baetis* female dry mass to the number of eggs per female and egg volume were determined using regression analyses (Sweeney & Vannote, 1981).

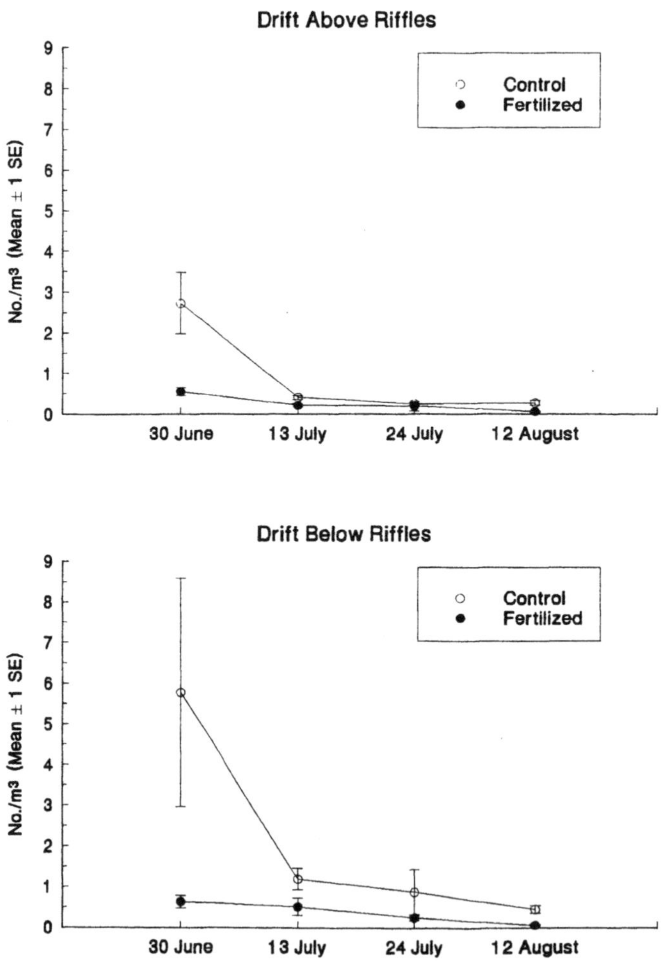

Fig. 1. *Baetis* drift density (no./m^3) above (top panel) and below (bottom panel) riffles in the control and fertilized sections of the Kuparuk River during 1988.

Results

Response of Baetis *drift patterns to river fertilization*

Drift was significantly higher in the control section of the river than in the fertilized section for samples taken both above ($F_{1,24} = 26.23$, $P < 0.0001$) and below each riffle ($F_{1,24} = 32.34$, $P < 0.0001$). This trend was especially evident early in the season when drift was high in the control section (Fig. 1). For both comparisons, date of sampling was significantly different (above $F_{3,24} = 19.01$, $P < 0.0001$; below, $F_{3,24}$, $P < 0.0001$). The date of sampling and treatment interaction was not significant for either comparison (up $F_{3,24} = 1.064$, $P > 0.05$; down $F_{3,24} = 1.39$, $P > 0.05$). Lack of a significant interaction term suggests that drift was consistently higher in the control than fertilized section of the river throughout the season in spite of significant seasonal changes in drift.

Net drift was significantly higher in the control section compared to the fertilized section ($F_{1,22} = 6.72$, $P < 0.05$) and date ($F_{3,22} = 1.54$, $P > 0.05$) and date by treatment interaction ($F_{3,22} = 0.32$, $P > 0.05$) were not significant (Fig. 2). These results suggest that *Baetis* had a greater tendency to emigrate from the control riffles than from the fertilized riffles.

Baetis benthic density was significantly different between dates ($F_{3,24} = 98.45$, $P < 0.0001$). *Baetis* benthic density increased significantly from June 30 to July 8 and then significantly declined throughout the season (Fig. 3). The initial increase was probably due either to hatching of nymphs or to inadequate sampling of early instars, and the subsequent decline was probably due to drift, emergence, and mortality. In the early season (June 30 and July 8) density appeared more abundant in the control section (Fig. 3). In the late season (July 27) density appeared more abundant in the fertilized section (Fig. 3). By Aug. 8, most emergence was completed, thus density comparison between sections is not relevant. There was, however, no overall significant treatment effect for density and no significant date by treatment interaction (Treatment $F_{3,24} = 1.76$, $P > 0.05$; Treatment × date $F_{3,24} = 2.5$, $P = 0.8$).

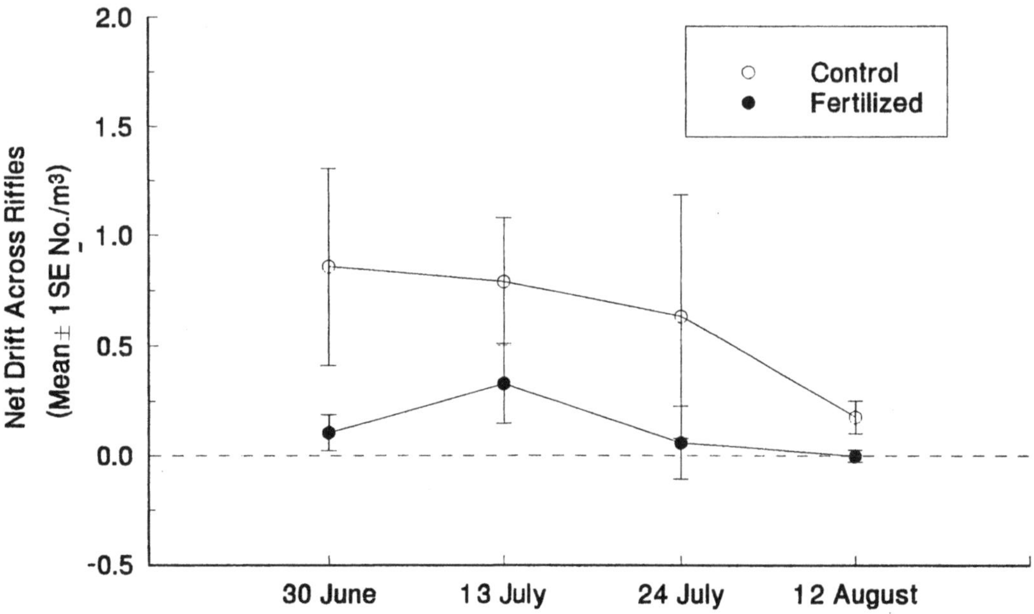

Fig. 2. Net drift across riffles (below-above) in the control and fertilized sections of the Kuparuk River. The figure shows that *Baetis* net drift was higher in the control section indicating that nymphs tended to leave these riffles compared to fertilized riffles.

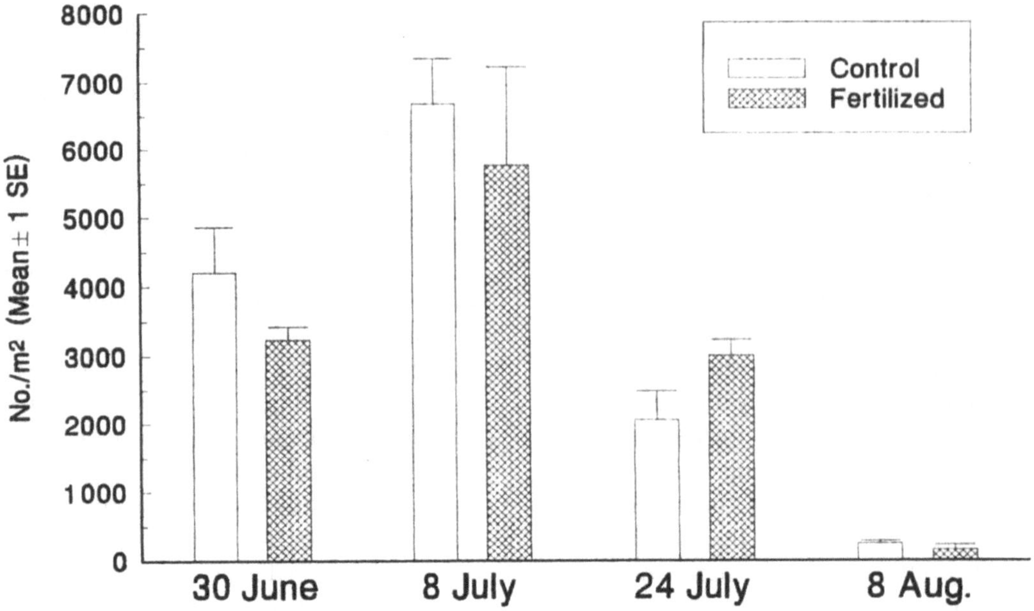

Fig. 3. Baetis benthic densities in the control and fertilized sections of the Kuparuk River during 1988.

An exponential regression model was slightly more satisfactory for describing the relationship of drift to benthic density than the linear model, but neither were very satisfactory (Table 1, Fig. 4). In the control section, the exponential model was significant and the lack of fit test indicated that this model was adequate, however, only 23% of the variation in drift was due to benthic density. In the fertilized section, the exponential model gave a significant lack of fit. The linear model was not significant in either the fertilized or the control section (Table 1).

Baetis *diel drift pattern*

Drift was not significantly different between the five sampling times for samples taken above ($F_{4,10} = 2.14$, $P = 0.15$) and below ($F_{4,10} = 0.27$, $P = 0.89$) each riffle suggesting no diel periodicity in the arctic summer drift patterns (Fig. 5). Treatment effect was also non-significant above riffles ($F_{1,10} = 3.12$,·$P = 0.11$), but was significant below riffles ($F_{1,10} = 9.7$, $P = 0.01$). Despite the overall non-significant treatment effect above riffles, *a priori* comparisons indicated a significant treatment

Table 1. Results of regression analysis used to describe the relationship between *Baetis* drift density (no./m^3) and *Baetis* benthic density (no./m^2). Linear model: $D = B_0 + B_1$ (Density). Exponential model: $Log_eD = B_0 + B_1$ (Density).

Model	n	Sig. of Reg. F(p)	R^2	Lack of Fit F(p)
Control:				
$D = 0.51 + 0.0001$ (Density)	24	2.66 (0.12)	0.11	1.05 (>0.05)
$Log_eD = -1.02 + 0.0002$ (Density)	24	6.75 (0.02)	0.23	1.40 (>0.05)
Fertilized:				
$D = 0.17 + 0.00004$ (Density)	24	2.56 (0.12)	0.10	7.04 (<0.01)
$Log_eD = -2.64 + 0.0003$ (Density)	24	12.48 (0.002)	0.37	4.67 (<0.05)

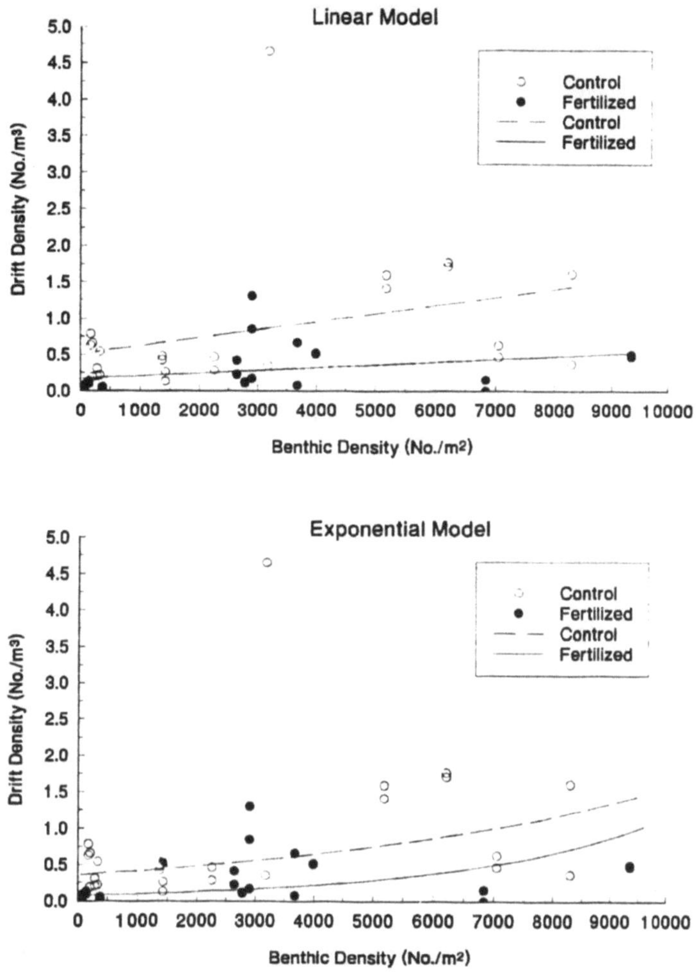

Fig. 4. Linear and exponential fits of *Baetis* drift density as a function of benthic density in control and fertilized sections of the Kuparuk River. Neither model gave a satisfactory fit to the data.

effect in the first sampling period for samples taken above each riffle (Fig. 5). Multiple comparisons also showed a significant treatment effect from the fourth sampling period for samples taken below each riffle ($P < 0.05$, Fig. 5). In all the below riffles; drift appeared higher in the control than in the fertilized section. These results are consistent with the generally higher drift observed in the control section of the river (Fig. 1). The transect run and treatment interaction was not significant for samples taken above or below riffles, suggesting that the effect of fertilization on drift did not vary with time of day.

Life history parameters for Baetis *in arctic Alaska*

The no./m^3 of adults collected as a measure of oviposition was not significantly different between the control and fertilized sections of the river. In general, the number of adults collected was not significantly different on different sampling dates except that on June 30 no adults were found. The date of sampling and treatment interaction was not significant. This result indicates that the adult density was not significantly different between the control and fertilized sections of the river throughout the season and supports the conclu-

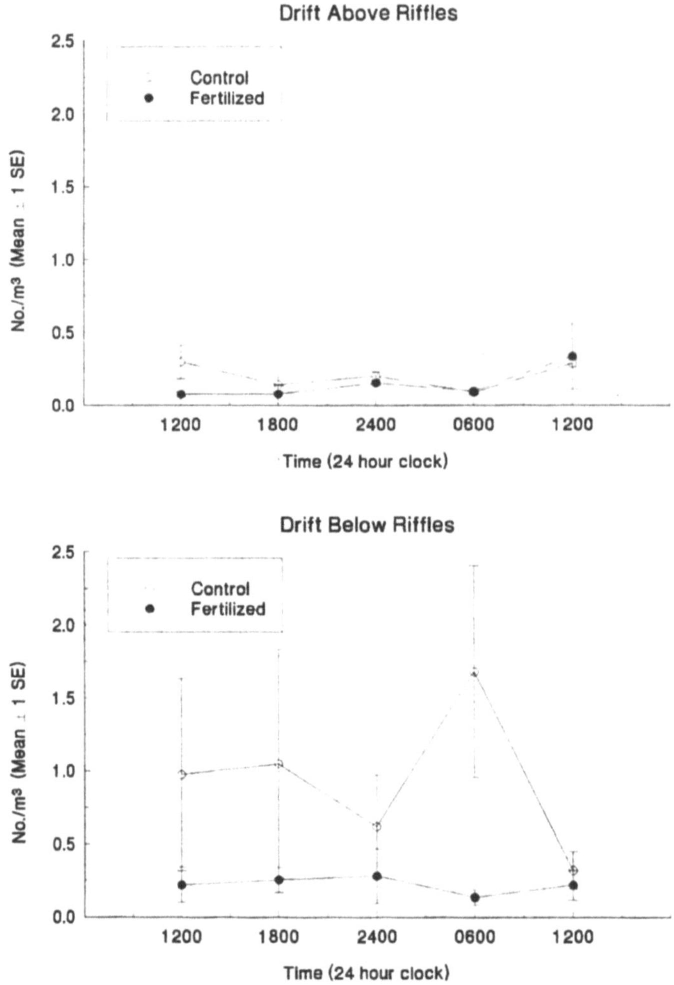

Fig. 5. Diel drift of *baetis* nymphs above (top panel) and below (bottom panel) riffles. The figure shows no strong diel drift pattern during this 24-light diel period of the arctic summer.

sion that these were ovipositing adults and had not necessarily emerged from the section of the river where they were collected. An average egg length/width ratio of 1.471 ± 0.144 was computed to back calculate individual egg widths for egg volume estimates. One data point, determined to be an extreme outlier at 13.5 times the interquartile range, was excluded (Devore & Peck, 1986).

The number of eggs per female was positively correlated with adult mass up to about 2.0 mg. Females larger than 2.0 mg tended to produce fewer eggs; therefore, there was no linear relationship ($r^2 = 0.085$) between female dry mass and the number of eggs per female. To get a better estimate of this relationship a quadratic equation was used:

$$Y = 64.5 + 112.7x - 226.7x^2.$$

The quadratic gave a slightly better fit ($r^2 = 0.212$), although neither model fit the data well. There was, however, a strong linear relationship ($r^2 = 0.746$) between egg volume and female dry mass (Fig. 6) suggesting that larger females tended to produce larger eggs rather than more eggs.

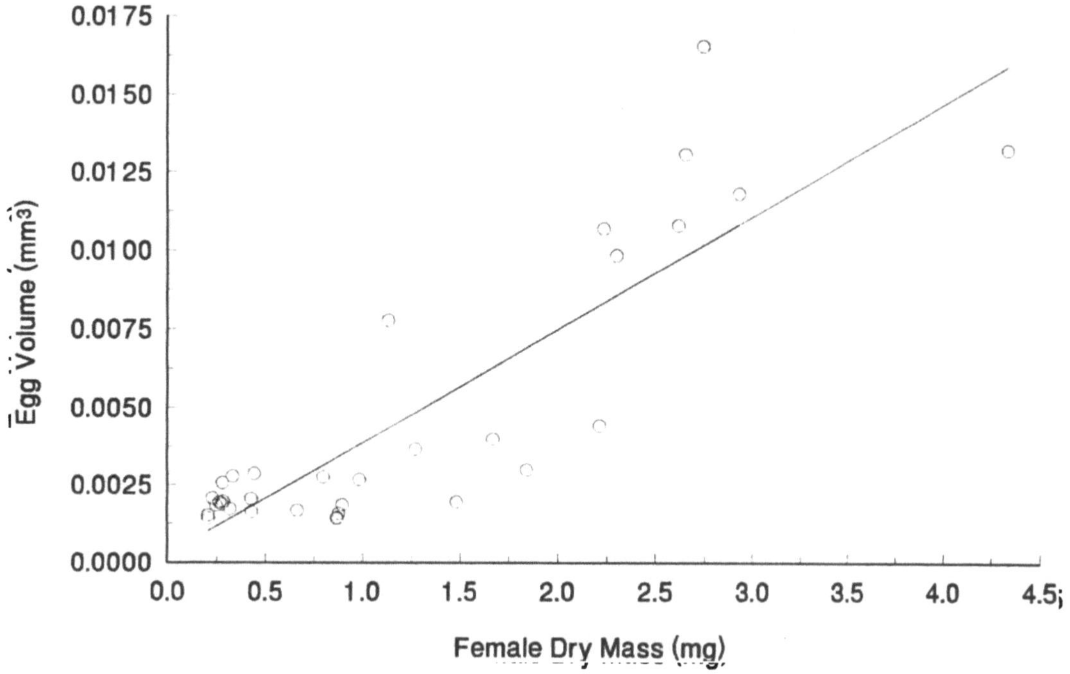

Fig. 6. Baetis egg volume (mm^3) as a function of female dry mass, indicating that egg volume increases linearly with female dry mass.

Discussion

Response of Baetis *drift patterns to river fertilization*

The results of this study show that drift density was significantly higher both upstream and downstream of control riffles than fertilized riffles, and that net drift out of riffles was higher in the control section. This suggests that nymphs were emigrating from the control riffles more than the fertilized riffles. Since epilithic algal biomass has increased significantly in the fertilized section of the river since the start of the fertilization in 1983 (Peterson *et al.*, 1985; Peterson, personal communication) the significantly higher drift of *Baetis* in the control section of the Kuparuk indicates a strong relationship between food abundance and drift.

The generally poor relationship between drift and benthic density suggests that using linear and exponential models to describe density independent versus density dependent drift is not adequate for these data. However, the comparisons of drift in control and fertilized sections of the river are more consistent with the conclusion that drift is a function of per capita food supply rather than interference or density of individuals. That conclusion is consistent with the findings of Hildebrand (1974), Bohle (1978) and Kohler (1984, 1985) in artificial streams. They found that drift was largely active, and density independent. Hildebrand (1974) concluded that increases in drift behavior result from increased activity in searching for food at low food levels, and Kohler (1985) concluded that individuals tend to abandon habitats when their estimate of habitat quality falls below some threshold level. Thus, a density-independent factor was the major force regulating drift in this population; the hypothesis that drift is a density dependent agent of population regulation (Muller, 1954, 1982; Waters, 1966) is not supported for this *Baetis* population. Dudley *et al.* (1986) and Richards & Minshall (1988) found higher *Baetis* densities in the presence of macroalgae and abundant periphyton. The late season trend toward higher *Baetis* density in the fertilized section is consistent with their findings.

The patterns of *Baetis* drift and benthic density are consistent with Muller's hypothesis (1954, 1982) that adult aquatic insects compensate for the gradual downstream movements of the larvae by flying upstream to oviposit. The significantly higher no./m^3 *Baetis* drifting in the control section in the early season (June 30) may be due not only to low food abundance but also to strong recruitment upstream the previous year. Allan (1989) found that small *Baetis* drift farther than large *Baetis* and that they drift farther at higher current velocity. In the Kuparuk, early season discharge and current velocity was high and *Baetis* were smaller than in the late season. This combined with the low per capita food abundance in the control section might account for the higher *Baetis* drift density in the control section early in the season. Small size and high current were clearly not the overriding factors, however, since early season drift in the fertilized section was not elevated. In the late season when *Baetis* was more abundant in the fertilized section, drift in the fertilized section did not increase. Some other factor was influencing drift more than density; our results suggest that the other factor is food abundance.

Baetis *diel drift pattern*

The results of this study indicate that there is no diel periodicity in arctic summer drift. Drift was consistent throughout the sampling period with no optimal time at which drift occurred. It is of interest to point out that during the third sampling period (2400 hours) the sky was cloudy, overcast and darker yet drift was not significantly different from any other time of the day. This may suggest that short-term changes in light availability are not enough to significantly influence drift patterns. These results support those of Elliot (1968) who found that over 24 hours of continuous light in an artificial stream the activity level of *Baetis* nymphs did not change. Activity increased, however, the first night after the light was turned off. In the arctic, once there are definite periods of darkness in the fall, *Baetis* nocturnal drift may increase. In Imnavait Creek, near the Kuparuk, marked diurnal drift of stoneflies and nocturnal drift of chironomids has been observed, but only in September, once there were definite periods of darkness (Miller & Stout, 1986).

Kohler (1985) suggested that selection directly influences drift behavior and that the minimization of day drift may be adaptive. It has been suggested that nocturnal drift is an adaptation to minimize mortality due to fish predation (Allan, 1984) or invertebrate predation (Corkum & Pointing, 1979). Turcotte & Harper (1982) examined drift in a fishless mountain stream and found no evidence for consistent diel periodicities with nocturnal maxima. Prolonged summer photoperiod at high latitudes might partially compensate for the reduced feeding rates associated with the low temperatures of more northern habitats because an increase in daylength increases the feeding period for day-active species (Sweeney, 1984). This would also suggest that there are no optimal times for *Baetis* feeding in the arctic summer since the threats from predation would be continuous.

Life history parameters for Baetis *in the Kuparuk River*

In the Kuparuk, there was no linear relationship between *Baetis* female dry mass and the number of eggs/female. This result is in contrast to the linear relationships previously found between fecundity and adult dry weight for other Ephemeroptera species (Sweeney & Vannote, 1981, review by Sweeney, 1984). There was, however, a strong linear relationship between egg volume and female dry mass suggesting that larger females tend to produce larger eggs rather than more eggs. As these results show, larger adult female *Baetis* produce larger and presumably better quality eggs (Richards & Kolderie, 1957); egg volume may be a better indicator of reproductive fitness than egg number. This is interesting because Hershey (unpublished data) for the years 1984 to 1986 found that there were significantly larger *Baetis* nymphs in the fertilized section of the river than in the control. Since the fertilization resulted in larger nymphs which presumably became larger adults, these adults should have had a higher reproductive success.

Summary

Environmental factors have been shown to affect many specific life-history parameters in aquatic insects. It has been shown that fertilization, which resulted in increased food abundance, influences drift density and reproductive fitness of an arctic *Baetis* population. *Baetis* had a significantly greater tendency to emigrate from the control riffles and drift was generally higher under control than fertilized conditions. Drift was a function of food supply and not dependent on the density of individuals.

Baetis diel drift behavior in the arctic summer showed no diel periodicity with no optimal time at which drift occurs. This would suggest that there are no optimal times for feeding since the threats from predation should be continuous.

References

Allan, D. J., 1984. The size composition of invertebrate drift in a Rocky Mountain stream. Oikos 43: 68–76.

Allan, D. J. & B. P. Feifarek, 1989. Distances travelled by drifting mayfly nymphs: factors influencing return to the substrate. J. N. Am. Benthol. Soc. 8: 322–330.

Bohle, H. W., 1978. Beiziehungen zwischen dem Nahrungsangebot, der Drift und der raumlichen Verteilung bei Larven von *Baetis rhodani* (Pictet) (Ephemeroptera: Baetidae). Arch. Hydrobiol. 84: 500–525.

Borror, D. J., D. M. DeLong & C. A. Triplehorn, 1981. (Fifth edition). An Introduction to the Study of Insects. Saunders College Publishing, Philadelphia, Pa: 186–196.

Butler, M. G., 1984. Life histories of aquatic insects. In V. H. Resh & D. M. Rosenberg, (eds) The ecology of aquatic insects. Praeger Scientific, N.Y., USA: 24–55.

Clifford, H. F. & H. Boerger, 1974. Fecundity of mayflies (Ephemeroptera) with special reference to mayflies of a brown-water stream of Alberta, Canada. Can. Ent. 106: 1111–1119.

Corkum, L. D., 1978. The influence of density and behavioural type on the active entry of two mayfly species (Ephemeroptera) into the water column. Can. J. Zool. 56: 1201–1206.

Corkum, L. D. & P. J. Pointing, 1979. Nymphal development of *Baetis vagans* McDunnough (Ephemeroptera: Baetidae) and drift habits of large nymphs. Can. J. Zool. 57: 2348–2354.

Devore, J. & R. Peck, 1986. Statistics, The Exploration and Analysis of Data. West Publishing Co., St. Paul, Mn: 92–96.

Dudley, T. L., S. D. Cooper & N. Hemphill, 1986. Effects of macroalgae on a stream invertebrate community. J. N. Am. Benthol. Soc. 5: 93–106.

Elliot, J. M., 1968. The daily activity patterns of mayfly nymphs (Ephemeroptera). J. Zool. Lond. 155: 201–221.

Hershey, A. E., A. L. Hiltner, M. A. J. Huller, M. C. Miller, J. R. Vestal, M. A. Lock, S. Rundle & B. J. Peterson, 1988. Nutrient influence of a stream grazer: *Orthocladius* microcommunities respond to nutrient input. Ecology 69: 1383–1392.

Hershey, A. E. & A. L. Hiltner, 1988. Effect of a caddisfly on black fly density: interspecific interactions limit black flies in an arctic river. J. N. Am. Benthol. Soc. 7: 188–196.

Hildebrand, S. G., 1974. The relation of drift to benthos density and food level in an artificial stream. Limnol. Oceanogr. 19: 951–957.

Humpesch, U. H., 1979. Life cycles and growth rates of *Baetis* spp. (Ephemeroptera: Baetidae) in the laboratory and in two stony streams in Austria. Freshwat. Biol. 9: 467–479.

Kohler, S. L., 1984. Search mechanism of a stream grazer in patchy environments: the role of food abundance. Oecologia (Berlin) 62: 209–218.

Kohler, S. L., 1985. Identification of stream drift mechanisms: An experimental and observational approach. Ecology 66: 1749–1761.

Miller, M. C. & J. R. Stout, 1986. Effects of a controlled under-ice oil spill on invertebrates of an arctic and a subarctic stream. Envir. Pollut. Ser. A, 42: 99–132.

Merritt, R. W. & K. W. Cummins, 1984 (second edition). An Introduction To The Aquatic Insects. Kendall/Hant Publishing Co. Dubuque, Iowa: 94–125.

McClure, R. G. & K. W. Stewart, 1976. Life cycle and production of the mayfly *Choroterpes* (Neochoroterpes) *mexicanus* Allen (Ephemeroptera: Leptophlebiidae). Ann. Ent. Soc. Am. 69: 134–144.

Muller, K., 1954. Investigations on the organic drift in North Swedish streams. Inst. Freshwater Res. Drottningholm, Rep. 34: 133–148.

Muller, K., 1982. The colonization cycle of freshwater insects. Oecologia (Berlin) 52: 202–207.

Pennak, R. W., 1978 (2nd edn). Freshwater invertebrates of the United States. John Wiley and Sons Publishing.

Peterson, B. J., J. E. Hobbie, A. E. Hershey, M. A. Lock, T. E. Ford, J. R. Vestal, V. L. McKinley, M. A. . Huller, M. C. Miller, R. M. Ventullo & G. S. Volk, 1985. Transformation of a tundra river from heterotrophy to autotrophy by addition of phosphorus. Science 229: 1383–1386.

Richards, A. G. & M. Q. Kolderie, 1957. Variation in weight, developmental rate, and hatchability of *Oncopeltus* eggs as a function of the mother's age. Ent. News. 68: 57–64.

Richards, C. & G. W. Minshall, 1988. The influence of periphyton abundance on *Baetis bicaudatus* distribution and colonization in a small stream. J. N. Am. Benthol. Soc. 7: 77–86.

Shiozawa, D. K., 1983. Density independence versus density dependence in streams. In J. R. Barnes & G. W. Minshall, (eds), Stream Ecology, Application and testing of general ecological theory. Plenum Press, New York, USA: 55–73.

Smock, L. A., 1980. Relationships between body size and biomass of aquatic insects. Freshwat. Biol. 10: 375–383.

Sweeney, B. W., 1984. Factors influencing life-history patterns of aquatic insects. Pages 56–100 in V. H. Resh and D. M. Rosenberg (eds), The ecology of aquatic insects. Praeger Scientific, New York, New York, USA.

Sweeney, B. W. & R. L. Vannote, 1981. *Ephemerella* mayflies of White Clay Creek: bioenergetic and ecological relationships among six coexisting species. Ecology 62: 1353–1369.

Tanaka, H., 1960. On the daily change of the drifting of benthic animals in streams, especially on the types of daily change observed in taxonomic groups of insects. Bull. Freshwat. Fish. Res. Lab. Tokyo 9: 13–24.

Waters, T. F., 1962. Diurnal periodicity in the drift of stream invertebrates. Ecology 43: 316–320.

Waters, T. F., 1965. Interpretation of invertebrate drift in streams. Ecology 46: 327–334.

Waters, T. F., 1966. Production rate, population density, and drift of a stream invertebrate. Ecology 47: 595–604.

Waters, T. F., 1972. The drift of stream insects. Annu. Rev. Ent. 17: 253–272.

Walton, O. E., Jr., S. R. Reice & R. W. Andrews, 1977. The effects of density, sediment, particle size and velocity on drift of *Acroneuris abnormis* (Plecoptera). Oikos 28: 291–298.

Wiley, M. J., 1981. Interacting influences of density and preference on the emigration rates of some lotic chironomid larvae (Diptera: Chironomidae). Ecology 62: 426–438.

Wiley, M. J. & S. L. Kohler, 1984. Behavioral adaptations of aquatic insects. In V. H. Resh & D. M. Rosenberg (eds), The ecology of aquatic insects. Praeger Scientific, New York, New York, USA: 101–133.

Wise, E. J., 1980. Seasonal distribution and life histories of Ephemeroptera in a Norhumbrian River. Freshwat. Biol. 10: 101–111.

Hydrobiologia **240**: 259–265, 1992.
W.J. O'Brien (ed.), Toolik Lake – Ecology of an Aquatic Ecosystem in Arctic Alaska.

Black fly (*Diptera*: *Simuliidae*) response to phosphorus enrichment of an arctic tundra stream

Allison L. Hiltner[1] & Anne E. Hershey[2,*]
[1]*Department of Zoology, University of Wisconsin, Madison, WI 53706, USA (present address: U.S. Environmental Protection Agency 5HR-11, 230 S. Dearborn, Chicago, IL 60604, USA);* [2]*Department of Biology, University of Minnesota-Duluth, Duluth, MN 55812, USA (* author for correspondence)*

Abstract

This study reports on the density, growth, and production response of the dominant black fly, *Prosimulium martini*, to whole river fertilization of the Kuparuk River in arctic Alaska during the summer of 1984. Beginning in 1983, a long term study of fertilization effects was initiated in the Kuparuk River. Increased nutrient supply stimulated algal and microbial biomass and microbial activity, which in turn affected the larval growth and abundance of *Prosimulium*. This experiment allowed us to isolate the effects of nutrient supply from other factors in determining black fly growth and abundance.

Phosphorus addition had the following indirect effects on *Prosimulium*: growth was higher, but abundance decreased in the enriched section, leading to a net decrease in secondary production from 2.62 g m^{-2} yr^{-1} to 0.77 g m^{-2} yr^{-1}. *Prosimulium* emergence rates were not measureably affected. The decrease in abundance and production appears to be a result of competitive displacement by the caddisfly *Brachycentrus americanus* which increased in abundance in the fertilized section of the river.

Introduction

Inorganic nutrient addition is known to have profound effects on the productivity of lake ecosystems (Edmondson, 1970; Schindler, 1974, 1977). Similar responses in stream ecosystems have recently been documented. Studies have shown increased algal biomass due to nutrient addition to streams (Stockner & Shortreed, 1978; Peterson *et al.*, 1985) and stream microcosms (Grimm & Fisher, 1986; Pringle, 1987; Hart & Robinson, 1990). Less attention has been paid to the response of benthic macroinvertebrates to nutrient enrichment, but evidence is accumulating that invertebrates may respond in terms of either density or biomass or both (Elwood *et al.*, 1981; Hershey *et al.*, 1988; Hart & Robinson, 1990).

Increases in food quantity (Colbo & Porter, 1978; Gallepp, 1977) and quality (Anderson & Cummins, 1979; Ward & Cummins, 1979) can stimulate somatic growth, increase survivorship, and accelerate developmental rates of stream macroinvertebrates in the laboratory (see Sweeny 1984 for literature review). Similarly, field populations of filter-feeding black flies and caddisflies have higher densities, larger larvae, and earlier emergence below lake outlets and impoundments compared to populations where food quality (estimated as the ratio of organic to inorganic matter in the seston) is lower (Richardson, 1984; Wotton, 1977, 1979; Parker & Voshell, 1982; Carlsson *et al.*, 1977; Oswood, 1979; Mackay & Waters, 1986). In some cases, increases in periphyton production may reduce black fly abundance by decreasing the surface area available for their attachment (Gersabek & Merritt, 1979; Chutter, 1968; Zahar, 1951).

The measurement of secondary production

should integrate the effect of food enrichment on growth, development, and abundance. Secondary production of stream insects has been measured often (e.g., Smock *et al.*, 1985; Waters & Hokenstrom, 1980; Freeman & Wallace, 1984; Parker & Voshell, 1983; Merritt *et al.*, 1982), but few studies have experimentally assessed whether secondary production in streams is limited by food or nutrient supply. Parker & Voshell (1983) compared production of filter feeding caddisflies in an impoundment where food quality of seston was high to production of the same species in a river with poor quality food. They concluded that higher food quality increased the production of these filter feeders. Merritt *et al.* (1982) concluded that increased stream temperature and food supply enhanced the production of black flies in Michigan, but were not able to separate the two effects. Krueger & Waters (1983) measured secondary production in three streams of varying water quality. They found a positive correlation between alkalinity and nitrate concentration and production of macroinvertebrates, lending support to the idea that in some cases nutrient supply may ultimately limit the production of these organisms.

The Kuparuk River in arctic Alaska was enriched with phosphorus during the summers of 1983–1985 to assess the importance of phosphorus in a stream where natural levels were very low. The Kuparuk River originates in the Brooks Range in northern Alaska and flows northward into the Arctic Ocean. At the study site (68° 38′ N, 149° 24′ W), it is a fourth order stream, with a substrate of boulders and cobbles. A more complete description of the study site is given in Peterson *et al.* (1986). Due to heavy rains during the summer of 1984 discharge was unusually high, ranging from low flows of 1.5–2 $m^3 sec^{-1}$ to 13 $m^3 sec^{-1}$ during spates. Temperature ranged from 2–14 °C during the experimental period.

During the months of July and August in 1983–1985 phosphoric acid was added to the river, increasing the concentration of PO_4^{3-} in the water from approximately 1 $\mu g l^{-1}$ to an average of 10 $\mu g l^{-1}$. After the 1983 nutrient addition, biomass of epilithic algae increased up to 10 times and the activity of the epilithic microbial community increased proportionately in the enriched section when compared to the upstream control section (Peterson *et al.*, 1985). A similar response was seen in 1984 (Peterson & Hobbie, unpublished data). Since the major source of particulate organic carbon in this relatively short section of the Kuparuk prior to the nutrient addition was peat eroded from the banks (Peterson *et al.*, 1986), this represents an increase in both the quantity and quality of food available to benthic invertebrates, but no change in other factors affecting growth, life history, and abundance of macroinvertebrates.

This study documents the 1984 response of the filter feeding black fly *Prosimulium martini*, to the phosphorus enrichment. *Prosimulium* was in its first instar when the fertilization began and it emerged from early to mid-August when the fertilization was terminated for the season. Peterson *et al.* (1985) showed that 23 days after the 1983 enrichment began, *Prosimulium* larvae 340 m below the site of P enrichment were significantly larger in total length than upstream controls, but they did not evaluate growth and life history effects.

In this study, *Prosimulium* was sampled in enriched and upstream unfertilized sections of the Kuparuk River from 25 June to 14 August 1984 to answer the following questions: 1) Do these larvae grow to a larger size when primary productivity is stimulated by nutrient addition? 2) Is developmental time accelerated? 3) Are the larvae more abundant in the enriched section? 4) Is secondary production increased in the enriched section?

Methods

Benthic macroinvertebrates were sampled every 10 days from 25 June to 14 August 1984. Three samples of four cobbles each with an average surface area of 363 ± 18 cm^2 were collected at locations 290 and 500 meters upstream from the nutrient addition site and 140, 340, 1005, 2015 and 2950 m downstream. Additional stations 10 m upstream and 590 m downstream were sampled during early summer. Samples were collected in riffles with fastest flow. Mean surface area of samples was determined by digitizing photographs of several representative samples with an Apple IIe graphics tablet. The rocks were scrubbed with a nylon brush in stream water, then the water was filtered through a 60 μm mesh net. Insects were preserved in 95% ethanol. Forty black flies from each sample (or as many as were present if less than 40) were mounted for identification and their total body lengths were measured with an ocular micrometer. Pupae were measured when present.

Density of all black flies was estimated from count data. Other species were included in the density estimate because that approach provides a better seasonal picture of an important group of filter feeders. Comparisons were made between total numbers of black flies in samples from upstream versus fertilized section of the river using t-tests between river sections for each date. Tests were performed on the $\log_e$ transformed data when necessary to stabilize variances. Black fly abundance prior to the first year of P enrichment was estimated from nine stored Surber samples collected in 1978 (curtesy of B. J. Peterson, The Ecosystems Center, MBL) from approximately 750, 1000, and 1500 m below the future P addition site. However, since these samples were taken using a different sampling technique, data are not strictly comparable.

Growth of *Prosimulium* was inferred from differences in length between upstream and fertilized sections of the river. For length comparisons, including black fly species other than *Prosimulium* would not have been appropriate. Mean lengths were compared between upstream and fertilized sections using the same analyses as used for density comparisons.

Emergence of *Prosimulium* was estimated by dividing the number of empty pupal cases by the number of larvae + pupae + empty pupal cases collected at each site on the three late season sampling dates. Two way analysis of variance by station and date was used to compare emergence.

Prosimulium secondary production was estimated using the size-frequency method (see Benke, 1984). Biomass needed for this estimation was calculated from length using the regression equation of Smock (1980):

$$\log_e \text{Weight (g)} = -5.339 + 2.55 \log_e \text{Length (mm)}.$$

Benke (1979, 1984) proposed that annual production estimates made by the size frequency method be multiplied by 365/CPI. CPI represents the cohort production interval or length of the larval stage in days. *Prosimulium* is univoltine and the length of its larval stage corresponded roughly to the length of the sampling period (very few larvae were collected on the first and last sampling dates), therefore, the CPI was assumed to equal the length of the sampling period. Since production was calculated only for this period no CPI correction was needed. A single production estimate was obtained for each sampling site, and stations were treated as replicates within a river section. A t-test was used to compare upstream to fertilized sections of the river.

Results

Black fly density was negatively affected by fertilization (Table 1). This pattern was significant on five of the seven sampling dates, and the control means were also higher than fertilized means on the remaining two sampling dates (Table 1). Black fly density ranged from 134 ± 24 to $2484 \pm 1030/m^2$ (mean $\pm$ SE), but was high in the control section until *Prosimulium* began its emergence in early August (Table 1).

Black fly density in July 1978 was 6682 ± 354, which was approximately 9 times higher than at similar stations in July 1984. Differences were not significant among the three sites sampled in 1978 (Kruskal-Wallis H = 0.089, *df* = 2, NS).

Prosimulium larval length was significantly greater in the fertilized section of the river on July 15, but not on the other sampling dates (Table 2). Lack of a difference in length on the earlier sampling dates indicates that there was no carryover

Table 1. Mean $\pm$ SE density/m² of black flies in upstream and fertilized sections of the Kuparuk River for each sampling date in 1984. Number of samples used in each comparison is given in parentheses. Direction of statistically significane differences ($p < 0.05$, *t*-tests) is indicated by inequalities between treatment abreviations.

Date	Mean ± SE		
	Upstream	Fertilized	*t*-test result
25 June 84	2408 ± 619 (5)	2195 ± 536 (16)	F = U
6 July 1984	2408 ± 303 (10)	661 ± 210 (6)	F < U
14 July 84	1680 ± 550 (6)	644 ± 137 (18)	F < U
24 July 84	2484 ± 1030 (6)	959 ± 215 (15)	F < U
2 August 84	1602 ± 384 (6)	594 ± 105 (9)	F < U
7 August 84	799 ± 225 (3)	236 ± 78 (9)	F < U
13 August 84	225 ± 54 (3)	134 ± 24 (9)	F = U

Table 2. Mean $\pm$ SE length (mm) of *Prosimulium martini* summarized for each sampling data at upstream control and fertilized sections of the Kuparuk River in 1984. N, which is given in parentheses, indicates the number of samples in each section of the river; mean black fly length from each sample are the observations used in the calculations. Pupal lengths and widths are mean values for samples containing pupae, regardless of date collected. Results of statistical comparisons between sections (*t*-tests, $p < 0.05$) are indicated by inequalities.

Date	Mean ± SE		
	Upstream	Fertilized	*t*-test result
28 June '84	3.2 ± 1.4 (3)	2.4 ± 0.3 (10)	F = U
8 July '84	3.0 ± 0.1 (8)	3.1 ± 0.2 (8)	F = U
15 July '84	3.4 ± 0.1 (7)	4.0 ± 0.1 (13)	F > U
25 July '84	4.8 ± 0.2 (6)	5.0 ± 0.1 (13)	F = U
2 August '84	5.8 ± 0.1 (6)	5.8 ± 0.3 (10)	F = U
8 August '84	5.8 ± 0.1 (10)	6.2 ± 0.2 (8)	F = U
13 August '84	5.8 ± 0.4 (6)	5.9 ± 0.2 (13)	F = U
Pupal length	3.2 ± 0.003 (3)	3.6 ± 0.09 (8)	F > U
Pupal width[1]	1.2 ± 0.003 (3)	1.3 ± 0.01 (8)	F > U

[1] (maximum dimension)

from the previous year's fertilization. Lack of a length difference on later dates, probably reflects developmental changes prior to emergence, such as more robust larvae undergoing early pupation.

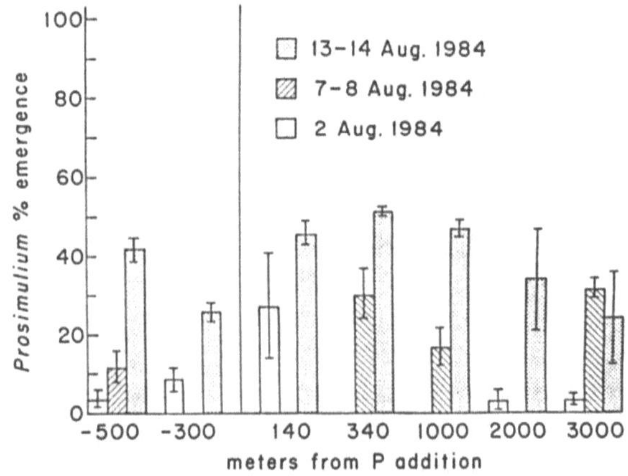

Fig. 1. Emergence rates of *Prosimulium* estimated by the number of empty pupal cases divided by the number of larvae + pupae + empty pupal cases collected at each sampling station from 2 August to 14 August. Error bars represent $\pm$ 1 SE.

Prosimulium pupae were 12.7% longer and 10.3% wider in the fertilized section (Table 2). These differences were both significant ($t_{length} = 2.636$, $p < 0.01$, $t_{width} = 4.32$, $p < 0.005$, $df = 9$). This increase in pupal size combined with the increase in larval size prior to pupation in the enriched section is strong evidence that growth was enhanced by fertilization and suggests that growth enhancement should result in larger adults.

Nutrient addition did not seem to affect time to emergence for *Prosimulium* (Fig. 1). Emergence estimates among stations on the 2–14 August sampling dates were not significantly different.

The net effect of the increase in individual biomass but decrease in abundance of *Prosimulium* was a decrease in the production of this species. *Prosimulium* production was 2.62 g/m² in the upstream section, compared to 0.77 g/m² in the fertilized section during the summer of 1984.

Discussion

The increase in length of individual black flies in the enriched section of the Kuparuk indicates that P addition did indirectly stimulate somatic growth, due to an increase in food quality and/or quantity. Fine detritus often makes up a large part of the seston in streams and the quality of this seston as a food source to filter feeders is dependent on the amount of bacteria and algae associated with the detritus (Lamberti & Moore, 1984).

The increase in larval length prior to the onset of pupation probably reflects a difference in the amount of fats stored in preparation for pupation. The increase in *Prosimulium* pupal lengths and widths with P addition supports this conclusion.

Although adults were not collected in this study, laboratory studies have shown that increased larval food supply at constant temperatures results in larger adults (Colbo & Porter 1979 and 1981). Chutter (1970) found a significant positive correlation between adult size and fecundity (number of oocytes/female) in 3 species of black flies reared without food in the laboratory. The feeding habits of adult female *Prosimulium martini* are not known, but most arctic black fly species are not dependent on a blood meal as adults (Downes, 1965). If this is the case for this species, larval nutrition may have a great effect on adult fecundity. [However, they clearly can take a blood meal, (AEH pers. obs.)].

Increased food supply in the Kuparuk River did not decrease the time to emergence for *Prosimulium*. The lack of evidence for reduced emergence times is surprising because several other studies demonstrate the importance of food supply in determining the life histories of aquatic insects (see Anderson & Cummins 1979 for literature review). Colbo & Porter (1981) have shown in the laboratory that increasing food supply at constant temperatures decreased time to emergence in two black fly species. Several authors have observed that black flies found near lake outlets emerged before those occurring where food was less abundant (Wotton, 1979; Carlsson *et al.*, 1977). Earlier emergence with food enrichment has been observed for other filter feeding species such as the caddisflies *Neureclipsis bimaculata* (Richardson, 1984) and *Clistoronia magnifica* (Parker & Voshell, 1982).

It is possible that our sampling regime was inadequate to detect small changes in emergence rates between sampling stations. *Prosimulium* densities were extremely low in the enriched section by the end of the summer (an average of 2.2 larvae and 6.7 pupae per sample were found at the enriched sites on the last sampling date).

The results of this study are in contrast to a well documented positive correlation between food supply and abundance of filter feeders (Oswood, 1979; Carlsson *et al.*, 1977; Wotton, 1978; Sheldon & Oswood, 1979). Current speed has also been found to be highly correlated with black fly abundance (Brown & Brown, 1984; Gersabek & Merritt, 1979). In this study, current speeds were measured above sampling sites in 1985 and there was no correlation between current speed and distance from the P addition site ($r^2 = 0.066$, $df = 46$). Food supply had an apparent negative effect on larval abundances, indicating that other factors must have been more important in determining their densities.

The two most likely explanations for the decreased black fly abundance in the enriched section are competition for space with insect species which increased in abundance as a result of the nutrient addition, or with the periphyton itself. Concurrent experimental studies suggest that behavioral interactions with the caddisfly *Brachycentrus americanus* were responsible for the reduced black fly density. *Brachycentrus* density was enhanced by fertilization, and black flies drifted in response to contact with *Brachycentrus* (Hershey & Hiltner, 1988). Thus, the decrease in black fly abundance with nutrient enrichment reflects higher order trophic interactions. Stream sections downstream of lake outlets, where enhanced black fly abundance has been observed, may be a simpler trophic climate.

The reduced density of black flies in the Kuparuk may have reduced their role in the cycling and downstream movement of P in the river. As filter feeders, black flies have the potential to hold P associated with fine particulate organic matter which would otherwise settle further downstream in a section of the river (Merritt *et al.*, 1984). Maciolek & Tunzi (1968) reported a 60% loss of suspended algae over 0.25 km in a California stream heavily populated by black flies compared to a 10% loss in an area where black flies were rare, although they did not quantify black fly abundance. McCullough *et al.* (1979) reported that the filter feeders *Hydropsyche occidentalis* (Trichoptera) and *Simulium* spp. (Diptera) together filtered approximately 1% of the seston in an Idaho stream at population densities of 1786 m^{-2} and 1699 m^{-2} respectively. In this study, black fly population densities averaged 2116 m^{-2} upstream of the P addition site and 1010 m^{-2} in the enriched section until emergence began in August. Since black flies decreased in abundance in the enriched section while seston quantity presumably increased due to sloughing of algae from rocks, the role of black flies in nutrient cycling appears to have declined with P enrichment.

Even though the individual growth of *Prosimulium* in the Kuparuk was enhanced by the addition of phosphorus, overall production of *Prosimulium* decreased by 71% in the enriched section. Although *Prosimulium martini* is univoltine, annual production falls near the middle of the annual production range for black flies of 0.06–3.56 $g\,m^{-2}\,yr^{-1}$ noted by Waters (1977). Morgan *et al.* (1980) concluded that food quantity and quality are the main factors limiting production of benthic invertebrates in lentic systems. Benke (1984) concluded that food quantity and quality, temperature, habitat complexity and biological interactions can all influence benthic production. In this case behavioral avoidance of *Brachycentrus* appears to have limited population size and low population density was more important than food limitation of growth in limiting secondary production of this species.

Acknowledgements

We thank Stanley Dodson for valuable support and editorial comments. Suggestions of an anonymous reviewer are also greatly appreciated. Bruce Peterson, John Hobbie, Robie Vestal, Mike Miller, Maurice Lock, Meredith Hullar, Bernard Moller, John Helfrich, Linda Deegan, David Socha, Tim Ford, and all others associated with the Kuparuk River fertilization project provided information essential to this project and assistance in the field. We are especially grateful to Bruce Peterson for providing 1978 insect samples. This work was supported by NSF grant DPP-8320544.

References

Anderson, N. H. & K. W. Cummins, 1979. Influence of diet on the life histories of aquatic insects. J. Fish. Res. Bd Can. 36: 335–342.

Brown, A. V. & K. B. Brown, 1984. Distribution of insects within riffles of a stream. Freshwat. Invert. Biol. 3: 2–11.

Benke, A. C., 1984. Secondary production of aquatic insects. In Resh, V. H. & D. M. Rosenberg (eds), The ecology of aquatic insects. Praeger Publishers, NY: 289–322.

Carlsson, M., L. M. Nilsson, B. Svensson & S. Ulfstrand, 1977. Lacustrine seston and other factors influencing the black flies (Diptera: Simuliidae) inhabiting lake outlets in Swedish Lapland. Oikos 29: 229–238.

Chutter, F. M., 1968. On the ecology of the fauna of stones in the current in a South African river supporting a very large *Simulium* (Diptera) population. J. appl. Ecol. 5: 531–561.

Chutter, F. M., 1970. A preliminary study of the factors influencing the number of oocytes present in newly emerged blackflies (Diptera: Simuliidae) in Ontario. Can. J. Zool. 48: 1389–1400.

Colbo, M. H. & G. N. Porter, 1978. Effects of food supply on the life history of Simuliidae (Diptera). Can. J. Zool. 57: 301–306.

Colbo, M. H. & G. N. Porter, 1981. The interaction of rearing temperature and food supply on the life history of two species of Simuliidae (Diptera). Can. J. Zool. 59: 158–163.

Downes, J. A., 1965. Adaptations of insects in the arctic. Ann. Rev. Ent. 10: 257–274.

Edmondson, W. T., 1970. Phosphorus, nitrogen, and algae in Lake Washington after diversion of sewage. Science 169: 690–691.

Elwood, J. W., J. D. Newbold, A. F. Trimble & R. W. Stark, 1981. The limiting role of phosphorus in a woodland stream ecosystem: effects of P enrichment on leaf decomposition and primary producers. Ecology 62: 146–158.

Freeman, M. J. & J. B. Wallace, 1984. Production of net-spinning caddisflies (Hydropsychidae) and black flies (Simuliidae) on rock outcrop substrate in a small southeastern Piedmont stream. Hydrobiologia 112: 3–15.

Gallepp, G. W., 1977. Response of caddisfly larvae (*Brachycentrus* spp.) to temperature, food availability, and current velocity. Am. Midl. Nat. 98: 59–84.

Gersabeck, E. F. & R. W. Merritt, 1979. The effect of physical factors on the colonization and relocation behavior of immature black flies. Envir. Ent. 8: 34–39.

Grimm, N. B. & S. G. Fisher, 1986. Nitrogen limitation in a sonoran Desert stream. J. N. Am. Benthol. Soc. 5: 2–15.

Hart, D. D. & C. T. Robinson, 1990. Resource limitation in a stream community: phosphorus enrichment effects on periphyton grazers. Ecology 71: 1494–1502.

Hershey, A. E., 1988. Effect of a caddisfly on black fly density: interspecific interactions limit black flies in an arctic river. J. N. Am. Benthol. Soc. 7: 188–196.

Hershey, A. E., A. L. Hiltner, M. A. J. Hullar, M. C. Miller, J. R. Vestal, M. A. Lock, S. Rundle & B. J. Peterson, year? Nutrient influence on a stream grazer: *Orthocladius* microcommunities respond to nutrient input. Ecology 69: 1983–1992.

Krueger, C. C. & T. F. Waters, 1983. Annual production of macroinvertebrates in three streams of different water quality. Ecology 64: 840–850.

Lamberti, G. A. & J. W. Moore, 1984. Aquatic insects as primary consumers. In Resh, V. H. & D. M. Rosenberg (eds), The Ecology of Aquatic Insects. Praeger Publishers, New York, pp. 165–195.

Maciolek, J. A. & M. G. Tunzi, 1968. Microseston dynamics in a simple Sierra Nevada stream system. Ecology 49: 60–75.

Mackay, R. J. & T. F. Waters, 1986. Effects of small impoundments on hydropsychid caddisfly production in Valley Creek, Minnesota. Ecology 67: 1680–1686.

McAuliffe, J. R., 1984. Competition for space, disturbance, and the structure of a benthic stream community. Ecology 65: 894–908.

McCullough, D. A., G. W. Minshall & C. E. Cushing, 1979. Bioenergetics of lotic filter feeding insects *Simulium* spp. (Diptera) and *Hydropsyche occidentalis* (Trichoptera) and their function in controlling organic transport in streams. Ecology 60: 585–596.

Merritt, R. W., D. H. Ross & G. J. Larson, 1982. Influence of stream temperature and seston on the growth and production of overwintering larval blackflies (Diptera: Simuliidae). Ecology 63: 1322–1331.

Morgan, N. C., T. Backiel, G. Bretschko, A. Duncan, A. Hilbricht-Ilkowska, Z. Kajak, J. F. Kitchell, P. Larsson, C. Leveque, A. Nauwerck, F. Schiemer & J. E. Thorpe, 1980. Secondary production. In E. D. LeCren & R. H. Lowe-McConnell (eds), The functioning of freshwater ecosystems. Cambridge University Press, Cambridge, England.

Oswood, M. W., 1979. Abundance patterns of filter feeding caddisflies (Trichoptera: Hydropsychidae) and seston in a Montana (USA) lake outlet. Hydrobiologia 63: 177–183.

Parker, C. R. & J. R. Voshell, Jr., 1982. Life histories of some filter-feeding Trichoptera in Virginia. Can. J. Zool. 60: 1732–1742.

Peterson, B. J., J. E. Hobbie & T. L. Corliss, 1986. Carbon flow in a tundra stream ecosystem. Can. J. Fish. aquat. Sci. In press.

Peterson, B. J., J. E. Hobbie, A. E. Hershey, M. A. Lock, T. E. Ford, J. R. Vestal, V. L. McKinley, M. A. J., M. C. Miller, R. M. Ventullo & G. S. Volk, 1985. Transformation of a tundra river from heterotrophy to autotrophy by addition of phosphorus. Science 229: 1383–1386.

Pringle, C. M., 1987. Effects of water and substratum nutrient supplies on lotic periphyton growth: an integrated bioassay. Can. J. Fish. aquat. Sci. 44: 619–629.

Richardson, J. S., 1984. Effects of seston quality on the growth of a lake outlet filter feeder. Oikos 43: 386–390.

Ross, D. H. & D. A. Craig, 1979. The seven larval instars of *Prosimulium mixtum* Syme and Davies and *P. fuscum* Syme and Davies (Diptera: Simuliidae). Can. J. Zool. 57: 290–300.

Sheldon, A. L. & M. W. Oswood, 1977. Blackfly (Diptera: Simuliidae) abundance in a lake outlet: test of a predictive model. Hydrobiologia 56: 113–120.

Schindler, D. W., 1974. Eutrophication and recovery in experimental lakes: implications for lake management. Science 184: 897–899.

Schindler, D. W., 1977. Evolution of phosphorus limitation in lakes. Science 195: 260–262.

Smock, L. A., 1980. Relationships between biomass and body size of aquatic insects. Freshwat. Biol. 10: 375–383.

Smock, L. A., E. Gilinsky & D. L. Stoneburner, 1985. Macroinvertebrate production in a southeastern United States blackwater stream. Ecology 66: 1491–1503.

Stockner, J. G. & K. R. S. Shortreed, 1978. Enhancement of autotrophic production by nutrient addition in a coastal rainforest stream on Vancouver Island. J. Fish. Res. Bd Can. 35: 28–34.

Sweeny, B. W., 1984. Factors influencing life-history patterns of aquatic insects. In Resh, V. H. & D. M. Rosenberg (eds), The ecology of aquatic insects. Praeger Publishers, New York, pp. 56–100.

Ward, G. M. & K. W. Cummins, 1979. Effects of food quality on growth of a stream detritivore, *Paratendipes albimanus* (Meigen) (Diptera: Chironomidae). Ecology 60: 57–64.

Waters, T. F., 1977. Secondary production in inland waters. Adv. Ecol. Res. 10: 91–164.

Waters, T. F. & J. C. Hokenstrom, 1980. Annual production and drift of the stream amphipod *Gammarus pseudolimnaeus* in Valley Creek, Minnesota. Limnol. Oceanogr. 25: 700–710.

Wotton, R. S., 1978. Growth, respiration, and assimilation of blackfly larvae (Diptera: Simuliidae) in a lake outlet in Finland. Oecologia 33: 279–290.

Wotton, R. S., 1979. The influence of a lake on the distribution of blackfly species (Diptera: Simuliidae) along a river. Oikos 32: 368–372.

Zahar, A. R., 1951. The ecology and distribution of black-flies (Simuliidae) in south-east Scotland. J. anim. Ecol. 20: 33–62.

Hydrobiologia **240**: 267–269, 1992.
W.J. O'Brien (ed.), Toolik Lake – Ecology of an Aquatic Ecosystem in Arctic Alaska.

Access pipes for sampling through thick ice

George W. Kipphut[1] & Stephen C. Whalen
Institute of Marine Science, University of Alaska Fairbanks, Fairbanks, AK 99775-1080, USA; [1] *Present address: Center for Reservoir Research, Murray State University, Murray, KY 42071, USA*

Key words: arctic limnology, ice

Abstract

Access pipes are described which permit water sampling beneath thick ice. The pipes prevent water overflow conditions which are sometimes encountered when sampling arctic lakes. Simple modifications to our design permit broader application to a number of winter sampling conditions.

Winter sampling of lakes in the arctic presents a number of challenges. Cutting access holes through very thick ice is the most common difficulty encountered. We experienced an additional problem while sampling limnological enclosures (limnocorrals) in Toolik Lake, Alaska (68° N, 149° W). Water flowed for several days from holes drilled for under-ice sampling of the limnocorrals, destroying the integrity of the enclosures. The overflow apparently results from differential pressures on the ice surface. Strong winds limit snow accumulation over most of Toolik Lake's surface (150 ha) to about 10 cm. However, in the sheltered cove where the limnocorrals are located, snow collects in drifts more than 100 cm thick. As a result of the weight of the drifted snow, the ice surface within the cove is depressed below the ice surface in other areas of the lake. We became aware of this condition when sampling the enclosures for the first time in May 1985, and did not attempt collection of samples through the ice in succeeding years. However, experimental work begun in the enclosures in the summer of 1989 necessitated winter sampling. The access pipes which are described here were developed to allow winter sampling without water overflow and with minimal disturbance of the enclosed water.

It is not clear if the overflow problem we encountered is common in the arctic or other areas. Cracking, seepage of water through ice, and formation of slush and refrozen ice relieve differential pressures on the frozen surface of lakes with heavy snow cover, but moderate ice thickness, forming 'white ice'. This is the usual condition at the Experimental Lakes Area, Ontario, Canada (Brunskill & Schindler, 1971). Welch *et al.* (1987) report that white ice development decreases with increasing latitude, and is rare in the high Canadian arctic. This presumably is the result of small snow accumulations on ice sufficiently thick to avoid cracking due to differential snow loading. We have not encountered overflow problems when sampling several of the small lakes (1–5 ha) in the vicinity of Toolik Lake, although none had nearly the variability in snow accumulation as Toolik Lake. We would expect overflow problems to be encountered most frequently on larger arctic lakes when sampling in sheltered covers with greater than average snow accumulation.

The access pipes which we designed were based

in part on the concept of Patterson *et al.* (1978) and Baird *et al.* (1985), who described methods for implanting access tubes and pipes through ice. Water forced from the access pipes of Baird *et al.* (1985) with air pressure prevented freezing between samplings. This method was not directly useful to us because it requires cutting through the ice to implant the pipes, and because our pipes were to be left unattended for a period of more than eight months.

Our access pipe is shown schematically in Fig. 1. The pipe consists of 10 cm diameter by 3 m long ABS pipe. In choosing the pipe length, we presumed an ice thickness of approximately 1.5 m. We required that the pipe extend 1 m beneath the ice bottom, and 0.5 m above the ice surface. We used large diameter pipe to enable access by small pumps and other sampling apparatus. The bottom of the pipe was sealed with an oversized, 5 cm thick plug of rigid styrofoam. The plug was secured to the pipe with waterproof, flexible, silicone sealant. Sufficient weight was attached just above the plug with stainless steel hose clamps to make the pipe nearly neutrally buoyant when 50 cm of pipe extended above

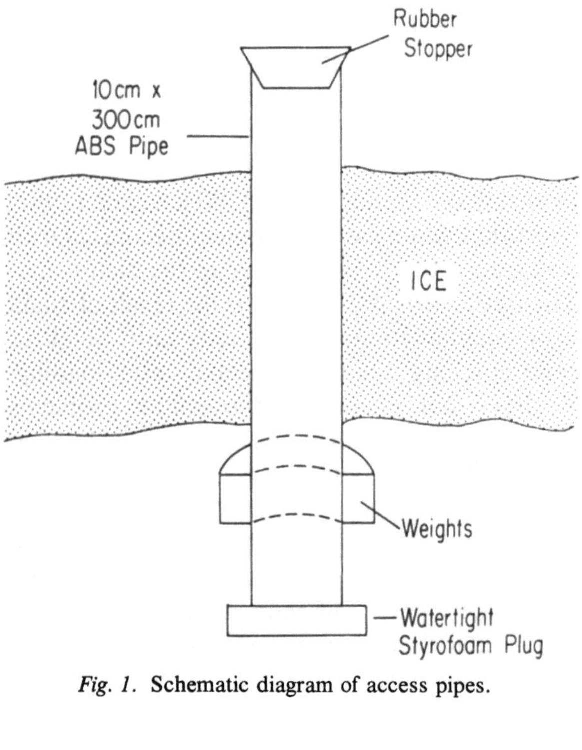

Fig. 1. Schematic diagram of access pipes.

the water surface. An initial estimate of the weight required was calculated from the dimensions and weight of the pipe (approximately 20 kg here); the exact weight was determined by flotation tests of the pipe. We used steel chain as pipe weights, but plastic-coated SCUBA weights would be preferable. The top most 1 m of the pipe was painted white to minimize solar heating.

Three access pipes were deployed on 28 August 1989 in the Toolik Lake limnocorrals (see 'O'Brien *et al.* (1992) for limnocorral description). The limnocorrals were designed and anchored to remain in place during lake freeze up (late September) and ice-out (June). The pipes were clamped vertically to the wooden frames of the limnocorrals with the plug end under water and the white, painted end extending above the water surface. Chain weights near the bottom of the pipe not only maintained neutral buoyancy, but also minimized the possibility of overturn caused by wave or wind action. Large rubber stoppers were fitted loosely into pipe tops and secured with silicone sealant to prevent entrance of precipitation. Water pressure against the bottom plug aided in maintaining a water-tight seal. The pipes were left unattended until 9 May 1990, when ice thickness at the pipe site was 121 cm, and snow accumulation on the ice was approximately 110 cm. The pipes had been previously marked with tall stakes to permit relocation beneath the snow. The rubber stoppers were removed from the pipe tops, and the bottom plugs were detached by pushing a pole through the pipe. When the bottom plugs had been displaced, water filled the pipes and sampling commenced. All three pipes functioned as planned. The water level in the pipes after filling was approximately 33 cm above the ice surface. This is nearly the depression of the ice surface at the pipe site (approximately 40 cm) which we calculated from the variability in snow mass over the surface of Toolik Lake. It is critical that the height of the pipes above the ice surface be greater than the expected equilibrium water level or overflow will reduce the usefulness of the pipes.

These access pipes overcame a difficult sampling problem in a remote arctic lake, but modifications to our methodology would permit broader application. We secured the pipes to large wooden frames permanently anchored in Toolik Lake, but such structures are not usually available. Alternatively, the pipes could be deployed in early winter by freezing the pipes into holes cut through thin ice, or the neutrally buoyant pipes could be anchored in open water just before freeze-up. Sufficient styrofoam flotation could be added to the top of the pipe to permit recovery after ice-out. We designed our pipes for a single sampling trip over a short time period. If the pipes are to be used repeatedly throughout the winter, then it will be necessary to prevent freezing of water in the pipes. Our pipes could be fitted with airtight caps (e.g., Baird *et al.*, 1985), which allow the use of air pressure to force water from the pipes between samplings.

This methodology may also be useful in ice-covered lakes in more temperate climatic regions. Water overflow from sample holes can be caused by varying lake depth as a result of changing inflow or outflow rates in lakes hydrologically connected to major stream networks (D. Huggins, pers. com.). Once frozen, loss or gain of water in the lake may create differential pressures on the ice cover or water under the ice.

Acknowledgements

This work was supported by National Science Foundation grants DPP-8722015 and BSR-8702328. P. Tinsley helped with deployment of the pipes, and B. Wimpee drafted the figure. We thank D. Huggins for his helpful review of the manuscript. Institute of Marine Science contribution 900.

References

Baird, F., D. C. Buso & J. W. Hornbeck, 1985. Access pipes for multiple sampling under ice. Limnol. Oceanogr. 30: 1129–1130.

Brunskill, G. J. & D. W. Schindler, 1971. Geography and bathymemtry of selected lakes, Experimental Lakes Area, northwestern Ontario. J. Fish Res. Bd Can. 28: 139–155.

O'Brien, W. J., A. E. Hershey, J. E. Hobbie, M. A. Hullar, G. W. Kipphut, M. C. Miller, B. Moller & J. R. Vestal, 1992. Control mechanisms of arctic lake ecosystems: a limnocorral experiment. Hydrobiologia 240: 143–188.

Patterson, R. J., L. Dykes, S. Frape, R. A. McLeod & J. Wickens, 1978. A new method for collecting water samples from beneath the ice. Limnol. Oceanogr. 23: 1029–1030.

Welch, H. E., J. A. Legault & M. A. Bergmann, 1987. Effects of snow and ice on annual cycles of heat and light in Saqvaqjuac Lakes. Can. J. Fish. aquat. Sci. 44: 1451–1461.

MIX
Papier aus verantwortungsvollen Quellen
Paper from responsible sources
FSC® C105338

If you have any concerns about our products,
you can contact us on
ProductSafety@springernature.com

In case Publisher is established outside the EU,
the EU authorized representative is:
Springer Nature Customer Service Center GmbH
Europaplatz 3, 69115 Heidelberg, Germany

Printed by Libri Plureos GmbH
in Hamburg, Germany